Impact of Digitalization on Modern Visual Media

Serhat Erdem
Atatürk Üniversitesi, Turkey

Zuhal Akmeşe
Dicle University, Turkey

IGI Global
Scientific Publishing
Publishing Tomorrow's Research Today

Vice President of Editorial Melissa Wagner
Director of Acquisitions Mikaela Felty
Director of Book Development Jocelynn Hessler
Production Manager Mike Brehm
Cover Design Jose Rosado

Published in the United States of America by
IGI Global Scientific Publishing
701 East Chocolate Avenue
Hershey, PA, 17033, USA
Tel: 717-533-8845
Fax: 717-533-7115
Website: https://www.igi-global.com E-mail: cust@igi-global.com

Library of Congress Cataloging-in-Publication Data

Names: Erdem, Serhat, 1982- editor | Akmeşe, Zuhal, 1982- editor
Title: Impact of digitalization on modern visual media / edited by Serhat Erdem, Zuhal Akmeşe.
Description: Hershey, PA : IGI Global Scientific Publishing, 2026. | Includes bibliographical references and index. | Summary: "This book aims to examine the transformation process of visual media through digitalization and evaluate the effects of this transformation on society, culture, and communication"-- Provided by publisher.
Identifiers: LCCN 2025022202 (print) | LCCN 2025022203 (ebook) | ISBN 9798337319674 hardcover | ISBN 9798337319681 paperback | ISBN 9798337319698 ebook
Subjects: LCSH: Visual communication--Technological innovations | Digital media--Social aspects
Classification: LCC P93.5 .I4778 2026 (print) | LCC P93.5 (ebook)
LC record available at https://lccn.loc.gov/2025022202
LC ebook record available at https://lccn.loc.gov/2025022203

British Cataloguing in Publication Data
A Cataloguing in Publication record for this book is available from the British Library.

Editorial Advisory Board

Table of Contents

Detailed Table of Contents

The rapid development of new media technologies has led to a profound transformation in the production of art and photography. This book chapter aims to examine the effects of new media dynamics on art and photography through aesthetic, ethical, and social dimensions. It offers a comprehensive analysis of how digitalization influences creative processes, the aesthetic values of these processes, and the transformative roles of online platforms in shaping creative practices. Prepared with both theoretical and practical approaches, this study draws on art theory, media studies, and visual culture to provide an in-depth exploration of its subject matter. Supported by qualitative analysis methods and case studies, this book serves as a significant resource for academics, students, and creative professionals seeking to understand how traditional paradigms of art and photography are being reshaped.

This chapter aims to examine how technologies create new dimensions in media production at the intersection of virtual reality and visual media. The rise of virtual reality (VR) technologies is transforming visual media experiences, offering innovative interaction spaces for both content creators and viewers. The goal of the study is to explore how these technologies are changing the effects on visual media, media perception, and audience engagement, and to discuss VR's potential impact on the future of media. The methodology will involve a review based on technology-based content analyses and media theories, along with a literature review on the effects of VR experiences on media consumers. The study will focus on the use of virtual reality applications in the media industry, with limitations that only certain VR platforms and media formats will be analyzed, excluding other media forms that reach broader audiences.

This study explores the profound impact of digital transformation on contemporary documentary cinema, particularly focusing on how digital technologies have reconfigured traditional narrative structures, production methods, and modes of audience engagement. In an era where media platforms are constantly evolving, documentary film has undergone significant conceptual and structural changes. Once rooted in linear storytelling and a singular authorial voice, the documentary genre is now increasingly characterized by interactivity, hybridity, and nonlinearity. These changes are not merely aesthetic or technological; they reflect a broader cultural and epistemological shift driven by the integration of digital tools into the filmmaking process.

With the increasing influence of technology, narratives are gradually transforming into grander stories. Undoubtedly, one of the most critical stages in a film's storytelling and creation process is the construction of space. The concept of space-time and spatiality, empowered by technological possibilities, gains a multidimensional significance, enabling creators in cinema to produce "surreal" spaces. In cinema, surreal creations emerge through the linkage of narrative time and narrative space via editing. This study will focus on space creation and the concept of constructed space in cinema. Based on Kuleshov's notion of "Creative Geography," it will also explore Eisenstein's theory of montage and J. Arnheim's concepts of aesthetics and cinematic reality, aiming to articulate the process of fictional space construction.

The integration of artificial intelligence into cinema has brought profound changes to creative processes and audience perception. This chapter examines how AI becomes part of the cinematic message and its cultural implications through the lenses of McLuhan's "the medium is the message" and Gerbner's theories of resonance and mainstreaming. The ethical implications of AI technologies used in scriptwriting, visual production, sound design, and editing are critically discussed. The chapter proposes an "Intelligent Marking Model" to transparently indicate the extent of AI involvement in film production. This model aims to inform both audiences and industry stakeholders, promoting ethical and responsible use. Ultimately, transparency and ethical approaches are essential for the sustainable integration of AI in cinema.

This study examines the effects of digital technologies on non-linear narrative structures in cinema. With the rise of digital cinema, the flexibility provided in areas such as editing, camera use, visual effects and sound design in filmmaking processes has transformed traditional narrative patterns. Non-linear narrative is characterized by flashbacks, parallel plots and structures where events are not presented in chronological order, and where time is broken. Digital technologies play a decisive role in the creation of such narratives; they support the development of new narrative forms that question the perception of time and space and direct the audience through memory, perception and emotions. The study aims to present a theoretical framework through post-2000 digital film examples and academic sources with a descriptive literature review method.

With the digital age, the ways of production, sharing and consumption of visual culture have radically changed. The perception of space in traditional arts has been reshaped with digital technologies, and representations of art and space form a more dynamic, interactive and multi-layered structure, especially with the influence of audio-visual media such as television. While space is generally represented with a static composition in traditional works of art, in the digital age the perception of space has become dynamic, variable and interactive with the viewer. Works of art move space from being just a physical phenomenon to digital environments with virtual reality, augmented reality or computer-based designs on digital platforms. Television has transformed works of art into visual experiences that merge with time and movement, rather than a static object. This transformation has allowed the work of art to be reproduced on digital screens, no longer dependent on physical space. Television is one of the most important tools that enable art to reach large audiences.

Foreword

In a media landscape increasingly shaped by the algorithmic gaze and platform-based aesthetics, the edited volume titled The Impact of Digitalization on Modern Visual Media, under the distinguished editorial leadership of Serhat Erdem and Zuhal Akmeşe, emerges as a timely and essential contribution to contemporary media studies.

This volume does not merely chronicle technological evolution; it critically interrogates how digitalization is transforming the fundamental structures of visual communication, narrative design, and viewer perception. Drawing on interdisciplinary perspectives that span visual culture, artificial intelligence, cinema studies, media psychology, and communication theory, the book offers a nuanced, layered, and academically rigorous roadmap for understanding the complex intersections between digital technologies and visual media practices.

What sets this book apart is the editorial precision and thematic cohesion achieved by Erdem and Akmeşe. Through their scholarly discernment, they have curated a series of chapters that not only reflect the depth and diversity of global media discourse but also anticipate future directions in both theoretical and applied domains. Topics ranging from AI-generated aesthetics and virtual reality storytelling to binge-watching behaviors and digital burnout are examined with both critical sensitivity and methodological innovation.

The contributors—academics and researchers from esteemed institutions—collectively advance our understanding of how digitalization reshapes authorship, agency, representation, and audience engagement. The volume succeeds in presenting digital transformation not just as a technical phenomenon, but as a cultural, ethical, and even existential shift.

In a time when visual media saturates every aspect of human life—from education to entertainment, from activism to identity—the insights offered in this book are not only informative but vital. I commend Professors Erdem and Akmeşe for assembling a work that will undoubtedly serve as a significant reference point for scholars, educators, practitioners, and students navigating the expanding horizons of visual media.

Nazife Güngör
Üsküdar University, Turkey

Preface

INTRODUCTION

In an era where digitalization has become the defining force behind the transformation of media, the book titled *The Impact of Digitalization on Modern Visual Media* provides a timely and critical examination of how visual communication, audience behavior, creative production, and technological ethics have been reshaped by digital processes. Curated by editors Serhat Erdem and Zuhal Akmeşe, this edited volume brings together scholars from Türkiye, Portugal, Malaysia, Czech Republic, and Indonesia, whose diverse perspectives span across disciplines including media studies, communication, psychology, artificial intelligence, and visual culture.

Through an interdisciplinary and cross-cultural framework, this book explores how digital technologies, ranging from artificial intelligence to streaming platforms and immersive virtual media, are altering the ways visual narratives are constructed, disseminated, and experienced. It is not only a testament to the global scope of digital media's influence but also an effort to map out the challenges and opportunities that arise in this evolving landscape.

CHAPTER OVERVIEW

Chapter 1 by Murat Han Er examines how digital media technologies have transformed the aesthetic, social, and ethical dimensions of photography and art production.

Chapter 2 by Serhat Erdem and Gülgin Altuğ discusses the impact of virtual reality on visual media perception and explore its transformative potential for immersive storytelling.

Chapter 3 by Yusuf Yurdigül and Mustafa Gülsün analyzes how digital tools have reshaped the narrative structures and production practices in contemporary documentary filmmaking.

Chapter 4 by Yelda Özkoçak focuses on spatial construction in cinema and explores the concept of fictional space through montage and narrative theory.

Chapter 5 by Elif Çanğa Bayer and Gökhan Toptepe proposes an ethical marking model for AI integration in cinema, highlighting concerns around transparency and viewer awareness.

Chapter 6 by Semra Kotan examines non-linear narrative structures in digital cinema, focusing on time manipulation and audience memory.

Chapter 7 by Nursel Bolat discusses the changing perception of space in art and how visual media such as television mediate these representations.

Chapter 8 by Recep Bayraktar analyzes the impact of streaming platforms on traditional TV viewing and audience engagement.

Chapter 9 by Aslı Yurdigül and Yusuf Ayaz compares traditional television broadcasting with digital video platforms in terms of content consumption habits.

Chapter 10 by Gözde Sunal, Gamze Nil Arkan and Gül Gürsoy interprets color, spatial symbolism, and childhood games in Netflix's Squid Game as metaphors for social critique.

Chapter 11 by Dilar Diken Yücel and Ömer Faruk Yücel studies the binge-watching phenomenon and its psychological and physical implications for modern viewers.

CONCLUSION

This edited volume offers a valuable resource for academics, media professionals, and students aiming to understand the socio-cultural, technological, and aesthetic transformations emerging from digitalization. Each chapter contributes a unique lens to this multifaceted terrain, presenting current debates as well as critical foresight into the near future of media.

As we continue to navigate a rapidly digitalizing society, this book stands as both a scholarly contribution and a reflective guide to those seeking to understand how media evolves alongside humanity.

Serhat Erdem
Atatürk University, Turkey

Zuhal Akmeşe
Dicle University, Turkey

Chapter 1
The Transformative Role of New Media Technologies on Art and Photography

Murat Han Er
https://orcid.org/0000-0002-3896-2796
Atatürk University, Turkey

ABSTRACT

The rapid development of new media technologies has led to a profound transformation in the production of art and photography. This book chapter aims to examine the effects of new media dynamics on art and photography through aesthetic, ethical, and social dimensions. It offers a comprehensive analysis of how digitalization influences creative processes, the aesthetic values of these processes, and the transformative roles of online platforms in shaping creative practices. Prepared with both theoretical and practical approaches, this study draws on art theory, media studies, and visual culture to provide an in-depth exploration of its subject matter. Supported by qualitative analysis methods and case studies, this book serves as a significant resource for academics, students, and creative professionals seeking to understand how traditional paradigms of art and photography are being reshaped.

INTRODUCTION

In today's world, where digital technologies are advancing rapidly, the transformation observed in the fields of art and visual culture profoundly affects not only technical processes but also the social, aesthetic, and conceptual frameworks

DOI: 10.4018/979-8-3373-1967-4.ch001

of production. In particular, the emergence of new media technologies has led to significant ruptures across a wide spectrum—from artistic creation to audience experience, from the role of the artist to the meaning of visual representations. This transformation redefines the relationship between visual production methods and traditional tools and techniques, repositioning practices such as art and photography within a multilayered, interactive, and pluralistic context. New media, through the flexibility and generative potential offered by digitalization, provides artists not only with new tools but also with alternative modes of expression and experiential domains. In this context, the evolution of photography from its analog phase to digital platforms can be interpreted as a significant rupture in the domains of knowledge production, identity construction, and social participation. This study aims to provide a comprehensive evaluation of the transformative effects of new media on art and photography through a historical, aesthetic, and critical lens, contributing to a deeper understanding of the dynamic nature of contemporary artistic production.

What is New Media?

The most commonly used term in the literature for *yeni medya* is "New Media." Depending on its areas of use and contextual references, it may also appear under different names such as "Digital Media," "Interactive Media," "Social Media," "Online Media," or "Internet Media." New Media has emerged alongside the widespread adoption and advancement of communication and digital technologies. One of the primary features distinguishing New Media from traditional media tools is its digital foundation and its structure oriented toward individual users. The internet, which led to radical changes during the transition to the digital age, has played a crucial role in shaping New Media. With the emergence of the internet in the 1990s—a vast network of data—it became possible to produce, distribute, and access visual content much more rapidly compared to traditional methods. Moreover, New Media has evolved into a platform that enables individual users to share and generate content, thus redefining the nature of media interaction and participation.

Lev Manovich provides a detailed account of its development in his seminal work *The Language of New Media*. According to Manovich, the *Analytical Engine* and the *Daguerreotype* represent two foundational innovations in the genealogy of new media. The *Daguerreotype* is announced as the first usable photographic technique (Manovich, 2001, p. 21). Arago proclaimed the extraordinary developments this major invention would bring with the following statement: "When observers begin to use a new device in the study of nature, the results obtained and the multitude of discoveries made possible through its application turn out to be vastly greater than initially anticipated" (Freund, 2008, p. 29). Arago's remarks regarding the connection between natural observation and technological tools signal, even at an early stage, the transformative impact of technological instruments on the production of knowledge.

Parallel to this development, in 1833, Charles Babbage began designing the *Analytical Engine*, an apparatus considered a precursor to the modern digital computer. Babbage's design incorporated many elements that constitute the basic architecture of contemporary computers. This system allowed for both data and command input via punched cards and enabled the storage of inputted information in the machine's memory. On one hand, the mechanization and mass distribution of visual representation, and on the other, the automation of calculation and the increasing capacity for abstraction—these two developments together laid the foundational groundwork for the digitalization of media in the centuries to come. While the *Daguerreotype* and the *Analytical Engine* each represent a revolutionary breakthrough in their respective domains, Manovich argues that they should be considered together in order to fully understand the historical origins of what is today referred to as digital culture or new media (Manovich, 2001, p. 21). However, as the example of the daguerreotype shows, the device quickly generated widespread technical and social effects, whereas the *Analytical Engine* never reached the stage of a functional prototype. For this reason, while the daguerreotype—recognized as the first modern media tool capable of reproducing reality—created an immediate cultural and social impact, the transformative potential of the computer was only realized after a significant delay (Manovich, 2001, p. 22).

Charles Babbage derived the idea of using punched cards as a means of data storage from a system developed in the early 19th century by Joseph Marie Jacquard, which was employed to program weaving looms. The Jacquard Loom enabled the automatic weaving of intricate patterns through punched paper cards, even producing figurative images such as Jacquard's own portrait. In this regard, the device functioned as a programmable machine for the production of visual images. The concept of generating images via programmed machines thus predates the development of numerical computation, indicating a profound interconnection between the histories of media and computing. While computer historians often regard the Jacquard Loom merely as an intriguing precursor, for scholars of new media, this connection holds significant meaning. It exemplifies not merely a coincidental parallel between two technical devices but the complementary evolution of two fundamental functions—media production and data processing—that have been instrumental in shaping modern society. In the operation of modern mass society, the dissemination of specific ideological content to broad populations has been as vital as the recording and management of individuals' birth, health, education, employment, and legal records. Therefore, mass communication tools and data processing systems represent not only technical advancements but also the dual axes along which the ideological, cultural, and administrative structures of modern society have evolved concurrently (Manovich, 2001, p. 22). This illustrates how visual production and data processing have historically been intertwined. The example of the Jacquard Loom demonstrates that the programmability of images via machines was conceived long before the advent of computer technologies. This suggests that new media should be understood not merely as a technical evolution but as a societal transformation wherein media and information management have co-developed.

In the late 19th and early 20th centuries, various mechanical and electrical calculators and tabulating devices were developed, becoming increasingly efficient in terms of speed and accessibility. This progression paralleled the growing need for systematic data organization among individuals and institutions, leading to the creation of tools essential for information processing (Manovich, 2001, p. 23). During this period, significant transformations occurred not only in computational systems but also in media technologies. Devices that allowed for the storage of visual and auditory information on physical media—such as photographic plates, film reels, and gramophone records—ushered in a new era of cultural production and transmission. This concurrent historical development reached a critical juncture in the 1890s. With the animation of still photographs, cinema was born, dramatically expanding the capacity for visual representation in modern media. In 1895, the Lumière brothers introduced their Cinématographe—a device combining camera and projection functions—to scientists and subsequently to the general public, rapidly transforming cinema into an international phenomenon (Manovich, 2001, p. 23). The fundamental idea behind the Lumière brothers' invention was the visual representation of movement on a surface. This innovation marked the inception of a new regime of representation that profoundly altered human perception and observation. The introduction of cinematography into human life extended beyond its functional use; by enabling the reproduction and viewing of movement on a surface, it facilitated the emergence of a novel human experience. Thus, the Lumière brothers pioneered one of the key epistemological shifts that shaped visual culture (Kılıç, 2012, p. 203). Additionally, in 1894, the Lumière brothers produced what is considered the first professional film in cinema history, capturing workers leaving their factory in Lyon for lunch (Acar, 2017, p. 22). The advent of cinema should not be viewed solely as a technical advancement; rather, it represents a cultural rupture that transformed visual perception. By enabling the representation of movement on a surface, the Lumière brothers' Cinématographe laid the foundation for modern visual culture. In this context, cinema can be seen as a representational tool that redefined how humans perceive and interpret the world.

With the advancement of cinematic techniques, practices such as staging reality before the camera, structuring captured material through editing, and mass-producing films on an industrial scale emerged. Film copies were distributed worldwide (Manovich, 2001, p. 23). The images presented on the silver screen largely comprised elements familiar to audiences from their daily lives. These images, reflecting the physical realities observed in viewers' own environments, structured the viewing experience through a sense of familiarity. In this context, the visual order constructed in early cinema gained meaning not only through the content represented but also through its definitional and perceptual resonance with the audience. Therefore, the images seen on the cinema screen transcended mere visuals, embodying a form of visuality that reconstructed the experiential connection between the viewer and the existing world (Kılıç, 2012, p. 209).

From the perspective of new media, another significant development is the advancement of technological systems, particularly the evolution of electronic technology. Electronic technology is regarded as the fundamental technological paradigm that shaped the 20th century. The construction of the societal, economic, and cultural structures of the 21st century is directly influenced by systems developed in line with the capabilities of this technology. The progression of electronic technology has not only transformed modes of production but also redefined meaning-making processes, enabling the emergence of multilayered visual structures open to diverse interpretations. In this context, digital systems developed in integration with electronic technology have become key drivers of globalization in areas such as production, consumption, control, communication, and information exchange. This technological transformation has also led to the establishment of new connections in human-machine relationships, bringing forth discussions on the digitization of human intelligence, perception, and memory structures. In this framework, the computer, which became widespread globally from the second half of the 20th century, stands out as a product of electronic technology. The computer is directly related to mathematical thinking as an extension of humanity's historical endeavor to question, discover, and systematize scientific truths. This is because computational methods and mathematical systems play a central role in the construction of scientific knowledge (Tuğal, 2018, p. 65). The electronic paradigm that particularly took shape in the 20th century has profoundly influenced meaning-making and cultural representations. The development of electronic technology is also pivotal in the digitalization of social structures and the redefinition of human-machine relationships. In this process, the computer has become a fundamental tool in knowledge production and meaning-making practices, integrated with mathematical thinking. Therefore, new media should be considered alongside content, structural, and cognitive transformations

The convergence of computers and media technologies has led to a profound transformation in the identities of both fields. From this point onward, the computer has evolved beyond being merely a tool for numerical calculations, a control mechanism, or a communication device; it has become a media processor capable of handling visual, auditory, and textual content. The computer has transformed into a media producer capable of generating new content and making aesthetic decisions. It has now become a media synthesizer. This transformation is one of the fundamental characteristics of the digital age, where the boundaries between media and information technologies have become increasingly blurred (Manovich, 2001, p. 26). This ongoing transformation has deeply affected communication forms, altering individuals' processes of meaning-making, interpretation, and participation. Consequently, a multilayered and dynamic structure has emerged in communication. In traditional media environments—such as newspapers, magazines, radio, and television—communication was typically conducted within a linear model directed from a specific center to a mass audience, often on a local or national scale. However, with the impact of digitalization, these structures have been replaced by new communication environments that transcend geographical boundaries, allow simultaneous access, operate on a multi-centered basis, and function on a global scale. Individuals can now actively participate in communication processes as content producers and distributors, rather than being passive recipients. In this context, digital transformation is considered a multifaceted and structural rupture with sociocultural dimensions (Sönmezer & Büyükbaykal Ilgaz, 2024, p. 757).

Ultimately, media technologies and computer sciences, which had long followed parallel but separate paths, have converged on the same axis. Daguerre's *Daguerreotype* and Babbage's *Analytical Engine*, along with the Lumière brothers' *Cinématographe* device and Hollerith's tabulating machine, have all merged within a common digital paradigm. As a result of this convergence, all existing forms of media—graphics, moving images, sounds, shapes, texts, and spaces—are transformed into digital data and become processable by computers. Each of these elements is now redefined as a computable data set. Media thus acquires a new ontological status: this structure, referred to as new media, represents a cultural and epistemological rupture (Manovich, 2001, p. 25).

When considered as a concept, Lev Manovich (2001, p.11) defines new media as the digital "medium" itself—its material and logical structure. He notes that the term is often used in reference to digital technologies such as the internet, websites, computer-based multimedia applications, digital games, CD-ROMs, DVDs, and virtual reality. However, he questions whether this list is sufficient to define the scope of new media. For instance, should television programs recorded digitally and edited with computer software, or films enriched with 3D animation and digital compositing, be included? Or should visuals, photographs, illustrations, and text-image compositions that are produced on computers but printed on paper be excluded? These questions point to the ambiguity surrounding where new media begins and ends. Popular understandings often limit new media to content that is distributed or exhibited using a computer. Thus, digital texts (such as e-books and web content) are accepted as new media, whereas their printed counterparts are not. Similarly, images stored on CD-ROMs and accessed via computer are counted as new media, but the same images in printed books are not. However, this approach is insufficient, as it focuses solely on the distribution process. Reducing the role of the computer to a display or transmission device neglects its significance in the production and storage of media. In fact, all these processes—production, processing, storage, and distribution—are equally influential in reshaping cultural narratives. Each stage of digital mediation holds the potential to transform existing cultural codes, making it highly unlikely for this transformation to occur without leaving a mark on cultural structures. When evaluated historically, it becomes evident that the digital age represents a profound cultural rupture—perhaps even deeper than those caused by the invention of the printing press in the 15th century or photography in the 19th century. While the printing press revolutionized distribution and photography transformed the production of still images, computer-assisted media touches every phase of communication—from data collection and processing to storage and dissemination—and encompasses all media types, including text, image, moving image, sound, and spatial design (Manovich, 2001, p.19). Understanding this transformation requires a detailed analysis of the computer's role in media production, editing, storage, and dissemination, and how these functions diverge from traditional media forms. This also raises the question of what is truly "new" in new media: the multilayered relationship between computer technologies and media reflects not merely a technical evolution, but an epistemological and cultural shift. As Manovich emphasizes, computer-based media is not only a technological innovation but a rupture in how cultural meanings are constructed. Thus, new media should be understood not just as a set of digital tools, but as a complex structure that induces radical changes in systems of cultural production and representation.

At the conceptual level, digital media is often referred to as "new media" or "digital new media." This terminological tendency is closely linked to the continuous nature of technological transformation in the field of media. Digital technologies evolve rapidly, and each new technological product or application is frequently categorized under the label of "new media." In this context, the notion of "newness" does not solely serve as a temporal indicator but also refers to the expansive and open-ended structure of technological potential. Manovich explains this "constant renewability" characteristic of digital media in terms of its structural flexibility. According to Manovich, the reason why digital media is persistently considered "new" lies in its inherent capacity to be reconfigured, transformed, and extended with new functions. Therefore, digital media should be understood as a dynamic cultural form with the potential to continually redefine itself (Boschele, 2023 cited in Sönmezer & Büyükbaykal Ilgaz, 2024, p. 758). In communication technologies, the concept of "new" is mostly defined through a relationship of succession, whereby each new medium positions its predecessor within the category of the "old." Within this dynamic and uninterrupted cycle of renewal, the designation of today's communication environment as "digital media" seems to be conceptually more accurate. However, the rise of digital media does not imply the complete functional disappearance of traditional mass media—such as newspapers, radio, or television. On the contrary, these media forms continue to address particular social segments, fulfill diverse communication needs, and maintain their presence within the broader media ecosystem. This indicates the emergence of a hybrid communication environment in which old and new media forms operate simultaneously. Such hybridity carries the characteristics of a multilayered media regime, simultaneously embodying historical continuity and technological transformation (Sönmezer & Büyükbaykal Ilgaz, 2024, p. 758). What is often referred to as "old media," i.e., twentieth-century media, transmits its messages via analog signals. Although analog communication systems theoretically allow for two-way interaction, their limited frequency and wave capacity have often resulted in a practically one-way communication structure. In contrast, twenty-first-century digital media, by increasing communication opportunities and enabling bidirectional interaction, has facilitated the mutual engagement of producers and consumers, gradually replacing traditional forms (Laughey, 2010 cited in Yaşar, 2020, p. 190). On traditional platforms, the audience was positioned as a passive receiver. However, when the same content is transferred to digital platforms, it becomes possible for the audience to comment, express opinions, and share evaluations, thus shifting from a passive to an active role (Yaşar, 2020, p. 192). New media environments have also transformed reading practices that are directly associated with media organization. One of the primary limitations of traditional media is its obligation to present content in a linear form and a predetermined sequence. For instance, in printed media such as books, magazines, or newspapers, content is structured and presented to the reader in a coherent whole during the production stage. This structure directs the reading process in the way the author intends; the reader is expected to decode and interpret the meaning in a fixed order. In this respect, traditional media imposes a unidirectional and hierarchical framework for meaning-making. The author holds a central role as the creator and transmitter of meaning, while the reader is positioned as a passive consumer of that meaning, which creates normative expectations such as "correct reading" or "correct interpretation" (Özdoyran, 2019, p. 197). Furthermore, since their inception, mass communication tools have consistently employed what we now refer to as "propaganda" as a communicative strategy. The very nature and emergence of mass communication seem to necessitate this form of communication. Mass communication is a one-way process, appealing more to emotions than to rational thought, in which messages are serially produced and disseminated to large audiences through a single source. The target of propaganda is the "mass"; it functions unidirectionally, aiming to influence the subconscious and emotional processes in order to serve the interests of specific institutions, individuals, or groups. It is plausible to assert that individuals and communities are exposed to propaganda in proportion to the time they spend engaging with mass media tools (Altun, 2022, p. 9). Consequently, it is evident that digital media constitutes a cultural, structural, and conceptual transformation. While its structural flexibility allows new media to be continuously redefined, it also forms a hybrid communication environment through its interactions with traditional forms. This enables the audience to transition from a passive consumer to an active producer, thereby shifting the locus of meaning-making from the center toward the individual. As such, media can evolve into a "field of propaganda," in which meaning and power relations are restructured.

With the process of digitalization, one of the most prominent concepts in the field of communication has been "interaction," which essentially signifies the transformation of the traditional communication model—based on one-way information transfer—into a bidirectional and reciprocal mechanism. This transformation represents one of the fundamental distinctions between digital mass media tools and traditional forms of media. Thanks to the hardware and software provided by digital technologies, users who were previously passive participants in the communication process have now become active agents. In this new structure, the user is no longer merely a content consumer; they have become an actor who produces, shares, and often reinterprets content through commentary. In this way, digital media environments have laid the groundwork for a user-centered and participatory communication ecosystem (Sönmezer & Büyükbaykal Ilgaz, 2024, p.758). The emergence of new media has led to profound transformations in the processes of production, consumption, and reproduction. This transformation has manifested in the shift of individuals from passive receivers to users, and from consumers to "prosumers"—individuals who both produce and consume content (Çakır, 2014, p.382). This transformation is not confined to media usage but has also had an impact on the structures of social labor. As Hardt and Negri note, digitalization has increasingly blurred the boundaries between labor and life, and between production and reproduction (Çakır, 2014; Hardt & Negri, 2011, cited in Özdemir & Şeyma, 2020, p.509).

Moreover, although mass media tools classified today as "traditional media" generally function according to a one-way communication model, they have had to take into account the demands and interests of the masses in the construction of popular culture. This condition laid the groundwork for the development of theoretical approaches that view the relationship between media and audiences as more complex and multidirectional rather than limited to passive reception. One of the first comprehensive analyses of this transformation in academic literature appears in Stuart Hall's seminal 1980 essay, *Encoding/Decoding*. In this work, Hall critiques the classical "sender-message-receiver" model and its associated linear communication framework, which had long been the foundation for explaining mass communication processes. According to him, the meaning of media texts is not solely determined by the intentions encoded during production; meaning is also reconstructed during the reception phase by viewers or readers. Thus, the communication process is repositioned as a multilayered and negotiable field of interaction rather than a one-way transmission (Ercansungur & Çetin, 2023, p.569). One of Hall's most significant insights is that the message encoded by the sender and the message decoded by the receiver do not necessarily coincide or form a unified meaning. Particularly in audiovisual media such as television, different social positions, cultural contexts, and interpretive strategies play roles in both the production and reception of messages. This allows for unexpected interpretations by the audience. Therefore, communication should be understood not as a unidirectional and passive information transfer, but as a dynamic and multilayered field where meaning is continuously reproduced (Hall, 1980, cited in Ercansungur & Çetin, 2023, p.569). Hall's theoretical framework enables a radical reassessment of the receiver's role in the communication process. According to this approach, the masses are not passive, homogeneous receivers who unquestioningly accept presented messages. Rather, individuals and social groups interpret and construct meanings from media messages differently, depending on their social, cultural, and economic contexts. This demonstrates that meaning production is not solely the domain of the sender, but that the receiver also acts as an active agent in the process. This perspective resonates with Jean Baudrillard's critiques of mass culture. According to Baudrillard, portraying the masses as merely manipulable, passive, and constantly deceived is a reductive generalization. Individuals may resist, reinterpret, or transform the presented content based on their own experiential frameworks. These pluralistic forms of reception transform communication from a mere information transfer into a dynamic, multidirectional field in which meaning is negotiated among various actors (Ercansungur & Çetin, 2023, p.569). It thus becomes evident that the communication process cannot be explained through a simple linear model of message transmission from sender to receiver. As seen in Hall's *Encoding/Decoding* model cited by Ercansungur & Çetin (2023), meaning is not fixed but varies depending on social and cultural contexts, thereby shifting the viewer from a passive consumer to a co-producer of meaning. This approach makes it possible to conceptualize the relationship between media and audiences within a layered, negotiable, and pluralistic framework.

In digital new media environments, however, reading practices differ significantly. Thanks to affordances such as hypertextuality, interactive interfaces, and multiple reading paths, the text is no longer a fixed and singular bearer of meaning. The user becomes an active participant who navigates among contents, makes choices, and constructs their own meaning by traversing complex layers of information. This transformation not only multiplies the reading experience but also shifts the authority over meaning from a writer-centered to a reader-centered structure. As a result, the digital text assumes a structure that is open-ended, negotiable, and conducive to multiple interpretations (Özdoyran, 2019, p.197). Consequently, the organization of content, presentation styles, and narrative strategies in new media environments have diverged from linear, hierarchical, and one-way models, acquiring a pluralistic, interactive, and participatory character. The direction of reading texts has evolved into a dynamic structure shaped by user preferences rather than following a fixed path. This structural transformation has led to a significant departure from traditional media forms and has been conceptualized by some theorists as "post-traditional media." However, this distinction is not merely a binary between "traditional" and "new." While new media tools clearly differ from traditional mediums in technical and structural terms, a more nuanced evaluation is needed concerning the nature of communication processes. Looking at human history, the dominant form of communication has often been dialogical—based on reciprocal interaction and bidirectionality. From this perspective, despite representing a "new" technological format, new media, with its user-to-user interaction, feedback openness, real-time communication, and participatory structure, signifies a return to an older, interaction-based communication model. In contrast, media tools typically classified as "traditional," such as television, radio, and newspapers, are defined by a "monologic" structure where messages are transmitted from a single center with little to no feedback mechanisms. In this context, new media should be interpreted as an indicator of a shift toward a more participatory and polyphonic paradigm in communication practices (Thompson, 2008, cited in Özdoyran, 2019, pp.197–198).

One of the key aspects that sharply distinguishes digital media from traditional media is the difference in content production and distribution. In traditional media environments, content production takes place under the supervision of certain institutional structures. At this point, the concept of "gatekeeping" becomes prominent. Editorial boards that decide which content will be published and which discourse will be emphasized also determine the ideological orientation of the media outlet and its strategy toward the target audience. However, this structure has undergone a significant transformation in digital media environments. Content production and distribution have largely shifted into the hands of individual users. In this context, users are not merely consumers; they also assume the role of producers, gaining considerable control and influence over media content. Additionally, the ability to circulate current content instantly and rapidly is one of the key innovations that digitalization has brought to communication processes (Sönmezer & Büyükbaykal Ilgaz, 2024, p. 759).

Another major development, which began to be used on a limited scale in the late 1980s and has significantly impacted all areas of media, is the Internet. The term "internet," which became increasingly widespread in the 1990s, is derived from the abbreviation of "Interconnected Networks" (Gülmez, 2015, cited in Yaşar, 2020, p. 192). With advancements in computer technologies, the innovations introduced by the web paved the way for the Internet to become a global information and communication technology. The Internet and globalization, as mutually reinforcing phenomena, began rapidly constructing what McLuhan had termed the "global village" (McLuhan & Powers, 2001, cited in Yaşar, 2020, p. 192). On a global scale, the Internet has created new and original forms of news and media, fundamentally disrupting and transforming the traditional functioning of mass communication tools (Yaşar, 2020, p. 192). As the Internet became embedded in daily life routines and social networks became widely used across the world, transformation in social structures and relationships became inevitable. Patterns of interaction within social structures, as well as the personal life practices and habits of individuals, have also been reshaped in parallel with this transformation (McLuhan, 2005, cited in Yaşar, 2020, p. 199).

The rapid developments in digitalization and Internet technologies have continued to expand both the usage areas and functional characteristics of web-based applications. Parallel to this technological transformation, the current era has been described in the literature by various terms, including the digital age, the age of informationalism, and the network society. In this era, communication—especially in areas such as e-commerce, culture, art, and education—has undergone a significant transformation under the influence of digital media tools and technological innovations (Castells, 2008; Koçyiğit, 2022 cited in Koçyiğit & Darı, 2023, p. 428). This transformation signifies a self-renewing structure. Even as digital platforms such as social media, the Web, and so-called new media—centered around user interaction—have yet to become obsolete, they are already being replaced by AI-based applications. As the role of artificial intelligence-supported systems increases in everyday life, many domains—from communication forms to decision-making processes—are shifting from human-centered structures to systems based on machine learning, algorithmic analysis, and automation (Koçyiğit & Darı, 2023, p. 428). Moreover, developments in the 2000s have significantly altered the nature of the web from its initial form. Since the advent of the Internet, the functions and usage purposes of Internet-based platforms have continually evolved. Within the context of digital environments referred to as new media, one of the most significant structural innovations has been the explosion of social networks that enable networking, information sharing, and new forms of structured content (Yaşar, 2020, p. 194). The digital age is marked by a dynamic and ever-evolving structure, which clearly demonstrates a transformative effect on social structures. Following the rise of user-centered platforms such as social media, the emergence of AI-based systems indicates a shift from human-centered communication to algorithmic interaction. As such, digital media now plays a decisive role in decision-making and social interaction processes.

With the integration of web technologies into our lives, social media platforms have provided individuals with the opportunity to acquire new identities in the digital environment and reconstruct themselves in different forms. This development has enabled the emergence of a flexible and multilayered subject position, where the individual can continuously transform their identity and exist through diverse representations in the digital realm. However, this identity plurality has also triggered significant debates regarding the perception of reality in digital contexts. The increasing distance between physical identity and digital representations has led to disruptions in meaning, trust, and authenticity in social relationships. While traditional media texts typically exhibit a linear structure with a clear beginning and end, the hypertextual structure dominant in social media applications—where interconnected content allows transitions from one text to another—has rendered meaning plural, dynamic, and open to interpretation. This structural transformation has facilitated more active user participation in the content production process. However, this participation has considerably limited users' control over the meaning of the content they produce. Therefore, although users have risen to the position of content creators in new media environments, it is often not possible for them to exercise direct control over how these contents are interpreted, how they circulate, or what social impacts they generate. This situation reveals that meaning production in the digital age occurs in a decentralized, polyphonic, and largely indeterminate structure (Stevenson, 2008; Binark, 2014 cited in Temel & Önürmen, 2017, p. 297).

Social media platforms such as Facebook, Twitter, LinkedIn, and YouTube are digital interaction spaces where user interaction and content sharing occur intensively. These platforms go beyond being mere channels for personal sharing; through groups, event pages, and communities formed with various motivations, they offer multifaceted opportunities for digital socialization. In this context, users can establish direct interactions on an individual level while also engaging in collective communication through group dynamics. These platforms enable individuals to reach a large number of users simultaneously and allow for plural forms of communication. Therefore, social media presents a hybrid communication environment that encompasses both micro-level interpersonal relations and macro-level communication practices targeting broader audiences (Temel & Önürmen, 2017, p. 297). Thus, it becomes clear that social media functions as a space where personal identity is constructed and exhibited. Platforms like Facebook, Twitter, and Instagram offer users opportunities for interaction on both individual and collective levels, while also playing an active role in processes such as self-presentation, identity performance, and the attainment of social status. Consequently, social media has transformed into a multilayered medium at the center of digital socialization, bearing both technical and cultural significance.

As a platform, Instagram has become a focal point, particularly for users seeking popularity, due to its structure that allows for integration and collaboration with other social media channels. This multidimensional capacity for integration not only enhances individual user interactions but also enables the platform to adopt a commercial structure through advertising tools and a professional management panel. Thus, the platform has transformed into a dynamic digital marketplace centered on interaction and visibility, situated at the intersection of user-generated content and economic activity (Chapple & Cownie, 2017 cited in Ayaz, 2024, p. 306). Moreover, Instagram is not merely a space for social interaction; it also stands out as a platform where digital marketing strategies and consumer culture take shape. Considering that approximately 90% of the information processed by the human brain consists of visual content, Instagram has become one of the most preferred social media platforms today. The fact that communication through visual content requires less cognitive effort and can be perceived more quickly makes this platform both accessible and effective for users. This has led to Instagram being intensively used not only for personal sharing but also for commercial and cultural interactions. In this respect, Instagram serves as a medium that supports the dissemination of innovative ideas and digital initiatives. It is also regarded as a rapidly growing and transforming field of digital marketing (Chen, 2018; Weismann, 2022; Ahmad, 2017; Baker, 2014 cited in Ayaz, 2024, p. 308).

From another perspective, Metaverse platforms—virtual environments integrated with augmented reality technologies that gained prominence especially after 2020—have introduced a significant paradigm shift in the field of digital media. This digital transformation continues to be debated by experts in both technological and socio-cultural contexts. Today, web environments are no longer limited to news portals, blogs, and animated infographics; they are increasingly manifesting through commercial websites. Many practices—from banking operations to retail and wholesale shopping, from cultural and artistic events to daily routines—have migrated from the physical world to digital spaces. The most distinctive difference between contemporary virtual platforms and traditional communication environments lies in the inclusion of advertising and data-driven marketing strategies in content delivery. In this respect, digital media is no longer merely a medium of communication; it also functions as a dynamic economic domain. The content circulating in virtual environments reveals a digital economic system that guides user behavior and shapes consumption practices (Sönmezer & Büyükbaykal Ilgaz, 2024, p.759). It is evident that digital media has become a multilayered structure at the center of economic and cultural transformation. Next-generation platforms such as the Metaverse have surpassed traditional media models by integrating user experience with data-focused marketing strategies. Thus, digital media has evolved into an interaction-based economic sphere, reshaping consumption habits and societal behaviors.

All these developments have also brought the issue of "surveillance" to the forefront. Surveillance has existed in various forms throughout the pre-modern, modern, postmodern, and digi-modern periods. In each historical era, it has assumed different forms depending on the societal structure and technological conditions of the time. However, the current structure of surveillance has emerged as a result of a long evolutionary process. While pre-modern surveillance practices were limited and relied on physical monitoring, they took on institutional and centralized forms in the modern era. In the postmodern and particularly the digi-modern period, surveillance has become far more comprehensive, penetrating, and invisible, facilitated by technological tools. With the infrastructure provided by digital technologies, surveillance today is no longer confined to monitoring individuals' physical movements; it has evolved into a continuous, automatic, and algorithmic system that monitors online behaviors, digital interactions, and personal data. In this sense, surveillance in the digi-modern era is both more systematic and more deeply embedded in individuals' daily lives than previous forms (Kalaman, 2019, p.579). The scale of digital surveillance technologies today has eliminated the possibility of a private space where individuals can hide, escape, or remain alone. It has significantly restricted areas of personal freedom and privacy. Most of these surveillance systems operate via the internet and are integrated into various aspects of everyday life—from security and entertainment to online shopping, digital socialization, and instant communication. While new communication technologies offer users benefits such as speed, accessibility, and interactivity, they expect users to share their personal data with the system unconditionally in return. This dynamic forms the core of what Zuboff (2019) calls surveillance capitalism. Today's systems constantly monitor and analyze a wide range of user data, including location, communication habits, consumption preferences, and social relationships. As a result, while individuals voluntarily participate in digital platforms, they also consent to these platforms' surveillance mechanisms. Moreover, they become an invisible and normalized part of the surveillance network (Kalaman, 2019, p.580). As Kalaman also states, surveillance today has become an invisible but continuously functioning system enabled by digital technologies. Not only physical movements but also our online behaviors, preferences, and communication habits are monitored. This situation has severely narrowed the domain of personal privacy and transformed the use of personal data into a marketplace where consumers are unaware of how their data is exploited.

Another important topic is academic research on new media. Öztürk et al. (2024) note a significant increase in academic studies focusing on related concepts such as creativity, new media, social media, storytelling, and persuasion in advertising literature since the 2000s. Especially between 2008 and 2023, there has been a noticeable rise in publications exploring the relationship between these themes and the advertising discipline, compared to previous periods. This increase is attributed not only to a shift in academic interest but also to structural factors such as the development of digital databases and the systematic archiving of academic publications. Indeed, the increased visibility and accessibility of articles during this period has emerged as a factor that supports the production of literature both quantitatively and qualitatively. As a result, creative strategies, digital media opportunities, and interactive relationships with consumers have become increasingly central topics in advertising research (Öztürk et al., 2024, p.61). In conclusion, the relationship between new media and advertising research has significantly strengthened in recent years. Digital platforms have become central to both academic attention and creative strategies; this development has been further reinforced by the increasing accessibility of digital databases. As a result, advertising research, which concerns multiple disciplines, has been reshaped through the interactive possibilities offered by media technologies.

New Media and Art

In everyday life, the increasing prevalence of virtual practices has profoundly transformed the experiences of individuals and institutions. Digital technologies, in particular, have led to significant changes in the perception of time and space. Thanks to mobile devices, a wide range of activities—such as shopping, education, legal procedures, and cultural events—can now be carried out independently of physical boundaries; meanwhile, brands are able to establish highly effective commercial domains through digital platforms that can even influence fashion trends. In this new era where the boundaries between the real and the virtual are increasingly blurred, Michel Foucault's concept of the "epoch of space" has gained new relevance, positioning virtual space as a multilayered heterotopic domain that enables the transformation of both individual and institutional experiences (Kazan, 2023, p. 821). New technologies that affect all areas of life have also transformed art and the modes of artistic production. Historically, artists have embraced the technological possibilities of their respective eras and integrated them into their creative processes. From the 20th century onward, the evolution of electronic and subsequently digital technologies has provided artists with new forms of expression and experimental spaces (Erdem, 2023, p.9). Accordingly, the relationship between art and technology has undergone continuous transformation, with the systems that emerged alongside digitalization now playing a decisive role in contemporary art production (Tuğal, 2018, p. 66).

Since the 2000s in particular, the rapid spread of real-time technologies, the internet, and the Web has increasingly blurred the previously assumed modern-era boundaries between art, science, and technology. This process of indeterminacy has shown that these fields can no longer be treated as isolated domains, thereby encouraging interdisciplinary collaborations among artists, scientists, technology experts, and relevant institutions. Such collaborations, driven by a search for the undefined and the unexplored, pave the way for open-ended creative processes and contribute to their development and support. In this context, new media art has emerged as one of the most dynamic artistic domains of our time, distinguished by its flexibility and capacity for rapid adaptation to these transformations (Yetişkin, 2012). With these developments, the boundaries between art, science, and technology have visibly dissolved, and interdisciplinary cooperation has become increasingly significant. New media art, as one of the fields most responsive to this indeterminate structure, allows for exploratory and experience-based production processes, reflecting the evolving nature of contemporary art. This situation indicates that art is now positioned not only as an aesthetic but also as a technological and intellectual field of inquiry.

When approached from a historical perspective, new media art can be traced back to the Dada movement. Tuğal (2018, p. 38) emphasizes that Dada was one of the significant movements that initiated technological and virtual art. He also notes that the origins of technological and virtual art can be found in the Fluxus movement and in Allan Kaprow's "happenings." In another assessment, Manovich (2001, p. 14) underlines the historical significance of the term "object," tracing the roots of new media art to the 1920s through this conceptual framework. According to Manovich, the Russian avant-garde artists of the 1920s, particularly those associated with the constructivist and productivist traditions, referred to their works not as "artworks" but directly as "objects" (vesh, konstruktsiya, predmet). These artists rejected the role of the autonomous and original artist, instead seeking professional identities aligned with industrial designers, architects, graphic designers, or clothing designers, much like the Bauhaus (Özkemahlı & Gençcelep, 2024, p.101). The notion of "object" implies mass industrial production rather than workshop-based individual creation. It also encompasses rational labor organization and engineering-based principles of efficiency, which were integral to this artistic understanding. These historical connotations are echoed in today's new media production. In contemporary digital culture, the boundaries between art and design have become increasingly permeable. Many artists sustain themselves through commercial design work, while professional designers develop experimental approaches and establish new aesthetic standards, emerging as key figures in transforming the language of new media. Moreover, the production of new media projects often involves collective or multi-actor structures. Although this does not precisely replicate the large-scale production model of classical Hollywood studios, the involvement of large teams or specialized small groups suggests a continuation of the industrial production model in digital culture. Thus, the concept of the "New Media Object" carries multilayered meanings at the technical, historical, and cultural levels (Manovich, 2001, p. 14). Manovich argues that many new media objects—particularly popular computer games and software applications—reach millions of users and have thus become integral to the global culture industry. Accordingly, new media productions are not merely artistic or personal expressions, but also standardized cultural practices subject to mass production. The field's reliance on technical standards in both hardware and software further integrates new media with large-scale industry (Manovich, 2001, p. 15). Manovich's use of the term "New Media Object" is not limited to production and distribution scales; rather, it recalls the experimental laboratory-based approaches of 1920s Russian and German avant-garde artists and seeks to highlight them within a contemporary context. While many artists today turn to digital technologies, only a few adopt a systematic approach that treats new media as an experimental laboratory. However, institutions like Vkhutemas and the Bauhaus, which conducted structural studies on emerging media forms of their time—such as photography, film, print technologies, and telecommunications—constitute the historical precedents of this discipline. Instead of immediately producing an interactive CD-ROM or a feature-length digital film, artists who choose to investigate the new media equivalents of a plan, a sentence, a word, or even a single letter with patience and precision achieve significant results both formally and conceptually. This approach allows the exploration of the expressive capacities of new media at the level of its fundamental components rather than its surface (Manovich, 2001, p. 15). A conceptual distinction is established between representation-based technologies and those founded on real-time communication. Media forms like film, audio and video tapes, or digital storage formats produce cultural objects fixed in time and space, referencing external realities. In contrast, "tele"-prefixed technologies—such as telegraph, telephone, television, telex, and telepresence—establish synchronous communication without producing lasting objects. In this context, new media compels a rethinking of cultural production, shifting it from an object-centered understanding toward a process-oriented form of communication (Manovich, 2001, p. 17).

Another historical approach seeks the origins of New Media Art in the foundations of Digital Art. Tuğal (2018) traces the roots of New Media Art to computer and electronic technologies, presenting the historical process as follows: Computers were invented to address complex calculations and scientific problem-solving that required rapid and precise results. Initially used only in high-tech centers, computers became accessible for personal use due to advances in microelectronics and semiconductor technologies. To understand current artistic trends, it is essential to examine the early computer-assisted art practices of the 1960s, as these experimental endeavors represent the initial steps in the formation of digital art and contemporary digital culture. While initially a technological tool in the hands of scientists and a few artists, the computer became widely accessible from the 1980s onward. Inquiries into human-machine relations, the emergence of alternative perceptions of reality, and the digitization of knowledge brought to light various dimensions of interaction with the cyber world. The acceleration of digital information sharing after the 1990s can be regarded as one of the major turning points in human history. The internet, enabling the exchange of data such as text, sound, images, and film via computers, became the foundation of today's digital media. From this point on, the digitization of all forms of knowledge in electronic environments rapidly increased. In the first quarter of the 21st century, a new electronic-based Industrial Revolution—known as Industry 4.0 and expected to be fully internet-centered—emerged on the agenda. It is anticipated that all these transformations will impact human perception and reshape the relationship between individuals and art (Tuğal, 2018, p.65).

In the 21st century, concepts such as knowledge, space, speed, and time have been reshaped through human interaction with digital reality, which is perceived as a realm distinct from physical life. All forms of information have been transformed into digital codes. People now live in a virtual universe composed of zeros and ones, using this technology to create artworks and experiences. Christiane Paul, in her book *A Companion to Digital Art* (2016), highlights the difficulty in defining digital art, noting that its close integration with technology makes it challenging to establish clear boundaries and leads to its continuous evolution (Tuğal, 2018, p.67). This field, referred to as "Computer Art" in the 1960s, began to be known as "Multimedia Art" by the end of the 20th century. Multimedia art includes hybrid forms such as film, video, and sound. Following the emergence of the internet in the 1990s, "Net Art" appeared as a distinct branch within the broader category of Digital Art. Although terms like multimedia and new media are frequently used today instead of digital art, this is not entirely accurate; the multidisciplinary New Media (Multimedia) Art that employs digital technologies is, in fact, a subfield of Digital Art that developed within the framework of Electronic Art. The processing and reconfiguration of digitally acquired images lead to multilayered meanings, while compositions made of visual and auditory elements transform into hybrid forms, offering artists boundless expressive possibilities. Indeed, the integration of digital technologies into the artistic production process has become increasingly common (Paul, 2016, cited in Tuğal, 2018, p.67). The difficulty in defining digital art lies in its continuously evolving, multidisciplinary, and technology-infused nature. With the transformation of knowledge into digital codes in the 21st century, artistic production has also been reshaped within digital reality; through multimedia, net art, and hybrid forms, modes of expression have diversified. This transformation has not only provided artists with new technical and conceptual opportunities but also necessitated the constant redefinition of art's boundaries.

Tokdil (2018) also connects the relationship between the digital and art to the foundations of the 'Avant-garde' movement. In the 1990s, new media transformed artistic practices into digital data, and through information technologies, it brought about fundamental changes in the perception of time and space. This transformation, enabled by the new forms of interaction with information in technological societies, marked the decline of classical avant-garde and laid the groundwork for the digital age. In this new era, traditional approaches to form endured by undergoing metamorphosis within similar conceptual systems. Cultural production transcended two-dimensional representational surfaces and evolved into multilayered, dimensional structures; as thought transformed into artistic production through information, new techniques and forms of expression also emerged. In this environment, all previous modes of expression were reinterpreted; fundamental concepts such as aesthetics, reality, form, surface, and dimension took on new meanings under the influence of the digital age and globalized culture. In this context, Piet Mondrian's geometric compositions evolved into Kawano's algorithmic aesthetics; László Moholy-Nagy's photomontage and compositions were reimagined through contemporary computer interfaces; Dziga Vertov's cinema gave way to three-dimensional visual narratives; and Kandinsky's abstract universe was reconstructed through digital codes. Thus, the emergence of new aesthetic structures and forms of expression went beyond mere formal change. As a result, in the digital age, art has evolved from merely incorporating knowledge and technological data into its formal structure to existing directly within these data systems. In other words, art has reconstituted its ontological existence through media, becoming intertwined with contemporary cultural forms and taking part in a process that institutionalizes the aesthetic domain of the digital (Tokdil, 2018, p.187).

Özdoyran (2019) contextualizes this historical structuring through Adorno's assessment of form and content, as well as through various artistic forms and practitioners, such as hypertextuality, collage, assemblage, and photomontage. As becomes evident, Adorno strongly emphasizes that the strategy of creating a work of art determines its form, and that form, in turn, determines content (Özdoyran, 2019, p.196). As seen in Richard Hamilton's work *Just What is it That Makes Today's Homes So Different, So Appealing?*, the phenomenon of hypertextuality operates in a manner similar to modernist aesthetic strategies like collage, assemblage, and photomontage. In these techniques, elements derived from different media are cut, replicated, and recombined to form a new whole. In this creative process, the artist re-designs existing forms by detaching them from their original contexts. In doing so, the representational relationship is problematized, and the meaning of the given form is multiplied (Bürger, 2003, cited in Özdoyran, 2019, p.201). In this sense, production is not merely recombination; the resulting work also entails a displacement of meanings. Such interventions disrupt the "definitive" perspective the viewer assumes they hold and open the regime of meaning to question. This multiplicity of perspectives resembles the aesthetic logic of Cubism, which fractures singular viewpoints. Michael Joyce underscores that hypertextuality is essentially a practice of "displacement" and "relocation." According to Joyce, the World Wide Web experience is, in itself, a process of "relocation": "Print stays itself; digital replaces itself" (Grusin & Bolter, 2000, cited in Özdoyran, 2019, p.201). Slavoj Žižek (2006, cited in Özdoyran, 2019, p.201), referencing Eisenstein's notion of "intellectual montage," defines hypertextuality as a "new montage practice." Lev Manovich (2002, cited in

Özdoyran, 2019, p.201), drawing on Frederick Jameson's *Postmodernism, or, the Cultural Logic of Late Capitalism*, views new media production as a typical form of postmodern cultural production. According to Manovich, digital media clones, collages, and reproduces existing material in various forms. Moreover, unlike pre-digital cinematic experience, the digital screen offers the viewer the possibility of seeing from multiple perspectives rather than a single one. In this context, within cyberspace, hypertextuality allows every user to become a "director" who charts their own narrative trajectory (Özdoyran, 2019, p.201). Yetişkin (2012) also evaluates New Media Art in terms of form, content, and structure. Since New Media Art encompasses numerous political, scientific, cultural, economic, and social elements, it necessitates a multilayered and multidimensional interpretive approach. It is quite difficult to comprehend this art practice through fixed, static texts, as is often the case with traditional artworks. Instead, New Media Art must be assessed through its real-time, dynamic, and interactive structures. At times, New Media Art transforms into a performance, offering an experiential space that interacts with the viewer in real time. This scenario compels the viewer to transcend conventional meaning systems and to develop a multidirectional thinking model and a new conceptual framework. Whereas artworks produced until the 20th century were "readable" through established theories, concepts, and fixed meanings, today's New Media Art is based on experience and instantaneous interaction. This has introduced a new understanding of art shaped not solely by the creator but also through reciprocal interaction with the audience. Rather than linking meaning closely to content, in New Media Art, interpretation is often derived from how form is employed. Traditional approaches to understanding art—heavily reliant on established meanings associated with symbolic and representative figures—are increasingly insufficient for analyzing this artistic practice. This is because New Media Art is less concerned with grand narratives or predetermined meaning systems and more engaged with an open-ended, pluralistic, and constantly reconfiguring process of meaning production. This form of art does not claim completion; on the contrary, it aims to generate dynamic narratives based on experience that evolve alongside the viewer. Consequently, New Media works often appear as structures that resist fixed meaning, remain open to interpretation, and continuously transform through viewer participation (Yetişkin, 2012). As a result of these readings, it becomes clear that New Media Art possesses a dynamic, multilayered, and experience-based structure that defies reduction to fixed or singular meanings. Within the theoretical framework spanning from Adorno to Manovich, the mutual determinism of form and content comes to the forefront, while aesthetic strategies such as collage, hypertextuality, and montage suggest that meaning can be fragmented, restructured, and multiplied. In this context, the viewer is no longer merely a consumer of the artwork but becomes an active subject who reshapes its meaning. The open-ended structures offered by New Media Art propose an aesthetic experience shaped through interaction rather than the production of fixed meanings.

New Media and Photography

The relationship between new media and photography is inherently linked to the origins of photography itself. According to Manovich, new media emerged at the intersection of two distinct historical trajectories—information technologies and media technologies. This process began in the 1830s with the nearly simultaneous development of Charles Babbage's Analytical Engine and Louis Daguerre's Daguerreotype (Manovich, 2001, p.20). However, as seen in the case of the daguerreotype, while it had an immediate technical and social impact, such an effect was not true for the Analytical Engine, which was never completed as a functional prototype. For this reason, while the daguerreotype, as the first modern media device designed for the reproduction of reality, had an immediate societal resonance, the transformative potential of the computer came to light with a significant delay (Manovich, 2001, p.21). Photography is one of the earliest examples that laid the groundwork for what we now call new media. As Lev Manovich emphasizes, with the invention of the daguerreotype in the 1830s, photography emerged as the first powerful tool demonstrating that reality could be reproduced by technical means. This technology quickly transcended being a mere invention and found immediate social and cultural resonance, radically altering the ways in which visual information was produced and circulated.

After years of experimentation, Daguerre succeeded in 1837 in permanently fixing the image produced in the *camera obscura*. However, he quickly moved beyond photographing architectural structures and began capturing life itself (Hacking, 2015, p.23). On August 19, 1839, the Institut de France in Paris witnessed an extraordinary crowd composed of intellectuals and artists of the time. The reason for this overwhelming interest was the official public announcement of a new image production technique known as the *Daguerreotype*, developed by Louis Daguerre. Already known in artistic circles for his diorama work, Daguerre's invention was regarded as a transformative development in visual culture. Soon after the announcement, optical shops in Paris were flooded with amateurs trying to purchase daguerreotype devices. Cameras were pointed in all directions, and people lined up to capture views from their windows. Those who succeeded in capturing silhouettes of rooftops against the sky in their first attempts considered themselves successful adopters of this new technology. Thus began the first global wave of technical enthusiasm and social curiosity in the history of visual media. The daguerreotype technique, which originated in France, quickly garnered great interest in both Europe and America. Initially used to capture architectural structures and natural landscapes, public interest soon expanded; with technical advancements, portrait studios proliferated rapidly within two years, and the desire for self-representation through this new imaging technology spread across the masses (Manovich, 2001, p.22). An equally significant development as the daguerreotype was the invention of the negative-positive method, which made photography reproducible. In 1840, William Henry Fox Talbot used a chemical bath in his *calotype* process to obtain negatives—this process became one of the foundational stages of photography. Talbot's method also contributed to the advancement of printing techniques, turning visual information into a universal form of communication (Hacking, 2015, pp.21–25). The invention of photography was not merely a technical breakthrough but the beginning of a profound transformation in visual culture. Daguerre's ability to fix the image in 1837 and soon bring it into the streets and everyday life indicates that photography was embraced not just as a tool but as a form of representation. The excitement generated by the 1839 announcement and the rapid spread of the daguerreotype reveal society's intense interest in this new mode of image production. With Talbot's negative-positive technique enabling mass reproduction, this transformation reached the public at large. These developments clearly demonstrate that photography became a cornerstone of individual identity, collective memory, and cultural production.

Alongside these developments, it was assumed in the nineteenth century that painting initially preceded photography, later inspired it, and eventually perceived it as a threat. In the twenty-first century, the digital form of photography—wired, instantaneous, automated, manipulable, and a component of a broader multimedia environment—has developed a more distanced relationship with film-based and chemically processed photography that preceded it. When one considers all these developments, it becomes evident that no single, monolithic form of photography ever existed at any one time. From the earliest periods in history, a multitude of photographic strategies emerged from phantasmagoric "spirit" photographs to mugshots. Photography has been used in utilitarian ways for documentation, in exalted ways to create art, and as a vibrant hybrid of the two. Despite varying approaches, photography has paradoxically gained prestige as both a reliable source and a subjective, interpretive medium. Ultimately, photography has served as a mediator of personal and social experience (Ritchin, 2012, p.19).

As Manovich also notes, the avant-garde aesthetic techniques of the 1920s have regained functionality today through their integration with digital technologies and have become the foundational elements of new media aesthetics (Manovich, 2002 cited in Tokdil, 2018, p.182). In the 1920s, traditional rules of perspective, conventional modes of seeing, and systematic compositional elements were replaced by radical formal experiments. Among the key representatives of this formal transformation were Lászl6 Moholy-Nagy and Aleksandr Rodchenko. Particularly in his photomontage works, Moholy-Nagy developed an abstract and experimental visual language. These visuals created a sense of forward and backward motion, producing a transient sense of depth in the viewer. This approach presents a model of representation resembling the non-spatial yet infinitely expandable windows of contemporary computer screens (Tokdil, 2018, p.182). Rodchenko's photographs, on the other hand, featured compositions with unconventional framing that challenged visual perception. As Arnheim states, “the boundaries of a film or visual frame are quite limited when compared to the range of human vision” (Arnheim, 1997 cited in Tokdil, 2018, p.182). To produce a visual experience akin to human perception, Rodchenko organized sequential images and moved the camera axis. This technique laid the groundwork for the approaches adopted by avant-garde filmmakers of the 1920s (Tokdil, 2018, p.183). Moholy-Nagy and Rodchenko carried these techniques into the fields of photography and photomontage, creating a dynamic visuality that allowed viewers to perceive not only the single frame but also its potential continuation. Their interventions in compositional angles opened the door to an aesthetic that invited the viewer to mentally complete the entire scene, thus placing these artists among the pioneering figures of modern visual culture (Tokdil, 2018, p.183). By the 1990s, the image strategies underpinning the concept of the “new vision” had been transformed into standard software techniques through advancements in computer technology. While a figure situated in a physical space—such as a room—can only be observed from a certain angle, in virtual environments this figure can be rotated on its axis via a computer mouse, enabling it to be fully defined from all perspectives. In this context, the visual effects created by avant-garde artists such as Rodchenko have evolved into technical functions integrated into digital software (Tokdil, 2018, p.183).

In the mid-twentieth century, the development of digital computers capable of high-efficiency data processing began to replace earlier information-processing tools such as mechanical calculators and tabulating machines with more advanced systems. During the same period, media technologies that allowed for the recording of visual and auditory information—such as photographic plates, film reels, and audio recordings—became widespread. The point at which these two separate trajectories converged was the digitization of all media forms. In other words, images, moving visuals, sounds, texts, and even spatial structures have become computable, editable, and reproducible within digital environments. The resulting structure is what we now call new media. New media should not be understood solely as a technical development but rather as a multi-layered transformation that affects all dimensions of cultural communication. This structure, in which the computer functions not only as a tool but also as the very carrier of cultural production, constitutes a foundational reference point for understanding today’s media environment (Manovich, 2001, p.20).

Within the evolution of digital art—classified under electronic art and renamed in accordance with emerging technologies—Computer Art occupies a significant position. In the 1990s, the term "Digital Art" began to replace "Computer Art." By the 2000s, the terms "Digital Art" or "New Media Art" started to be used alongside "Computer Art" with increasing frequency. With the acceleration of digitalization and rapid change, the term "post-digital" has emerged in contemporary discourse (Taylor, 2014 cited in Tuğal, 2018, p.70). As Tribe also states, the term "new media art" today either replaces or offers an alternative framework to previously used terms such as "digital art," "computer art," "multimedia art," and "interactive art" (Tribe, 2006 cited in Özel Sağlamtimur, 2017, p.86). In addition, genres such as "net art," "software art," and "pixel art" are considered contemporary forms of new media art, reflecting the diversified modes of expression characteristic of this artistic approach (Özel Sağlamtimur, 2017, p.86). Post-Internet, meanwhile, refers to art created in relation to the conditions of the digital age. Its main themes include the role of photography in a networked culture and the ways in which the internet has shaped our understanding of what photography is (Lewis, 2018, p.140).

With the integration of new media into contemporary human life, it has increasingly come to be regarded as a medium for cultural and artistic production. Positioned at the intersection of technological developments and modes of artistic expression, new media art offers artists a space of freedom and experimentation. This space functions not only as an aesthetic platform but also as an interactive arena where direct communication with the audience is made possible. In this context, a dynamic and multi-layered network is established between the artist, the viewer, and the artwork; the viewer is transformed from a mere observer into an active subject participating in the production process (Özel Sağlamtimur, 2017, p.86). New media art is defined as a contemporary art form that draws from various disciplines and synthesizes their methodologies to generate current modes of artistic production. Through the use of diverse technological tools and techniques—including digital media, sound art, computer graphics, virtual reality, internet art, robotics, video, interactive ap plications, and computer animations—it clearly diverges from traditional visual art practices (Korkmaz, 2015 cited in Özel Sağlamtimur, 2017, p.86). The development of digital art has evolved not only technologically but also conceptually, moving from computer art to new media art. Today, terms such as "post-digital" and "post-internet" suggest that digitalization is not merely an instrument for artistic practice but also an intellectual foundation. This transformation has opened new avenues for expression within visual culture, especially altering how photography is perceived.

Witnessing, participating in, and academically studying the new media environment shaped by the rise of digital visuality entails a highly complex intellectual process. This process contains significant areas of tension in relation to visual culture, human interaction, social relations, and active participation in everyday life. The affordances of digital technologies have influenced camera-mediated sociality in far more complex and varied ways than previously imagined. In this context, the capturing of both still and moving images has evolved—from DSLR cameras to compact cameras, then to tablets and smartphones. Simultaneously, social media platforms such as Facebook, Twitter, and Instagram have emerged as the dominant channels through which user-generated and found images are globally circulated. Improvements in internet bandwidth and the proliferation of wireless connectivity technologies have not only enabled digital communication but have made it a routine component of daily life. Within this framework, the role of photography in global communication networks has been radically transformed (Scott, 2020, p.7).

With the digital revolution, the digitization of photography and the widespread adoption of this technology at a mass level have radically altered the role of visual imagery in everyday communication. Photography has evolved into a multi-layered mode of communication that generates meaning on individual, local, national, and global scales (Scott, 2020, p.18). This transformation is largely shaped by the historical progression of photographic technologies and the creative adaptation of these technologies by photographers and artists. Photographers have employed this medium not only to document what is visible but also to convey what is felt—capturing not only the personal but also the social and political dimensions of experience. The immediacy of production and sharing offered by photography has facilitated the emergence of new visual regimes both technically and aesthetically. Each new technological step has prompted photographers to rethink not only their tools but also their photographic gaze (Scott, 2020, p.18). In conclusion, the digital revolution has fundamentally altered the function of photography, transforming it into a multi-layered communicative tool that goes far beyond personal documentation. Photography now captures not only moments but also emotions, social issues, and political positions. With every technological advancement, photographers are compelled to redefine themselves not only technically but also aesthetically—revealing photography as a continuously evolving and living form of expression.

Sample Works

In the contemporary period, the use of technology within the art scene has, at times, transformed into a form of critique for artists. Just as postmodernism witnessed the appropriation of mass media imagery as a means of questioning its influence, Post-Internet artists adopt digital media aesthetics to expose the pervasiveness of these visual cultures. Stock images, corporate branding, emojis, GIFs, and memes; the personalization of Tumblr, Instagram, and Facebook pages—all of these play a role in shaping this approach (Lewis, 2018, p.141). Kate Steciw, Constant Dullaart, and the DIS collective have all appropriated stock images in their work. They draw attention to the omnipresence of such imagery in everyday life and investigate how these visuals, when recontextualized within the domain of fine art, acquire new cultural and economic values (Lewis, 2018, p.141). What is often appropriated is someone else's photograph. This constitutes one of the postmodern strategies that digital culture deems its right. Numerous photographers have raised concerns not only about the unauthorized use of their photos but also about the modifications made to them (Ritchin, 2012, p.32). Some artists reconstruct historically significant photographs by reconfiguring or embedding visuals within one another. For example, Polish artist Zbigniew Libera transforms iconic images of suffering—such as those of concentration camps, the napalm-scarred Vietnamese girl, or Che Guevara's death—into scenes depicting smiling people behind barbed wire (Ritchin, 2012, p.33).

Contemporary artists have also employed a strategy of highlighting the distinction between the ephemeral nature of online images and the physical photograph as a tangible, labor-intensive object. This has expanded the notion that photographic works must exist in flat, two-dimensional forms, leading to the prominence of photo-based sculptural practices as seen in the work of artists such as Artie Vierkant and Anouk Kruithof (Lewis, 2018, p.141). Other artists have preferred to work directly on the Internet. Among the pioneers in this field, Dutch artist Loan Heemskerk and Belgian artist Dirk Paesmans created the Jodi art collective in the 1990s. Works such as "http://globalmove.us/" (2008) exemplify guerrilla art, filled with humorous animations, witty and constantly changing URLs, and satirical images critiquing everything from the software industry to capitalism (King, 2017, p.549). This era marks a transition in photographic action from a practice that required a seer and a seen, the distillation and creation of aura, and the intuitive mind of the photographer—toward one of ubiquitous, instantaneous communication. It is a form now commonly produced by mobile phones, PDAs, webcams, and satellites (Ritchin, 2012, p.18). Furthermore, many Post-Internet works reflect anxiety about the overwhelming volume of images freely accessible online. Penelope Umbrico's photographic installations, sourced from the internet, serve as a prime example, emphasizing the usage and mass image production by online photo communities (Lewis, 2018, p.141).

Penelope Umbrico explains her project "Sunset" in detail in an interview with Jonathan Blaustein. She states that the project began when she found 541,795 sunset photographs on the image-hosting site Flickr by searching the word "sunset." Umbrico created her work by cropping only the suns from these photos and uploading them to Kodak to produce 4" x 6" machine prints. Her key selection criterion was that the sun had to be decontextualized, completely isolated from any Earth-bound interference—resulting in images devoid of subjects. In these cropped images, only the sun is depicted, and there are no references to the location or position of the photographer. She also prioritized iconic images of sunsets, reflecting collective ideas of what a sunset looks like. Specifically, Umbrico sought the familiar orange-yellow hues of the sun hovering just above the horizon. She explains her motivation behind the "Sunset" project as follows: "There is a single, life-giving, hot sun in the sky. In 2006, in this digital, electronic, blue-lit virtual space, I found 554,000 images of the Sun in a single search. That, in itself, was strange, uncanny, and something I couldn't ignore." The "why" behind this project, she notes, relates to a kind of exploration into what it means to be digital, to exist in the world as a number, to experience a sense of subjectless existence within the context of a consumer-oriented web. She observes that in the act of creating, sharing, and consuming images, the more one shares their own visuals, the less presence they seem to have in the world—suggesting that sharing is a manifestation of anxiety or fear of disappearance. For her, the project was about erasing the individual. While the photographs signify the photographer's intended position, Umbrico removes the subjectivity she finds in these images, thus marking the unintentional and unconscious participation of subjects in a collective process. In the second part of the project, where people appear in front of the sunsets, the notion of individual agency is further problematized. Here, because the camera exposes for the sun, the human subject is erased. As one witnesses thousands of silhouetted figures, the individual disappears. When encountering a large installation of this work, one may even identify with the lost figures—because one has, in fact, occupied that very position (Blaustein, 2016).

However, Umbrico states that she has never been interested in going outside to take photographs. Instead, she has always been concerned with the deceptive spaces we create for ourselves as a culture (Blaustein, 2016). As Ritchin (2012, p.181) notes, contemporary photographers are excited about discovering entirely new paths.

Since the 19th century, photography has served as both a tool for self-expression and as an instrument embedded within a complex social structure—a strategy not only for enlightenment and affirmation, but also for disruption and subversion. However, the medium has transformed significantly from what it once was. As Peter Plagens remarked in *Newsweek* at the end of 2007, "The next generation of great photographers will have to find a way to reassert photography's special connection with reality. And they'll have to do it in a whole new way" (Ritchin, 2012, p.181). Moreover, the photographic image today has become a widespread and democratic means of communication. While analog imaging techniques remain part of the past, digital image production has introduced an unprecedented intensity and velocity in both quantitative and qualitative terms within the historical evolution of the photographic medium. This transformation has progressed so rapidly that it has become increasingly difficult to reimagine the role of photography in the contemporary digital landscape (Scott, 2020, p.7). Yet, as previously discussed, Umbrico stands out as one of the artists who has critically engaged with this radical shift. She has transformed the innovations brought forth by technology into new forms of artistic expression, incorporating them into her work.

Umbrico believes that all art, to a certain extent, is or should be transgressive. She accepts this as a given and is unbothered by it. What compels her to produce work are the things that affect her—those that force her to question their presence. She believes that, regardless of whether the material is transgressive, she must engage with it on the internet, because the digital realm poses questions she cannot ignore. She describes this impulse as follows: "I think, 'What is going on here?' Then I have to look closer. And once I see that there's something going on with the images that wasn't their original intent, I absolutely feel entitled to use them" (Blaustein, 2016). While some photographers today continue to adhere to rigid and traditional practices, Umbrico views the data-rich world of digital media as a resource—perhaps even a right—to be accessed and utilized. Additionally, Ritchin's (2012, p.30) reflection—"When I was a university student, watching an exposed sheet of paper mysteriously transform into a photograph in a chemical bath seemed magical, like a form of alchemy. Now, in the digital age, the photograph doesn't emerge later; the magic lies in the transformation of what already exists"—is a particularly poignant and insightful observation. It underscores the vast and unprecedented opportunities introduced by the remarkable tools and applications of this new era.

One of the other notable names utilizing the tools of this new era is Erica Scourti. Like Umbrico, Scourti benefits from large data networks. She uses tools such as Google's "Search by Image" function as part of her artistic process to explore how one can preserve individuality amid the abundance of personal snapshots published online (Lewis, 2018, p.141). Another example, French internet artist Christophe Bruno, deals with issues of language. In his project titled "The Google Adwords Happening," Bruno purchased certain keywords from Google's search engine and ensured that users who searched those terms would encounter nonsensical poems in the keyword advertisement section. Although Google blocked the artist on the very day he began publishing the poems, Bruno recorded and shared the incident on his website. Today, the internet has become a powerful tool for artists (King, 2017, p.549).

Moreover, all alternative initiatives emerging from the social and cultural environment strongly support the formation of such ideas. The various stages of global circulation, communication, and interaction—whether individual or collective—effectively obstruct, or even eliminate, the persistence of oppressive and authoritarian ideologies in the international sphere. Additionally, a revitalized cultural consciousness aligned with contemporary times has already increased the use of new technologies and novel perspectives. For this reason, new media not only invades the public sphere but also adds numerous dimensions to the production and dissemination of contemporary art (Girgin, 2018, p.220). In this context, some artists have produced works that directly critique media culture. For instance, Cindy Sherman's "History Portrait" series reinterprets women commodified and objectified in

Western society by reversing the media's representational codes. The 35 photographs in "History Portrait" are notable for their postmodern reinterpretation of canonical works by European artists that have stood the test of time (Girgin, 2018, p.62). According to Manovich, new media objects imply that users' choices are unique and individual. These technologies, in a way, function as a form of compensation for their previous roles in standardizing individuals, now working to convince them of their uniqueness. The digital successors of Jacquard's loom, Hollerith's statistical tabulator, and Zuse's cinematic computer now operate to affirm our individual differences (Manovich, 2001, p.42). New media has become a powerful agent of cultural, artistic, and social transformation. While weakening authoritarian structures, global communication has expanded the space for individual expression. Artists have responded to this transformation with critical and creative interventions, such as Sherman's reversal of media imagery to question societal norms. As Manovich emphasizes, new media offers an alternative to the homogenizing systems of the past by promising users a sense of individuality and uniqueness. This demonstrates that new media provides a multilayered space of liberation, both in terms of personal identity and artistic production.

Gerard Rancinan also reflects Sherman's critique of media culture in his work. In his reinterpretation of "The Maids of Honor," Rancinan adheres to the painting's composition while transforming its meaning. This work, intended to entertain, is produced through imitation in the context of intertextuality. In his version, the noble family depicted in Velázquez's painting is replaced by elements of popular culture. By incorporating Marilyn Monroe—an icon of popular culture—the artist references the fetishized image of the female figure and contemporary social life. The work presents a complex structure where religion, taboos, entertainment, popular culture, nobility, branding, advertising, and media intertwine. A figure standing next to Monroe wearing a dress made of newspapers alludes to media manipulation, while a D&G belt to the right, Converse shoes on the left, and a dollar-sign necklace around Monroe's neck clearly point to consumer culture (Girgin, 2018, p.203).

Within this data-driven world, everyone has now become a content and data producer. Scott (2020) raises the critical question of our time: "Where exactly are we?" Often, the response is: "We are all photographers now." However, this statement warrants a more nuanced exploration: Today, we are not only photographers, but also image creators and publishers. To share, send, or publish an image on an online platform is essentially to engage in a form of personal publishing. This development not only transforms the function of the photographer, but also endows the visual producer with greater authority and responsibility. Key discussions in contemporary visual communication revolve around the context in which images are shared, the meanings ascribed by the platforms where they are disseminated, and the expectations of those who publish them. Accordingly, the skills demanded of today's photographers have diversified significantly; they are no longer limited to aesthetic sensitivity or technical expertise alone (Scott, 2020, p.18).

Finally, Ritchin's (2012, p.11) assessment of this new era is noteworthy. He argues that photography, by freezing and fragmenting the visible, has become a central player in depicting reality and— as many critics have noted—subtly distorting our perception of it. As one of the most universal elements of media, photography also serves as an expanding lens through which to examine the shift from analog to digital. Although the divide between pre-digital and post-digital media continues to grow, the moments where they radically diverge or intersect have given rise to new assumptions about the nature of existence itself. The history of media does not follow a clear, linear narrative of progression. Rather, it is a chaotic interplay of rising and fading forms, of possibilities that both converge and diverge. Each medium filters the world through its own unique characteristics, and in doing so, the multiplicity of meanings often becomes lost in the dominant vortex of the message. Digital media, however, is particularly well-suited for offering a cubist-like, multi-perspective outlook—an ability to see in all directions simultaneously. If developed appropriately, new imaging strategies emerging from the digital revolution could transcend many of analog photography's limitations while prompting novel approaches. The humanistic potentials of emerging media should be carefully and swiftly maximized—especially considering the extensive efforts by governmental and commercial interests to exploit or domesticate these technologies (Ritchin, 2012, p.11). According to Ritchin, in the digital age, photography has evolved into a powerful language—one that fragments, transforms, and even directs its subject. Digital innovations not only introduce new modes of expression but also offer a chance to amplify the humanistic and creative capacities of photography.

CONCLUSION

With the integration of new media technologies into artistic production processes, art and photography have transcended their classical boundaries, transforming into more flexible, interactive, and experience-oriented structures. The multiple perspectives offered by digitalization have repositioned visual production not merely as a form of representation, but also as a space for questioning and intervention. In this transformation, photography has moved beyond its function as a mere document, becoming a medium for cultural, political, and personal narratives; users, in turn, have become active agents in the process—not only as viewers, but also as producers and distributors. The aesthetics of new media necessitate a rethinking of traditional artistic paradigms, while contemporary art practices shaped by digital technologies provide a framework that supports interdisciplinary collaboration and experimental approaches. Moreover, concepts such as surveillance capitalism, data-driven production, and digital identity construction have rendered art a subject of socio-political interpretation, rather than solely an aesthetic concern. This study reveals the transformative role of new media technologies in artistic production and aims to foster a critical awareness of the creative possibilities that this transformation may enable in the future.

REFERENCES

Acar, A. (2017). *Sessiz Sinema Tarihi*. Doruk.

Altun, F. (2022) Geleneksel Medyadan Yeni Medyaya Propaganda: Süreklilik ve Değişim. Mevzu: Sosyal Bilimler Dergisi, 7 (Mart 2022): 1-22.)

Ayaz, F. (2024). Instagram'da Büyümek: Takipçi Etkinliklerine Yönelik Karma Bir Analiz. *Hitit Sosyal Bilimler Dergisi*, *17*(2), 303–330. DOI: 10.17218/hititsbd.1437098

Blaustein, J. (2016) Jonathan Blaustein, Penelope Umbrico Röportajı, aphotoeditor, https://www.aphotoeditor.com/2016/04/27/penelope-umbrico-interview/

Ercansungur, D. O. & Çetin, E. (2023). Popüler Kültür ve Dijitalleşme: Yeni Medya Ekosisteminde Yakınsama ve Kitle Üzerine Sosyolojik Bir Analiz . İnsan ve Toplum Bilimleri Araştırmaları Dergisi, 12 (2), 565-584 . https://doi.org/ DOI: 10.15869/itobiad.1215377

Erdem, S. (2023). Görsel Okuryazarlığının İşaret Ettiğine Dair: Görsel İletişim Çağında Bilinçli Medya Tüketimi. In Bayraktar, R. (Ed.), *Medya Okuryazarlığı Analizleri*. Eğitim Yayınevi.

Freund, G. (2008). *Fotoğraf ve Toplum*. Sel Yayıncılık.

Girgin, F. (2018). *Çağdaş Sanat ve Yeniden Üretim Alıntı, Öykünme, Kolaj, Taklit*. Hayalperest.

Hacking, J. (2015). *Photography The Whole Story*. Thames&Hudson.

Kalaman, S. (2019). Yeni Medya ve Dijital Gözetim: Türkiye'deki Sosyal Medya Kullanıcıları Üzerine Bir Araştırma, Yönetim ve Ekonomi, Yıl:2019, Cilt:26. *Sayı*, *2*, 575–594.

Kazan, F. (2023). İlk Yerli Dijital İnsan Örneği: Aypera. *RumeliDe Dil ve Edebiyat Arastirmalari Dergisi*, (34), 819–838. DOI: 10.29000/rumelide.1316233

Kılıç, L. (2012). *Fotoğraf ve Sinemanın Toplumsal Tarihi*. Dost Kitapevi.

King, C. (2017). In Farthing, S. (Ed.), *Dijital Sanat, Sanatın Tüm Öyküsü* (pp. 548–549).

Koçyiğit, A., & Darı, A. (2023). *"Yapay Zekâ İletişiminde ChatGPT: İnsanlaşan Dijitalleşmenin Geleceği", Stratejik ve Sosyal Araştırmalar Dergisi, C.7*. Temmuz.

Lewis, E. (2018). *İzmler Fotoğrafı Anlamak*. Hayalperest.

Manovich, L. (2001). *The Language of New Media.* MIT Press.

Özdemir, M., & Şeyma, Y. (2020, January). Medya ve Gerçeklik İlişkisini Küresel Biyopolitik Dijital Çağda Yeniden Düşünmek. *Erciyes İletişim Dergisi, 7*(1), 507–532. DOI: 10.17680/erciyesiletisim.541652

Özdoyran, G. (2019). Yeni Medya'ya Yönelik 'Yeni' Yaklaşımlar: Yeni Medya ve Post-Yapısalcılık/Postmodernizm. *Etkileşim., 4*(4), 186–218. DOI: 10.32739/etkilesim.2019.4.70

Özel Sağlamtimur, Z. (2017). Yeni Medya Sanatı ve Fotoğraf. *Sanat Ve Tasarım Dergisi, 7*(2), 82–100. DOI: 10.20488/www-std-anadolu-edu-tr.393543

Özkemahlı, K. K., & Gençcelep, B. (2024). Konstrüktivizmin Fotoğraf Sanatına Etkisi. In Bulat, M., & Daşkesen, H. (Eds.), *Sanatsal Araştırmalar II.* Detay Yayıncılık.

Öztürk, G., & Dayanç Kıyat, G. B., Yılmaz Altuntaş, E. (2024) Halkla İlişkiler ve Reklam Perspektifinden Yaratıcılığın Geleceği, Turkish Review of Communication Studies, Year: 2024, Issue: Special Issue, pp. e148-171 • ISSN: 2630-6220 • DOI: DOI: 10.17829/turcom.1360999

Ritchin, F. (2012). *Fotoğraftan Sonra.* Espas Yayınları.

Scott, G. (2020). *New Ways of Seeing, The Democratic Language of Photography.* Routledge. DOI: 10.4324/9781003103639

Sönmezer, Z., & Büyükbaykal Ilgaz, A. C. (2024). Medya ve Teknolojinin Kolektif Kullanmı. *The Turkish Online Journal of Design Art and Communication, 14*(3), 753–762. DOI: 10.7456/tojdac.1464676

Temel, F., & Önürmen, O. (2017, March).. . *Gelenekselin Dönüşümünde İletişim Akademisyenleri ve Sosyal Medya, E-Gifder, Volume, 5*(1), 292–313.

Tokdil, E. (2018) Yeni Medya, Temel Bileşenleri ve Sanatın Değişen Estetik Dili, İletişim Kuram ve Araştırma Dergisi - Sayı 47 /2018, pp.167-190

Tuğal, S. A. (2018). *Oluşum Süreci İçinde Dijital Sanat.* Hayalperest.

Yaşar, İ. H. (2020). s.192). İletişim tarihine kısa bir bakış: Analogdan dijitale iletişim araçlarının gelişimi ve bireye yansıması. *Ufkun Ötesi Bilim Dergisi, 20*(1), 183–207.

Yetişkin, E. (2012). *Sular Artık Tersine Akıyor: Yeni Medya Sanatı Nasıl Okunabilir? Art Unlimited.* Temmuz.

ADDITIONAL READING

Azoulay, A. (2008). *The civil contract of photography*. Zone Books.

Batchen, G. (1997). *Burning with desire: The conception of photography*. MIT Press.

Benjamin, W. (2008). *The work of art in the age of mechanical reproduction.*

Borden, I., Rendell, J., & Pivaro, A. (2000). *The Unknown City: Contesting Architecture and Social Space*. MIT Press. DOI: 10.7551/mitpress/7008.001.0001

Burgin, V. (1982). *Thinking photography*. Macmillan. DOI: 10.1007/978-1-349-16716-6

Kracauer, S. (1995). *The mass ornament: Weimar essays*. Harvard University Press.

Pink, S. (2007). *Doing visual ethnography: Images, media and representation in research*. SAGE Publications. DOI: 10.4135/9780857025029

Sekula, A. (1984). *Photography Against the Grain: Essays and Photo Works 1973–1983*. Nova Scotia College of Art and Design.

Sontag, S. (2001). *On photography*. Picador.

Wells, L. (Ed.). (2021). *Photography: A critical introduction* (6th ed.). Routledge. DOI: 10.4324/9780429274183

KEY TERMS AND DEFINITIONS

Aesthetic Representation: Describes the way photographers frame urban elements with artistic intent. This includes composition, perspective, and visual storytelling to convey the city's beauty or complexity.

Documentary Photography: A photographic genre dedicated to truthfully capturing social realities and urban life. It serves both as an archive and a critical commentary on societal conditions.

Ideological Function of Photography: Recognizes that photography is not neutral—it can reinforce, question, or subvert ideological structures. What is photographed and how it is framed carries implicit cultural and political meanings.

Modernity and Space: Addresses the transformation of urban environments in the context of modernization. Through photography, shifts in architecture, infrastructure, and spatial planning become visually traceable.

Social Critique: Photography's capacity to highlight urban inequality, marginalization, or transformation. It positions the photographer as a conscious observer who reflects or challenges dominant narratives.

Street Photography: Focuses on candid moments in public urban spaces. It emphasizes the dynamic interaction between people and their surroundings, capturing everyday life in its spontaneous form.

Urban Memory: Refers to the collective memory embedded in a city's historical structures, spatial transformations, and lived experiences. Photography plays a central role in preserving and visualizing this memory over time.

Chapter 2
Virtual Reality and the New Dimensions of Visual Media

Serhat Erdem
https://orcid.org/0000-0002-3782-0147
Atatürk Üniversitesi, Turkey

Gülgin Altuğ
https://orcid.org/0000-0002-7581-3036
Erzincan Binali Yıldırım Üniversitesi, Turkey

ABSTRACT

This chapter aims to examine how technologies create new dimensions in media production at the intersection of virtual reality and visual media. The rise of virtual reality (VR) technologies is transforming visual media experiences, offering innovative interaction spaces for both content creators and viewers. The goal of the study is to explore how these technologies are changing the effects on visual media, media perception, and audience engagement, and to discuss VR's potential impact on the future of media. The methodology will involve a review based on technology-based content analyses and media theories, along with a literature review on the effects of VR experiences on media consumers. The study will focus on the use of virtual reality applications in the media industry, with limitations that only certain VR platforms and media formats will be analyzed, excluding other media forms that reach broader audiences.

DOI: 10.4018/979-8-3373-1967-4.ch002

INTRODUCTION

In modern societies, people obtain almost all information about their environment and the world through the media, apart from basic daily life practices, and internalize information about facts and events as much as and in the way they appear in the media (Akmeşe, 2020, p.38). The media provide individuals with various experiences to rationally comprehend individual, social, national and international issues and to make the necessary decisions (Yurdigül & Yurdigül, 2014, p.122).

Transformations in visual media technologies have added a new dimension to traditional media such as cinema and television with virtual reality. Virtual reality is an environment that offers a 360° experience by immersing the viewer in the content, thus radically changing visual storytelling. Historically speaking, cinema's transition from silent to sound, from black-and-white to color, and from two to three dimensions has always been part of a quest to offer the audience a "more real" experience. Indeed, as Lev Manovich points out, from the invention of photography in the 1840s to the emergence of virtual reality in the 1980s, every new visual media technology has come up with the claim of representing reality more directly (Manovich, 2001). The current resurgence in popularity of virtual reality continues in this vein, but the interactivity and spatial depth offered by this technology creates new aesthetic and narrative possibilities that are different from traditional media (Manovich, 2001). Virtual Production is the digitalization of traditional production techniques. With this technology, physically built set environments are replaced by virtual set environments designed in a computer environment. Virtual set environments are created in predetermined virtual spaces. The basic components are LED walls and ceilings, unreal engine, motion capture, digital environment and 3D design (Çelik & Bolat, 2025).

Marshall McLuhan's famous expression "The medium is the message" gains a special meaning in the age of virtual reality. According to McLuhan, communication tools transform society not only with the content they transmit but also with the sensory/perceptual environment they create. As a matter of fact, virtual reality-based media, also called spiral media, completely integrate the audience with the message transmission environment, making communication highly personal and impressive (Arıkan, 2023). In McLuhan's distinction between "hot" and "cold" media, virtual reality can be considered an example *of "hot" media* with its high-resolution, multi-sensory structure that encompasses the perception of the participant (Ashley, 2015). On the other hand, visual culture theorists such as W.J.T. Mitchell discuss how images redefine the perception of reality in today's digital visual environments (Mitchell, 2005). The concept of *"pictorial turn"* put forward by Mitchell can be said to have reached its climax with virtual reality: Images are no longer merely representations, but are transformed into spaces that are experienced through virtual reality.

Media theorist Jean Baudrillard's simulation theory is also frequently used to analyze the phenomenon of virtual reality. Baudrillard argues that reality in modern society has been replaced by images and simulations and that today is a phase of *"hyperreality"*. According to Baudrillard, the age of simulation we are living in is an "age of virtual realities" and in this age, image and reality have become intertwined and the "real" has completely disappeared (Gençcelep & Özkemahlı, 2023, p. 3063; Kara, 2022). As a matter of fact, Fahrettin Altun, the Presidential Communications Director Fahrettin Altun emphasized at the TRT International Metaverse and Broadcasting Forum held in Istanbul in 2022 that "image and reality have disappeared in the universe of simulation" and that the representation of truth has become impossible (Kara, 2022). This approach points to the crisis of representation created by virtual reality technologies: The boundaries between the real and the virtual are blurred to such an extent that the possibility of "a reality in which reality is lost" as a result of virtual reality experiences comes up for discussion (Kara, 2022). Baudrillard's words of nearly forty years ago that "in this age, ideas no longer have the possibility of representing reality" remain relevant as a strong criticism of the world of virtual reality and the *metaverse*.

The effects of virtual reality in cinema and digital video are rewriting the rules of visual narrative. The new narrative form called cinematic virtual reality (cine-VR) allows the viewer to experience the story from the inside by moving the viewer's perspective beyond the fixed frame. In this case, camera angles, editing and framing, which are the basic elements of traditional cinematic language, are replaced by the viewer's 360° free gaze. The director now has to perform the task of directing attention not through methods such as cuts or close-ups, but indirectly through the light, sound and spatial arrangement of the virtual reality environment (Kobylinski & Pochwatko, 2020). This change transforms the viewer from a passive to an active participant; the viewer becomes the "center of the environment". Famous directors have not remained indifferent to this potential: For example, the Oscar-winning Alejandro G. Iñárritu, with his virtual reality project *Carne y Arena*, which he presented at the Cannes Film Festival in 2017, took the story of migrants to a dimension that the audience could experience first-hand (Ross & Munt, 2018). Similarly, Kathryn Bigelow transformed the experience of watching a documentary by positioning the viewer next to rangers traveling through an African national park in her 360° documentary-short film *The Protectors* (2016). These examples indicate that virtual reality is not only a new technique in cinema, but also a new narrative form (Ross & Munt, 2018). "Immersion" and "presence" become more important in VR than ever before in the traditional film experience; as the viewer's sensory presence in the space increases, their emotional engagement with the story deepens (Roth, 2016).

The television and digital publishing sectors have also begun to explore the possibilities offered by virtual reality. Especially in news and documentary broadcasting,

360° video and virtual reality applications give viewers the opportunity to get inside the events. For example, in 2015, the New York Times launched a virtual reality journalism experience by distributing Google Cardboard to its readers, presenting news topics ranging from war zones to refugee camps in virtual reality. In sports broadcasting, international networks are also experimenting with virtual reality applications that allow multi-directional viewing of major events (Olympics, World Cup, etc.). Thus, the viewer can, for example, experience a football match as if they were in a virtual stadium or feel as if they were standing side by side with the artist on the stage of a concert. Digital platforms have also started to adapt to virtual reality: 360° streaming video, interactive virtual documentaries and virtual short films are being added to online content libraries. These developments are transforming the traditional practice of watching television and evolving towards a more interactive and personalized media experience. In Turkey, there are also remarkable initiatives in this area. The academic journal *"Virtual Reality and Media"* published by TRT in 2023 brought theoretical and practical discussions to the agenda by scrutinizing the communication and broadcasting dimension of virtual reality (Arıkan, 2023). In addition, in international forums organized by TRT, the role that virtual reality will play in the future of broadcasting in the axis of Web 3.0 and metaverse is discussed (Kara, 2022). In this context, the rapid popularization of the metaverse shows that virtual reality technologies have started to be seen not only for entertainment or narrative purposes, but also as a social interaction and broadcasting infrastructure. As a matter of fact, Altun (2022) states that the metaverse offers a new vision of the future of the internet and is considered as a "new world vision" on a global scale (Kara, 2022). This situation reveals that broadcasting organizations have started to evaluate virtual reality as a strategic tool.

In the Turkish context, the use of virtual reality applications in the cultural sphere is also becoming increasingly widespread. For example, Müzeverse, which opened in Istanbul in November 2024, was introduced as Turkey's first virtual reality museum and offered visitors the opportunity to experience historical sites in the form of an unusual, interactive virtual tour (Limon, 2025). In Müzeverse, visitors can "virtually travel" to the Pyramids of Ancient Egypt or medieval castles with virtual reality glasses and experience the narratives of the past as if they were there. The founders of the project emphasized that this concept "blurs the boundary between reality and the virtual" and stated that virtual reality technology is a step into the future of storytelling (Limon, 2025). Similarly, universities and independent studios have begun to produce new content at the intersection of virtual reality and cinema-television. For example, virtual reality categories have been established in some short film festivals, and young Turkish producers have started to appear in the international arena with 360° video and interactive virtual reality films. These developments show that visual media in Turkey is adapting to virtual reality in tandem with the digital transformation of the world and that a creative ecosystem is emerging in this field.

Virtual reality technology expands the boundaries of visual media and introduces a new paradigm of experience. Cinema, television and digital publishing, together with virtual reality, have the potential to become media that are experienced, not just watched. In the following sections of the study, the historical background and theoretical framework of this transformation will be examined; the place of virtual reality in the media ecology will be discussed from the perspectives of thinkers such as McLuhan, Baudrillard, Manovich and Mitchell. In addition, how virtual reality reshapes cinematographic narrative, what kind of innovations it brings to television broadcasting and its current applications in Turkey will be discussed. Virtual reality and the new dimensions of visual media are not only a technological innovation but also a cultural and aesthetic transformation. In order to understand this transformation, it is necessary both to develop a historical perspective in the light of past experiences and to gain a critical perspective through current examples and theoretical approaches.

THEORETICAL BACKGROUND and LITERATURE REVIEW

Virtual Reality Concept and Historical Development

Virtual reality (VR) is generally defined as interactive three-dimensional artificial environments created by computers, where the user experiences the feeling of being in a real space (Ferhat, 2016). In these environments, special equipment (e.g. head-mounted displays, data gloves) is used to provide multi-sensory inputs and allow the user to interact with simulated scenes in real time. The term *"virtual reality"* was first used in the literature in the late 1980s by Jaron Lanier, one of the pioneers of virtual reality technology (Ferhat, 2016). This concept is also referred to by different terms such as artificial reality, cyberspace or virtual environment. Virtual reality is a simulation environment created with software and special hardware that gives people a perception of reality (Ferhat, 2016). This world that participants experience, witness and interact with is characterized as the virtual world (Çakın & Yaman, 2020).

The development of virtual reality technology is based on several important scientific and technical milestones. Pioneering initiatives include Morton Heilig's multi-sensory cinema experience Sensorama in 1962 and Ivan Sutherland's "Ultimate Display" concept in 1965 (Altınbay & Gümüş, 2020). Sutherland developed the first computer graphics-based head-mounted display prototype (HMD) with his student in 1968 (Basu, 2019). This system presented stereo images with a device worn on the user's head, realistically updating the visual scene according to head movement. In the literature, this device is referred to as the "Sword of Damocles" because it hangs from the ceiling and is considered the beginning of modern virtual reality (Basu, 2019). In the 1980s, VPL Research, founded by Jaron Lanier and his team, developed the first commercial virtual reality equipment (Data Glove, EyePhone headset, etc.), thus introducing the concept of "virtual reality" to the masses (Basu, 2019). Throughout the 1990s, the concept progressed from academic research and military simulations to interactive games and room-scale projection systems such as CAVE. For example, the CAVE system developed at the University of Illinois in 1992 offered a multi-user virtual space experience with 3D images projected onto walls. With the introduction of cost-effective headsets such as the Oculus Rift in the 2010s, virtual reality technology gained popularity again, and immersive experiences became widespread in many fields such as education, health, design and entertainment (Basu, 2019).

Transformation of the Concept of Visual Media

Visual media has undergone a transformation from analog technologies such as photography and cinema to digital imaging and interactive screens. In the analog era, images were captured and transmitted on physical film or printed material; in the digital age, images are stored and transmitted in digital codes (bits) (Özkemahlı & Er, 2024, p. 126). New media theorists emphasize that digitalization has radically changed visuality. For example, Lev Manovich (2001) states that the most fundamental feature of new media is *digital representation*; that is, all the contents of traditional media (text, image, sound, etc.) are recoded as a series of 0s and 1s in the computer environment (Bayer & Bulut Özek, 2021). Thanks to this digitization, media content can be flexibly processed in a modular structure, copied and instantly reproduced over networks. As early as the 1960s, Marshall McLuhan stated that electronic media expanded the central nervous system of human beings across the world, almost eliminating spatial and temporal barriers (McLuhan, 1964). McLuhan's thesis *that "the medium is an extension of man"* has been reaffirmed, especially in the age of the internet and smart devices, as visual communication has become instantaneous and interactive in a global village.

Digitalization has created a "visual explosion" in our culture. W.J.T. Mitchell identifies a sharp shift in modern society from a print-centered culture to a visual culture and explains this shift with the concept of *"pictorial turn"* (Aldemir, 2022). Mitchell (1994) emphasizes that, especially since the second half of the 20th century, visual simulation and illusion techniques have gained unprecedented power with the opportunities provided by technology (Aldemir, 2022). Digital images no longer merely mimic reality, but have become so effective that they create a world of meaning and impact in their own right. Indeed, today, social media streams, television, video games and augmented reality applications have placed the visual image at the center of life. These developments have also transformed the concept of visual perception (visuality): Images have ceased to be mere representations of reality and have become the very subject of everyday experience. As Mitchell puts it, in the digital age, images have acquired "unprecedented power" and have even begun to produce simulations powerful enough to replace reality (Aldemir, 2022).

Another important theorist thinking about new visual media is Vilém Flusser. Flusser characterizes the emergence of technical images such as photography and video as a revolution in the history of human communication. According to him, images produced by technical devices are qualitatively different *from* previous *traditional images* (e.g. painting, drawing) and give rise to a new form of consciousness in society (Kara, 2019). Flusser (1985) argues that the age of digital technical images symbolizes a break from the linear thinking system based on text and creates a brand-new cultural situation (Kara, 2019). In this new era, meanings are transmitted through images and codes on the screen instead of texts that are read line by line. According to Flusser, technical images are "reprogramming" culture by radically changing the way we experience and express the world. As a matter of fact, today, cameras, computers and smartphones have popularized individuals' practice of *"visual thinking"*; people now communicate through photographs and videos instead of words. In this context, McLuhan, Flusser and Mitchell share a common point: The form of media (e.g. digital visual media) is an active factor transforming the way we think and perceive (Canán, 2008). Lev Manovich, on the other hand, argues that new media assimilate traditional forms of visual media such as cinema and television and bring them into an interactive and software environment (Manovich, 2001). New media principles such as *modularity, variability and automation* identified by Manovich explain the flexible and adaptable nature of digital visuality (Bayer & Bulut Özek, 2021). The transition from analog to digital is not only a technical evolution but has also changed the theoretical framework of visual culture; media theorists have developed new concepts and approaches to make sense of this transformation.

Virtual reality and Media Theories

The media dimension of virtual reality can be analyzed in depth in the light of classical media theories. McLuhan's basic thesis that *"the medium is an extension of man"* is extremely accurate for virtual reality. Marshall McLuhan argues that each new media redefines the individual's relationship with the world by expanding human senses and perception capacity (McLuhan, 1964). As a matter of fact, virtual reality encompasses multiple senses, especially sight and hearing, and brings the user's body and senses into a digital space. In this respect, virtual reality is a step towards the "technological simulation of consciousness" as predicted by McLuhan (McLuhan, 1964). McLuhan stated that the final stage of electronic communication will be the simulation of consciousness where all humanity unites in a common experience (McLuhan, 1964); today, virtual reality technologies can be considered as a medium that brings this prophecy closer to reality. In other words, virtual reality offers the user a new space-time experience by embodying the concept of *"extension of the senses"* discussed in media theory.

On the other hand, Jean Baudrillard's theory of *simulation* provides a critical framework for understanding the social and philosophical implications of virtual reality. Baudrillard argues that in modern society, the media *"eviscerates reality"* by replacing truth with copies (Ferhat, 2016). According to him, our age lives in a simulation universe called *hyperreality*, where reality and image are replaced by each other (Baudrillard, 1981). Virtual reality is a technological reflection of this simulation universe described by Baudrillard: Real-world spaces and situations are reproduced with the contribution of computer graphics and artificial interactions and presented to the user *"as if they were real"*. According to Baudrillard, images and virtual signifiers can be powerful enough to eventually replace reality; they even run the risk of "killing" reality (Baudrillard, 1981). Indeed, in virtual reality experiences, one can forget the sensations of the physical world and immerse oneself in a completely artificial environment. Using the example of Disneyland, Baudrillard explains how simulacra (images that are not real but seem real) encompass all reference systems (Baudrillard, 1981). In this context, virtual reality, as Baudrillard warns, is one of the most extreme media extensions that can lead to the loss of *the principle of reality* (Baudrillard, 1981). From the perspective of simulation theory, virtual reality experiences may result in the loss of the sense of truth and the weakening of the ability to distinguish between the real and the virtual. As a result, hyperreality may become a part of everyday life through virtual reality and individuals may find fulfillment in simulated experiences without feeling the lack of *the "real thing"*.

The interactive nature of virtual reality and its transformation of the viewer-user position also overlaps with new media audience theories. While the audience is more passive in traditional mass communication, virtual reality makes the user an active participant. Therefore, virtual reality, as one of the most extreme examples of the concept of interactive media, embodies the phenomenon of *active participation* defined in audience theories. Communication scientist Henry Jenkins states that in the age of digital networks, audiences now form a participatory culture that participates in content production and meaning construction (Uyanık, 2023). In virtual reality experiences, the user is a subject who can partially determine the course of the story or content and interacts with the environment through physical actions.

This is explained in media theory with concepts such as two- way communication, player experience and audience participation. For example, presence, an important concept in virtual reality, defines the depth of users' interaction with the media. Research has shown that increasing the level of *presence* in VR increases the user's emotional engagement and attention (Interaction Design Foundation, 2023). The more intensely the user experiences the feeling of "being there" when wearing the virtual reality goggles, the more strongly they are involved in the experience and the immersion effect with the media is reinforced (Interaction Design Foundation, 2023). This phenomenon shows that the concept of *the audience* in the classical sense has gone beyond the concept of *the audience*, and that the individual is now a part of the experience. New media theorists such as Janet Murray emphasize that interactive environments allow users to actively participate in the narrative world by giving them a sense of *freedom* and *agency*. Virtual reality, on the other hand, maximizes this interaction and makes the viewer a kind of "co-creator". Virtual reality stands at a concrete intersection of theoretical debates ranging from McLuhan to Baudrillard: On the one hand, it shows how the media can expand the limits of human senses and perception of reality, and on the other hand, how it can blur the line between reality and simulation. For the viewer, virtual reality is a platform that accelerates the transition from passive reception to active participation and is a striking example that supports theories of interaction and participation. These multifaceted theoretical perspectives help us to situate the phenomenon of virtual reality within both technology and media ecology.

VIRTUAL REALITY APPLICATIONS and VISUAL MEDIA EXPERIENCE

Cinema and Virtual Reality

Virtual reality cinema transforms the viewer from a passive observer to an active participant in the story (FILMD, 2024). The narrative that is conveyed in one direction in traditional cinema is transformed into a 360-degree spatial narrative in virtual reality; the viewer can explore different aspects of the scene by turning his head or changing his position (C-Istudios, 2024). In this way, the flow of the story is not limited to a fixed frame, and a multifaceted narrative is possible. For example, the Oscar-winning virtual reality film *Carne y Arena* (2017) offered an in-depth experience in which the viewer could walk in the middle of the desert, putting them in the shoes of illegal migrants crossing the US-Mexico border (Oscars, 2017). This work by Alejandro G. Iñárritu was considered a milestone in the history of cinema as the first virtual reality film to take part in the Cannes Film Festival and was awarded a special prize by the Academy for its innovative narrative that "opens new doors in cinematic perception" (Oscars, 2017).

The use of physical space and 360° cameras in virtual reality cinema allows the viewer to physically experience the narrative. This immersive experience confirms Marshall McLuhan's thesis that "the medium is the message"; the media technology used radically changes the form of the narrative and its impact on the audience (Spampinato & Catricalà, 2021). Indeed, according to McLuhan, communication tools shape us as our extensions (Sunal, 2016). The virtual reality movie deepens the emotional impact by creating an environment where image and sound surround the viewer. The audience feels as if they are experiencing the events rather than watching a movie. In the context of Jean Baudrillard's simulation theory, virtual reality cinema can create a hyperreality experience by intertwining reality and fiction; when the image becomes effective enough to replace reality, it brings it closer to the situation described by the expression "reality has been expelled from reality" (Sunal, 2016). As W.J.T. Mitchell states in his "pictorial turn" theory, the central role of images has increased in modern culture; virtual reality films take this dominance of visual media to a new dimension and create a new cinematic language that takes the audience into the story (Spampinato & Catricalà, 2021). The combination of cinema and virtual reality transforms the viewer from a spectator sitting in his or her seat to a participant traveling through the space of the story, making narrative structures interactive and spatial (C-Istudios, 2024).

Use of Virtual Reality in Television and Broadcasting

In the field of visual broadcasting, virtual reality technologies are transforming the presentation of news, sports and documentary content. In television journalism, a trend called "immersive journalism" aims to involve the viewer in the news. For example, in a project supported by TRT in Turkey, the critical moments of the July 15, 2016 coup attempt were made experienceable through 360° video and virtual reality glasses. A special truck traveled to different provinces and offered citizens the opportunity to relive the night of the coup attempt through virtual reality (TRT, 2017). Wearing virtual reality goggles, viewers found themselves on the July 15th Martyrs' Bridge and witnessed the historical event by seeing what was happening around them in 3D with real images. Such applications transform news from a content that is merely watched into an event that is experienced. Similarly, important steps have been taken in international broadcasting. For example, in the UK, BT Sport broadcast Premier League football matches with multiple 360° camera systems with 8K resolution, giving viewers the feeling of being live in the stadium (Grotticelli, 2021). Through the mobile app, viewers were able to virtually visit normally inaccessible areas such as the locker room and tunnel by choosing between various angles. Thus, the narrative in sports broadcasting has evolved from the flow dictated by a single camera to a personalized experience shaped according to the viewer's preference (Grotticelli, 2021).

The use of virtual reality in documentaries and news programs requires the reconstruction of narrative structures. The time and space weave is designed to be flexible to fit the viewer's interactions; for example, in virtual reality documentaries, the flow of the story is designed to guide the user as they explore the scene. This means a shift from linear narrative to a multi-sequential and interactive structure. The traditional hierarchy between the narrator and the viewer is also broken in this way. The viewer is now like a co-narrator who partially controls the course of the news or program. As media theorist John V. Pavlik emphasizes with his concept of "experiential media", virtual reality-based broadcasts radically change the storytelling of journalism by focusing on making the story experienced rather than communicating it (Pavlik, 2020). This transformation reveals the power of media to transform the receiver as McLuhan predicted. From the perspective of Baudrillard's "universe of simulation", virtual real news and broadcasting practices construct worlds of representation that are convincing enough to replace real events. As a matter of fact, in a well- designed virtual real news experience, the viewer can forget that he/she is in front of the screen and feel himself/herself at the scene of the event; this creates a subjective experience that is completely different from the passive viewing attitude in traditional broadcasting. The use of virtual reality in television and broadcasting redefines the production and consumption of all visual content, from news to sports, and expands the narrative boundaries by transforming the viewer into an active experiencer.

Digital Games and Virtual Reality Technologies

The digital gaming industry is one of the most intense and innovative applications of virtual reality technologies. Modern virtual reality games are created with the help of high-performance game engines. Game engines such as Unity and Unreal Engine make it possible to create immersive virtual worlds with real-time 3D graphics and physics interactions. As a matter of fact, Izmir-based The Pack Studios from Turkey has attracted attention in this field with its game projects such as *Ephesus* and *Quicksand*, developed using Unreal Engine 5. Such local studios make use of advanced visualization tools to bring historical places and unique stories into VR-supported gaming experiences. In VR games, real-time interaction is essential: players are able to penetrate the game world with body movements as well as controllers, grasp and lift objects, and physically interact with virtual characters (Çakın & Yaman, 2022). For example, in a virtual reality adventure game, the player can illuminate a dark cave with a virtual lantern while simultaneously hearing the surrounding sounds directionally. The concept of avatar in games has gained a new dimension with virtual reality; players are embodied in the virtual environment through three-dimensional avatars representing themselves, which embodies the experience of identity and self in the virtual world. The interaction of multiple avatars in multi-user virtual reality games creates a social virtual reality experience, enabling players to perform tasks or compete together.

Space design is also an integral part of virtual reality games. While in traditional digital games the camera angles are determined by the game designer, in virtual reality the environment needs to be designed consistently and in detail from all angles as the user looks freely. The geometry, scale and visual style of the virtual environment are designed to support the player's sense of presence. In this context, the phenomenon of "navigable space" mentioned by Lev Manovich in his theory of digital media is fully realized in virtual reality games. Manovich stated that digital cinema has almost turned into a sub-genre of painting by approaching animation and interactive environments (Sunal, 2016). Indeed, the visual aesthetics seen in virtual reality games is a combination of cinematic details and interactive animation; the player not only encounters high quality graphics as if watching a movie but also can move freely in this world. The logic of visual aesthetics and immersive media is applied to these games through techniques that allow the player to remain in the world without distraction. For example, game interfaces are designed to be as diegetic as possible, with information such as a health bar or map displayed as a device on the player's wrist or as a screen hanging on the wall, so that the user retains the feeling that they are not seeing a real-world HUD (head-up display). Sound design is as important as visuals, and 3D audio techniques are used to create a spatial sound atmosphere that changes according to the player's position. All these elements aim to keep the player in the virtual world without damaging the sense of reality within an immersive media logic. McLuhan's proposition that "tools shape us" is also valid in this field; virtual reality gaming technologies create a new gaming culture by transforming both the expectations and gaming practices of players (Sunal, 2016). In addition, Virtual Reality/Augmented Reality laboratories established in technology centers such as METU Teknokent in Turkey support local initiatives in this field and develop a wide range of projects from education to design. For example, animation and game development (ATOM) centers and creative technopark companies at METU are working on innovative virtual reality applications ranging from wearable devices to metaverse platforms. All these developments show that digital games are taking the visual media experience to a new level through virtual reality. Virtual reality- enabled digital games offer an immersive gaming experience with real-time interaction, avatars, and rich audiovisual design elements that redefine traditional gaming formulas both aesthetically and narratively, placing the player at the center of the virtual world.

Recoding Visual Perception

Virtual reality radically alters traditional paradigms of visual representation and recodes visual perception. For example, the fixed and singular point of view of classical linear perspective, which has been dominant since the Renaissance, loses its validity in virtual environments; the user no longer looks from outside a frame, but plunges directly into the virtual world. William J. Mitchell defines this as "the erosion of the boundary drawn by the edge of the computer screen" and states that in virtual reality the viewer "can immerse himself in simulated environments instead of looking through a small window" (Mitchell, 2005). Thus, the user becomes a participant, an "insider"; no longer a passive spectator (Mitchell, 2005). The perception of depth also gains a new dimension in virtual reality environments. While in traditional 2D media, depth is created only through illusion, in these environments, stereoscopic double images and head tracking create a realistic sense of spatial depth. For this reason, Lev Manovich states that with the advent of computer vision, "perspective vision reached its apogee and simultaneously began to recede". Perspective is no longer the only method of representing digital reality, but only one of a myriad of visual tools. McLuhan's theory that the media is an extension of the human senses suggests that virtual reality extends the human perceptual experience with the new visual paradigms it offers, transforming the "message" of the media into a transformed perceptual experience (McLuhan, 1964); this view is in line with W. J. T. Mitchell's concept of the "visual turn" (Mitchell, 2005). Virtual reality expands the boundaries of visual communication by transforming the image from mere representation into direct experience.

The 360° immersive visuality of virtual environments requires different strategies for directing the viewer's attention than traditional cinematic aesthetics. While film directors can determine where the audience looks with framing and editing techniques, in virtual reality the user's field of view is completely free and there are visual elements in every direction. In order to ensure that critical details are noticed in this visual density, producers are developing new methods of directing attention. For example, experienced virtual reality storytellers draw attention to important elements with cues of sound, light, or movement, while spreading the visual composition across a frameless 3D space to provide the user with an experience that makes them feel like they are really "there" in the world. Janet

H. Murray emphasizes that the embodied visitor should be at the center of virtual reality design instead of the camera's frame, and states that non-narrative elements such as external voice narration or on-screen text should not be used, and that everything the user hears and sees should be part of the virtual world (Murray, 2016). As a matter of fact, spatial sound arrangement stands out as an important guidance tool in virtual reality: For example, when a sound comes from the left or from behind, it draws the user's gaze in that direction, functioning almost like an "invisible director" (Murray, 2016).

In addition, virtual reality allows the user to directly control the point of view; the viewer can experience the scene from any angle in 3D space. This liberation of the gaze ends the monopoly of the fixed camera perspective and offers the user an active exploration of the space. The viewer ceases to be an observer and becomes an actor in the experience, constantly reconstructing the scene they see by moving around the space and approaching objects. Virtual reality transforms visual perception from an external presentation into a dynamic experience shaped by user action.

Narrative Forms and Media Aesthetics

By breaking the linear structure of storytelling, virtual reality has created new narrative forms that can branch according to user interaction, as opposed to the single-stream stories of classical cinema. While in classical cinema, the story follows a single predetermined flow, in virtual reality experiences, narratives often have branching structures that are guided by user interaction. In such user-centered narratives, the story is experienced through interaction and exploration rather than a fixed sequence. Lev Manovich, with his concept of "navigable space" in digital media, emphasizes that moving around in space has become as important an element in narrative as time; digital games and virtual reality experiences, on the other hand, resembling the space-traveling hero stories of antiquity, continue the plot through spatial progression rather than character development. Therefore, virtual reality transforms the narrative from a single-line narrative to a multi-part experience that the user freely explores.

The user's guiding role is at the forefront in the narrative aesthetics of virtual reality. Rather than being fixed, the structure of the story world is designed to be flexible to adapt to the user's actions and choices. Janet H. Murray (2016) states that "the sense of agency experienced by the user is the most important design value" in interactive experiences in digital media design and argues that virtual reality should follow the game model rather than cinema in this respect (Murray, 2016).

Editing (montage), one of the basic elements of traditional cinema language, is also rethought with virtual reality. Since sudden scene cuts or harsh camera angle changes in virtual reality can create discomfort and disconnection in the user, softer and more immanent transitions are used in story transmission. As Murray states, scene changes that are "not triggered by the user's conscious movement" should be minimized in the virtual reality experience, and instead natural transitions that coincide with user actions should be preferred (Murray, 2016).

Empathy and Emotional Impact in Virtual Reality

Virtual reality experiences stand out for their potential to evoke empathy and generate intense emotional responses in the user. Virtual reality has been described as "the ultimate empathy machine" (The Guardian, 2016); it is thought that a carefully designed virtual reality experience can "teleport" the viewer into another person's point of view, creating a deep emotional connection. For example, *Clouds Over Sidra*, the United Nations' 2015 eight-minute virtual reality documentary on the Syrian refugee crisis, immersed the viewer in the life of a 12-year-old refugee girl and helped raise much more than expected (about 70% more than expected) in a subsequent fundraising campaign (Guardian, 2016).

On the other hand, the claim that virtual reality alone will automatically produce empathy has been criticized. Janet H. Murray emphasizes that the empathy effect in virtual reality does not occur spontaneously, but depends on skillful storytelling, just like in literature or cinema (Murray, 2016). To deeply move the user, the virtual experience itself is not enough; well-chosen human details and strong compositional techniques are required (Murray, 2016). In virtual reality-based social documentaries, the limits of presenting real suffering in a virtual environment and the oversimplifying aspects of the "empathy machine" discourse are debated. In line with the principles of "affective design", designers are trying to produce experiences that both evoke empathy and do not traumatize the user.

The emotional impact of VR is not limited to empathy; it can also offer experiences of unparalleled intensity. *Notes on Blindness* is a good example: This virtual reality movie, which conveys the world of a person who has lost his sight through his audio diaries, transports the viewer to a "world beyond sight". Using blurred, silhouetted images and rich spatial sound instead of sharp visual details, it creates a striking experience of how virtual reality can be perceived without sight. Designers who have analyzed this production have noted that VR's ability to turn off external stimuli and focus attention entirely on the story makes such a profound experience of empathy possible (Playfull, 2021). The physical envelopment of the virtual reality environment can also intensify the sensory experience and strengthen emotional responses: For example, in a height simulation, the user may feel a real sense of fear, or be unable to hold back tears at the end of a touching scene (Guardian, 2016). The unique power of virtual reality to mobilize emotions requires ethical responsibilities and careful consideration of long-term impacts.

ANALYSES ON PRACTICAL EXAMPLES

Analyzing Successful International Projects

Notes on Blindness: Into Darkness; This interactive virtual reality documentary (2016) conveys the experience of a visually impaired individual through a six-part narrative structure. *Notes on Blindness: Into Darkness* is based on the audio diaries of John Hull during the process of losing his sight, offering an immersion into a "world beyond sight" through visual minimalism and rich spatial sound design. In 2016, the work won the Tribeca Film Festival's Storyscapes award and the Sheffield Doc/Fest Virtual Reality Award, and was recognized by critics as one of the "best virtual reality works of 2016" (Filmmaker, 2016). Indeed, one critic described it as "an astonishing work of immersive cinema", noting that putting on the VR headset and "virtually falling into the world beyond sight" almost brought him to tears (Filmmaker, 2016). *Notes on Blindness* has thus become a striking example that goes beyond the traditional documentary and demonstrates the power of virtual reality technology to create empathy.

Clouds Over Sidra, an eight-minute virtual reality experience produced by the United Nations in collaboration with Chris Milk and Gabo Arora to draw attention to the Syrian refugee crisis in 2015, takes the viewer to the Zaatari Refugee Camp in Jordan with 12-year-old Sidra. *Clouds Over Sidra* uses 360-degree images of a real refugee camp to create a sense of being there. As the narrative conveys Sidra's daily life through her eyes, the idea that virtual reality can be "the ultimate empathy machine" has often been exemplified. Indeed, after showing this film at a donor conference in Davos, a total of USD 3.8 billion was raised, 70% more than expected (Robertson, 2016). According to UNICEF, one in six people who watched the movie pledged to donate to charity, more than twice the rate of traditional methods (Robertson, 2016). According to media reports, viewers were so moved by the story that some were moved to tears as soon as they saw the title and donated to help more refugee children (Lu, 2019). *Clouds Over Sidra* has been cited worldwide as a successful example of the potential of documentary to build empathy.

The Key; *The Key*, which won the Storyscapes Award at the Tribeca Film Festival in 2019, is a virtual reality narrative directed by Celine Tricart that draws attention with its metaphorical connection to reality. This fifteen-minute experience draws the viewer in with dream sequences and a mysterious narrative in the beginning and unexpectedly reveals a strong message about the refugee phenomenon in the finale (Ha, 2019). The unique structure of the narrative is based on the viewer first exploring the dream world of an anonymous narrator and then connecting it to a compelling situation in the real world (the refugee experience). What is unique about *The* Key is that it is not only limited to the virtual environment, but is also supported by a physical space: Viewers begin the experience by entering a room-sized fictional space and interacting with an actor (Ha, 2019). Participants are then immersed in the story through the Oculus virtual reality headset, and experience an immersive adventure enriched with interactive elements. This multi-layered structure of the work blurs the boundaries between narrative and experience and offers a successful analysis of how an impressive storytelling can be constructed in a virtual reality environment.

Virtual Reality Media Applications from Turkey

In Turkey, virtual reality technology has been used in various fields by both large institutions and independent producers in recent years. Although the national public broadcaster TRT has not yet produced extensive virtual reality content productions, it has taken developments in this field into its agenda and has started to support them. For example, the Communication Presidency of the Republic of Turkey produced a three-dimensional virtual reality video titled "The Victory of the Nation on July 15th" as part of the commemoration of the fourth anniversary of the July 15th coup attempt (Tosun, 2020). This 360° digital reality experience, produced in Turkish and English, was presented to the public with virtual reality glasses specially printed by the institution and was experienced with virtual reality cartons distributed to the audience in digital screening centers (Tosun, 2020). This study is an example of the use of virtual reality for public communication in Turkey. Similarly, TRT and other corporate media outlets broadcast programs that draw attention to the potential of virtual reality in educational and cultural themes and raise awareness about the use of technology. The use of virtual reality is increasing, especially in the field of education; in fact, for the first time in Turkey, an undergraduate interior architecture course at a university (Beykent University, 2022) started to be taught with virtual reality glasses. In this way, students can virtually navigate through their designs and the effectiveness of applied courses is increased. The Ministry of Education has also implemented pilot projects such as virtual reality-supported language classes and laboratory experiments in some high schools.

Looking at the private sector and independent studios, it is seen that virtual reality content production in Turkey has sprouted especially in the fields of documentary and cultural heritage. A new generation of directors such as Deniz Tortum have started to use virtual reality media as a creative narrative tool. For example, *Selyatağı VR* (2018), directed by Tortum, stands out as a local and independent 360° virtual reality film adapted from Emre Yeksan's *Yuva* and screened at many international festivals (Erkılıç & Dönmez, 2016). Set in a natural forest area and dealing with themes of ecological destruction, *Selyatağı VR* combines cinematographic narrative and virtual reality experience to present an innovative example of a virtual reality film from Turkey. On the cultural content side, museums and art institutions are embracing VR.

The virtual reality experience *Journey into Osman Hamdi Bey's World*, which Pera Museum offered to its visitors in 2017, made the famous painter's studio and the world of the Turtle Trainer painting virtually navigable. By digitizing a museum collection and transforming it into an interactive story, this project set an example for the interpretation of cultural heritage with new generation technologies.

More recently, the Müzeverse initiative, touted as Turkey's first virtual reality museum, has attracted attention. Founded in Istanbul, this museum offers multi-sensory history experiences to its visitors in an area of 900 square meters with thematic content such as *the Pyramids of Ancient Egypt* and *the Medieval Castle* (Aksoy, 2025). The founders of the Müzeverse project emphasized that they have welcomed 30,000 visitors since the opening and that the content they have produced has "aroused great curiosity" (Aksoy, 2025). This interest points to a rising demand for virtual reality-based cultural content in Turkey. As a result, both in public communication through TRT and official institutions and in educational, documentary and cultural projects of independent producers, virtual reality media is slowly finding its place in Turkey and continues to develop with exemplary practices.

User Experience and Audience Feedback

The user experience in virtual reality-based visual media is quite different from traditional screen- based content and often has a more intense level of interactivity. As viewers experience a sense of presence in VR, attention and emotional engagement levels can be heightened. In successful international examples, the depth of audience engagement is often emphasized. For example, many viewers who experienced *Clouds Over* Sidra reported feeling as if they were in the refugee camp with Sidra and felt a strong emotional connection to the story (Lu, 2019). As a result of this level of connection, many viewers were motivated to help the refugees after the film, which is evidence of the power of narrative to engage (Robertson, 2016). Similarly, *Notes on Blindness: Into Darkness* experience, especially in the section titled "Panic", a critic stated that "I experienced an indescribable anxiety when the voices stopped and you could only hear your own breath" and that the virtual reality environment pulled him into the dark world of John Hull. This experience simultaneously evoked feelings of helplessness and empathy in the viewer, with the author adding that "throughout the episodes, Hull's experience became my own, creating a highly emotional impact". Such qualitative feedback, through press and media commentary, reveals the deep impression that virtual reality content makes on the viewer.

Another point frequently mentioned in audience feedback is that the narrative structure and interactive nature of VR experiences make the duration of engagement remarkable. Although an average VR movie lasts between 5-15 minutes, the audience is involved in the story as an active participant throughout this time. In productions such as *The Key*, the audience's attention is directed not only to the audiovisual elements, but also to the interactive elements, where users become part of the experience through simple tasks such as picking up an object, opening a door, or engaging in physical interaction. This is described in audience feedback as "getting lost in the story" and temporarily disconnected from the real world.

Qualitative evaluations show that viewers who complete virtual reality experiences are often moved "as if they have come back from a much longer journey" and continue to reflect on the theme of the narrative after the experience. For example, after experiencing *The* Key at the Tribeca screening, a journalist emphasized that the surprise ending of the production stayed with him for days and that the message delivered in the virtual reality environment left a lasting impression (Ha, 2019). Likewise, Lauren Wissot, who reviewed *Notes on* Blindness, summarized the impact of this experience by saying that "it is rare to have an emotional connection with virtual reality, but this piece brought me to the point of tears" (Filmmaker, 2016). Press reviews and viewer feedback agree that VR content leaves a strong impression on the viewer and provides more intense feedback than traditional media.

The user experience in virtual reality media applications is characterized by high levels of empathy and engagement. When audience feedback is analyzed, it is seen that it increases emotional interaction thanks to the "feeling of being there" and that viewers often give strong reactions (excitement, fear, joy, curiosity) during the experience. Although quantitative survey data is still limited, qualitative observations and media reviews reveal that the impact of VR on the audience is significant on both emotional and intellectual levels. This shows that virtual reality narratives have the potential to create a new language in the future of visual media by deepening the bond with the audience.

LOOKING TO THE FUTURE POTENTIAL EFFECTS OF VIRTUAL REALITY ON VISUAL MEDIA

New Directions in Media Production

Virtual reality applications, which first emerged with military technology, paved the way for the development of augmented reality applications and today these types have emerged as two separate fields. The most popular augmented reality application in recent years is the game "Pokemon Go" (Yılmaz, 2017, p. 265). The maturation of virtual reality (VR) technology and advances in artificial intelligence (AI) pave the way for groundbreaking new directions in visual media production. AI-powered VR production is notable for its potential to reduce the costs and time of content creation. For example, researchers at Meta have developed a method that generates a realistic 3D VR scene from a single 2D photograph (Wolfenstein, 2025). Such approaches promise democratization by automating the previously highly specialized and labor-intensive design of VR environments. Indeed, IEEE Digital Reality reports that *"3D scene generation from individual photographs can dramatically reduce the cost of immersive experience creation"* (Digitalreality, 2022). AI and VR content is also being used to create intelligent characters and dynamic worlds, and algorithms that have been used in the gaming industry for years are now being integrated into VR experiences for automated world building, route finding and even designing game levels (Digitalreality, 2022).

AI-powered tools have become an integral part of VR production, providing content creators with automated visual storytelling systems. For example, recent research has developed director models that "automatically select the most compelling camera viewpoint using machine learning" in VR environments (Wu, 2024). In this way, during the recording or editing of a virtual reality experience, artificial intelligence can determine the most appropriate angles, alleviating the workload of human directors and contributing to the creation of a coherent visual language.

Another innovative trend is the use of Extended Reality (XR) and augmented content. XR combines VR, Augmented Reality (AR) and Mixed Reality (MR) technologies, enabling the integrated production of the physical and digital world. Especially in the movie and television industry, applications called virtual production create virtual backgrounds in real time with LED screens and game engines. In recent years, investments in this field have accelerated due to the COVID-19 pandemic (Apaydın, 2024). In Turkey, the first XR studios were established and started to be used in movie production (Apaydın, 2024). For example, Zero Density, a local initiative, has achieved international success by providing virtual studio services to more than 70 countries on 6 continents with its real-time graphics engine (Apaydın, 2024). XR-based virtual studio techniques increase production efficiency as they allow directors and cameramen to instantly see and interact with 3D virtual sets that appear in the background during shooting (Apaydın, 2024). In Turkey, companies such as MGX Film have established studios equipped with this technology, while industry professionals are generally satisfied with the high-quality visual effects and realistic set integration offered by XR (Apaydın, 2024). In short, storytelling automated with the support of artificial intelligence and XR-based virtual production techniques usher in a new era in visual media production.

Ethical, Social and Philosophical Debates

The proliferation of virtual reality technology raises important ethical and social debates about the distortion of perceptions of reality and the media's responsibility for accuracy. Research has shown that powerful virtual reality experiences can blur the line between virtual and reality in individuals' minds. For example, in an experiment in 2025, 20% of participants sat on a virtual chair they saw in virtual reality, even though there was no physical support in the real world, and 45% expected to find a virtual tablet placed in a drawer in virtual reality in the same place in the real room a week later (Wiesing & et al, 2025). Moreover, even small discourse differences of a virtual character measurably affected participants' attitudes (Wiesing & et al, 2025). These findings suggest that the illusion of presence created by virtual reality can fool users into thinking that virtual objects are real, and that the impressions gained in virtual reality can be reflected in real-life decisions. Virtual reality is a form of media that, when used in the wrong hands, can expose individuals to perceptual misconceptions and manipulation (Wiesing & et al, 2025). Indeed, philosophers Madary and Metzinger have suggested the development of an ethical code for the use of virtual reality, emphasizing that virtual reality is *"a powerful form of mental and behavioral manipulation"*, which can be particularly problematic if there are commercial, political or religious interests behind it.

In this context, media responsibility and the risk of manipulation becomes a critical agenda item. Since immersive media content evokes a strong sense of reality in the viewer, there is a serious danger of presenting VR fictions as "real" for misinformation or propaganda purposes. For example, in the field of *immersive* journalism, there are warnings *that "the biggest challenge of VR journalism is to prevent malicious individuals and organizations from producing fake VR content and circulating it as real news"*. For this reason, news media and visual content producers are updating their existing journalistic codes of ethics and redefining the principles of transparency, accuracy, and non- psychological harm when using VR. As emphasized in Frontiers in Robotics and AI, journalists should make it clear to the target audience *"what they are really seeing there"* and ethically manage the boundaries of virtual illusion. Experts even recommend further research on the effects of virtual reality content on children and vulnerable groups, and that users should be informed about this, *as "the effects of prolonged exposure to AI on behavior are poorly understood"*.

Another dimension of discussion brought to the agenda by virtual reality is the transformation of the body, identity and self in virtual environments. In virtual reality experiences, users can detach from their own bodies and take on the body of an avatar, or even assume a completely different identity or form. The impact of this "embodiment" experience on individuals' self-identity and behavior has been scientifically studied. Studies show that there can be significant changes in the attitudes and behaviors of people who assume a different body or role in virtual reality. For example, experiencing another ethnic identity or gender in virtual reality can increase empathy and reduce prejudice towards that group (Wiesing & et al, 2025). A narrative review reported that perspective- taking from a virtually embodied object can have a major impact on people's emotional and cognitive responses, even reducing the propensity for violence (Wiesing & et al, 2025). Such findings suggest that the concept of "self" can stretch in virtual reality environments and that digital experiences can shape real-life *self-perception.*

At the philosophical level, the possibility of adopting multiple and fluid identities in virtual worlds is discussed. As Sherry Turkle puts it, online virtual universes offer users the opportunity to experiment with alternative lives and fragmented identities that are not possible in the physical world (Çiftçi & Karakaş, 2019). As a matter of fact, in virtual worlds such as *Second Life*, which became popular in the early 2000s, many people have adopted their presence there as a *"free and alternative way of life"* rather than seeing it as a game (Çiftçi & Karakaş, 2019). This situation is interpreted as the experience of body and identity in virtual reality creating an intermediate space of existence that is disconnected from reality but not entirely fictional. The ability to redefine the body through digital representations challenges traditional assumptions about the question "who am I?" and points to a post-modern experience in which the self can be divided between virtual and real or new hybrid identities are formed (Çiftçi & Karakaş, 2019). Virtual reality triggers intense ethical-philosophical debates on body and identity both at the practical and theoretical levels; in this environment where reality and the virtual are intertwined, the way people perceive themselves and the other is re- evaluated.

CONCLUSION

Thanks to the integration of virtual reality (VR) technology into visual communication design, it is clear that visual media has gained new dimensions. Virtual reality transcends the traditional two- dimensional design language and offers a three-dimensional and even multi-dimensional field of expression, making design works more vivid and interactive. Through virtual reality, viewers can feel as if they are physically present in the content and connect with the visual content in a deeper way than ever before.

Storytelling in these environments is also undergoing a radical transformation. Instead of traditional linear narratives, multi-layered narrative experiences based on audience interaction come to the fore. This immersive form of storytelling, referred to as cinematic VR, takes the viewer out of a passive position and makes them a part of the story. However, this new narrative structure also brings challenges such as the 'narrative paradox'; this tension between the freedom granted to the user and the effort to preserve the integrity of the narrative has been identified as a critical issue that VR content designers must resolve. Issues such as the viewer's fear of missing important events (FOMO) and balancing the level of engagement are emerging as a priority area for future research.

In addition, virtual environments have also opened new horizons in terms of aesthetics; visual communication designers can design multi-sensory experiences that include not only visual but also auditory and tactile elements thanks to virtual reality, thus creating richer and innovative visual narratives by freely reflecting their imagination in three-dimensional spaces without being confined to two-dimensional planes. As a matter of fact, current studies reveal that this technology provides a deeper experience to the viewer by integrating abstract and dynamic content into the design scheme. In addition, virtual reality has become an interdisciplinary field of creation that brings together the disciplines of art, design and technology, bridging fine arts and popular culture. In this way, this new potential not only diversifies forms of aesthetic expression but also redefines the relationship between artist and audience.

All these developments have transformed the viewer experience. VR-based visual communication transforms the viewer from a passive recipient to an active participant, increasing interactivity and emotional engagement. For example, one study reported that VR experiences increased users' level of immersion by approximately 40% and overall satisfaction by 25%. These findings show that this technology has a groundbreaking impact not only in terms of aesthetic and narrative dimensions, but also in terms of audience interaction and satisfaction. These new dimensions offered by virtual reality in visual communication design radically change narrative forms and media experience, strengthening the bond with the audience.

On the other hand, there are important issues that need to be addressed for VR to reach its full potential in visual communication. Issues such as ethical design, accessibility, user privacy, technical standardization and environmental sustainability will play a critical role in the widespread adoption of virtual environments. The future should also consider the cultural and ethical contextualization of virtual content, ensuring user trust and training designers in these new media. Future academic work should focus on developing solutions to these challenges through interdisciplinary collaboration, as well as delving deeper into areas such as VR narrative structures, visual aesthetic norms and audience psychology. Thus, it will be possible to realize the transformational effect of virtual reality in visual communication design in a permanent, inclusive and socially beneficial way.

REFERENCES

Akmeşe, Z. (2020). Televizyon Ana Haber Bültenlerinde Çerçeveleme: Türk Telekom Reklam Filmi Örneği. *Middle Black Sea Journal of Communication Studies*, *5*(1), 37–49.

Aksoy, Ü. (2025). *Müzeverse'nin ikinci içeriği "Son Kale - Orta Çağ'ın Keşfi" ziyaretçilerle buluştu,* https://www.aa.com.tr/tr/kultur/muzeversenin-ikinci-icerigi-son-kale-orta-cagin-kesfi-ziyaretcilerle-bulustu, [Erişim tarihi: 02.06.2025].

Aldemir, A. (2022). Dijital Çağda Fotoğrafik İmge ve Sözün Düşüşü. Intermedia. *International Journal (Toronto, Ont.)*, *9*(17), 192–211. DOI: 10.56133/intermedia.1109495

Altınbay, R., & Gümüş, N. (2020). Sosyal bilgiler öğretmenlerinin sanal tur uygulamalarıyla ilgili görüşleri. *Journal of Innovative Research in Teacher Education*, *1*(1), 60–71. DOI: 10.29329/jirte.2020.321.5

Apaydın, C. (2024). *Film üretiminde gerçek zamanlı görsel efekt kullanımına ilişkin bir araştırma: Genişletilmiş gerçeklik (XR) stüdyoları örneği. (Doktora Tezi)*. Selçuk Üniversitesi, Sosyal Bilimler Enstitüsü.

Arıkan, B. A. (2023). Sanal Gerçeklik ve Medya. *TRT Akademi*, *8*(17), 5–7. DOI: 10.37679/trta.1245474

Ashley, J. (2015). Marshall McLuhan and Understanding Digital Reality, https://www.imaginativeuniversal.com/blog/2015/06/23/marshall-mcluhan-and-understanding-digital-reality, [Erişim Tarihi: 17.04.2025].

Basu, A. (2019). A brief chronology of Virtual Reality, Cornell University arxiv, 1, 2-18.https://doi.org//arXiv.1911.09605.DOI: 10.48550

Baudrillard, J. (1981). *Simülakrlar ve Simülasyon, 1981* (O. Adanır & A. K. Mollaoglu, Çev.). İstanbul: Yüksekistan Yayınları. (Orijinal eser yayımlanma yılı: 1981).

Bayer, H. ve Bulut Özek, M. (2021). Yeni Medya Eğitimi ve Önemi. *Journal of Communication Science Researchs*, *1*(2), 127–138. DOI: 10.7456/100102100/004

C-Istudios. (2024). 360-Degree Video: Immersive Experiences and Virtual Reality, https://c-istudios.com/360-degree-video-immersive-experiences-and-virtual-reality, [Erişim tarihi: 11.04.2025].

Çakın, Ö. ve Yaman, D. (2020). Fijital (Phygital) Pazarlama Uygulamaları Üzerine Bir İnceleme: Amazon Go Örneği. *Middle Black Sea Journal of Communication Studies*, *5*(1), 1–10.

Çakın, Ö., & Yaman, D. (2022). *Dijital Oyunlar- Fijital Deneyim Alanları, Gamevertising Dijital Çağda Oyunlaştırma ve Reklam İçinde* (Nur Erdem, M., Ed.). Eğitim Yayınları.

Canán, A. C. (2008). McLuhan, Flusser, and the Mediatic Approach to Mind. Flusser Studies 6.

Çelik, A. G., & Bolat, N. (2025). The Recreation of Space in Digital Broadcasting: Virtual Production. In Bolat, N. (Ed.), *Innovative Applications and Changing Framework in Digital Broadcasting* (pp. 513–539). IGI Global. DOI: 10.4018/979-8-3373-1957-5.ch019

Çiftçi, A., & Karakaş, Y. (2019). Dijitalleşen Zamanın İzdüşümünde: Kimliğin, Bedenin ve İletişimin Dönüşümü. AJIT-E. *Academic Journal of Information Technology*, *10*(37), 7–30. DOI: 10.5824/1309-1581.2019.2.001.x

Digitalreality, (2022). *Common uses for AI in virtual reality today,* https://digitalreality.ieee.org/publications/ai-in-virtual-reality, [Erişim tarihi: 17.01.2025].

Erkılıç, H., & Dönmez, S. C. (2016). Sanal Gerçeklik Anlatısının İzini Sürmek: Trinity VR ve Selyatağı VR Örnekleri. *SineFilozofi*318-344. DOI: 10.31122/sinefilozofi.674107

Ferhat, S. (2016). Dijital Dünyanın Gerçekliği, Gerçek Dünyanın Sanallığı Bir Dijital Medya Ürünü Olarak Sanal Gerçeklik. *TRT Akademi*, *1*(2), 724–746.

FILMD. (2024). Emerging Filmmaking Technologies: Shaping the Future of Cinema, https://www.filmd.co.uk/articles/emerging-filmmaking-technologies-shaping-the-future-of-cinema, [Erişim tarihi: 15.04.2025].

Filmmaker, (2016). *The Best VR of 2016: Arnaud Colinart and Amaury La Burthe on Notes on Blindness: Into Darkness,* https://filmmakermagazine.com/100779-the-best-vr-of-2016-arnaud-colinart-and-amaury-la-burthe-on-notes-on-blindness-into-darkness, [Erişim tarihi: 09.04.2025].

Gençcelep, B., & Özkemahlı, K. K. (2023). "Gerçek Olandan Simülasyona: Fotoğraf Stüdyosu Eğitiminde "Set.A.Light 3D" Örneği", International Academic Social Resources Journal, (e-ISSN: 2636-7637), Vol:8, Issue:51; pp:3062-3071. DOI: 10.29228/ASRJOURNAL.70969

Grotticelli, M. (2021). *BT Sport's Live VR 360 Coverage Of Premier League Brings Fans Closer To The Action*, https://www.thebroadcastbridge.com/content/entry/17429/bt-sports-live-vr-360-coverage-of-premier-league-brings-fans-closer-to-the, [Erişim tarihi: 17.04.2025].

Ha, A. (2019). *'The Key,' a VR story about dreams and refugees, wins Storyscapes prize at Tribeca*, https://techcrunch.com/2019/05/03/the-key-tribeca-immersive-storyscapes, [Erişim tarihi: 20.04.2025].

Interaction Design Foundation - IxDF. (2023). *What is Presence in Virtual Reality (VR)?* Interaction Design Foundation - IxDF. https://www.interaction-design.org/literature/topics/presence

Kara, K. (2022). *Cumhurbaşkanlığı İletişim Başkanı Altun: Türkiye gibi bir güç, "metaverse"nin inşa sürecinin parçası olmalı.* Anadolu Ajansı (AA) Kültür Haberleri. https://www.aa.com.tr/tr/gundem/cumhurbaskanligi-iletisim-baskani-altun-turkiye-gibi-bir-guc-metaversenin-insa-surecinin-parcasi-olmali. [Erişim tarihi: 19.05.2025].

Kara, O. Ç. (2019). Vilém Flusser'in Soyutlama Modeli ve Sinematografik İmge. *SineFilozofi*221-240. DOI: 10.31122/sinefilozofi.514984

Kobylinski, P., & Pochwatko, G. (2020). Detection of strong ´ and weak moments in cinematic virtual reality narration with the use of 3D eye tracking. *Proceedings of the 13th International Conference on Advances in Computer–Human Interaction*, pp. 280–284.

Limon, Ö. (2025). *Türkiye'nin ilk sanal gerçeklik müzesi "Müzeverse", tarihi alışılmışın dışında keşfetme imkanı sunuyor.* Anadolu Ajansı (AA) Kültür Haberleri. https://www.aa.com.tr/tr/kultur/turkiyenin-ilk-sanal-gerceklik-muzesi-muzeverse-tarihi-alisilmisin-disinda-kesfetme-imkani-sunuyor. [Erişim tarihi: 21.05.2025].

Lu, X. (2019). *"Clouds over Sidra": Empathy and Immersion in VR Storytelling,* https://medium.com/@irenelu728/clouds-over-sidra-empathy-and-immersion-in-vr-storytelling, [Erişim tarihi: 06.02.2025].

Manovich, L. (2001). Reality Media, https://manovich.net/index.php/projects/reality-media, [Erişim Tarihi: 13.04.2025].

McLuhan, M. (1964). *Understanding Media: The Extensions of Man.* McGraw-Hill.

Mitchell, W. J. T. (2005). *What Do Pictures Want?* University of Chicago Press. DOI: 10.7208/chicago/9780226245904.001.0001

Murray, J. H. (2016). *Not a Film and Not an Empathy Machine*, https://immerse.news/not-a-film-and-not-an-empathy-machine, [Erişim tarihi: 19.04.2025].

Oscars (2017). The academy's board of governors awards an Oscar to alejandro g. Iñárrıtu's "carne y arena" vırtual realıty ınstallatıon, https://www.oscars.org/news/academys-board-governors-awards-oscarr-alejandro-g-inarritus-carne-y-arena-virtual-reality, [Erişim tarihi: 26.04.2025].

Özkemahlı, K. K., & Er, M. H. (2024). "Yapay Zekâ Destekli Fotoğrafik Yaklaşım: Portre ve Mekân Temelinde Mustafa Kemal Atatürk Fotoğraflarının Yeniden Üretilmesi". (Ed.), Yağmur, Ö. & Şengünalp, C., Disiplinlerarası Sanat Pratikleri 2. Ankara: Detay Yayıncılık.

Pavlik, J. V. (2020). Drones, Augmented Reality and Virtual Reality Journalism: MappingTheir Role in Immersive News Content [Erişim Tarihi: 28.03.2025]. *Media and Communication, 8*(3), 137–146. DOI: 10.17645/mac.v8i3.3031

Playful, (2021). *Creating Empathy in Notes on Blindness VR,* https://www.playfulrealities.com/2021/01/creating-empathy-in-notes-on-blindness.html, [Erişim tarihi: 03.04.2025].

Robertson, A. (2016). *The UN wants to see how far VR empathy will go,* https://www.theverge.com/2016/9/19/12933874/unvr-clouds-over-sidra-film-app-launch, [Erişim tarihi: 19.05.2025].

Ross, M., & Munt, A. (2018). Cinematic virtual reality: Towards the spatialized screenplay. *Journal of Screenwriting, 9*(2), 191–209. DOI: 10.1386/josc.9.2.191_1

Roth, P. C. H. (2016). *Experiencing Interactive Storytelling.* [PhD-Thesis – Research external, graduation internal, Vrije Universiteit Amsterdam].

Spampinato, F. M., & Catricalà, V. (2021). Contemporary Art and Virtual Reality: New Conditions of Viewership. *Cinergie – Il Cinema E Le Altre Arti, 10*(19), 121–133. DOI: 10.6092/issn.2280-9481/12322

Sunal, G. (2016). Sanal Gerçeklik ve Dijital Sinemanın Olanakları. İnönü Üniversitesi İletişim Fakültesi Elektronik Dergisi (İNİF E-Dergi), 1(2), 294-309.

The Guardian. (2016). *Can tearjerker virtual reality movies tempt donors to give more aid?* https://www.theguardian.com/global-development/2015/dec/31/virtual-reality-movies-aid-humanitarian-assistance-united-nations, [Erişim tarihi: 09.03.2025].

Tosun, M. (2020). *15 Temmuz anma programlarında dijital çalışmalar öne çıktı,* https://www.aa.com.tr/tr/15-temmuz-darbe-girisimi/15-temmuz-anma-programlarinda-dijital-calismalar-one-cikti, [07.04.2025].

TRT. (2017). *"360 derece, karanlık gece" tırı Türkiye turunda,* https://www.trthaber.com/haber/gundem/360-derece-karanlik-gece-tiri-turkiye-turunda, [Erişim tarihi: 13.03.2025].

Uyanık, G. (2023). Dijital medya platformlarını Henry Jenkins'in kavramları üzerinden okumak. *Etkileşim, 11*(11), 432–444. DOI: 10.32739/etkilesim.2023.6.11.204

Wiesing, M., Comadran, G., & Slater, M. (2025). Confusing Virtual Reality with Reality – an Experimental Study. *iScience, 28*(6), 112655. DOI: 10.1016/j.isci.2025.112655 PMID: 40520086

Wolfenstein, K. (2025). *VR gelişiminde AI dönüşümü: Meta, 3D dünyaları tek görüntülerden nasıl çağırıyor,* https://xpert.digital/tr/vr-gelisiminde-ki-donusumu, [Erişim tarihi: 17.04.2025].

Wu, Z, (2024). "Automated Cinematographer for VR Viewing Experiences", *Dartmouth College Master's Theses*. 162.

Yılmaz, R. (2017). Türkiye'de Reklam Anlatısının Dijital Çağdaki Yapısal Dönüşümü. *1st International Conference on New Trends in Communication*. İstanbul Ticaret Universty.

Yurdigül, Y., & Yurdigül, A. (2014). Tv Haberlerinin Anlatı Yapısının Oluşturulması Sürecinde Özel Efekt Teknolojileri: Ntv ve Cnn Türk Ana Haber Bültenleri Üzerinden Bir İnceleme. *İstanbul Üniversitesi İletişim Fakültesi Dergisi | Istanbul University Faculty of Communication Journal*(*46*), 121-146

ADDITIONAL READING

Bolter, J. D., & Grusin, R. (2000). *Remediation: Understanding new media*. MIT Press.

Dorner, R.. (2018). *Virtual and augmented reality: Changing realities in a dynamic world*. Springer.

Flusser, V. (2000). *Towards a philosophy of photography* (Mathews, A., Trans.). Reaktion Books.

Geroimenko, V. (Ed.). (2022). *Augmented reality art: From an emerging technology to a novel creative medium*. Springer.

Grau, O. (2003). *Virtual art: From illusion to immersion*. MIT Press.

Manovich, L. (2001). *The language of new media*. MIT Press.

Murray, J. H. (2017). Hamlet on the holodeck: The future of narrative in cyberspace (Updated ed.). MIT Press.

Papagiannidis, S., & Bourlakis, M. (2021). Virtual reality in retail and service environments: A research agenda. *Journal of Business Research, 136*, 451–460. DOI: 10.1016/j.jbusres.2021.07.030

Rose, M., Taylor, C., & Hull, J. M. (2017). *Notes on blindness: A VR journey into a world beyond sight.* In A. Jones & D. Pennington (Eds.), *Digital immersion and empathy in visual storytelling* (pp. 112–129). Palgrave Macmillan.

Slater, M., & Sanchez-Vives, M. V. (2016). Enhancing our lives with immersive virtual reality. *Frontiers in Robotics and AI, 3*(74), 1–10. DOI: 10.3389/frobt.2016.00074

KEY TERMS AND DEFINITIONS

Aesthetic Experience: Aesthetic experience is not limited to visuality in VR environments; it deepens with sound, movement and spatial arrangements. The user experiences the content both visually and physically. This multi-sensory structure creates a depth that is not possible in traditional media forms. Aesthetic decisions are now integrated with space, movement and the user. This requires a new aesthetic approach in visual communication design.

Empathy: VR is a powerful tool with the potential to bring the user emotionally close to someone else's experience. Especially in documentary genres, the production of empathy becomes an ethical narrative strategy. Examples such as "Clouds Over Sidra" show that this potential can be used effectively. In VR, empathy is not only part of storytelling but also part of social responsibility. In this respect, VR offers a new paradigm in emotion-oriented media design.

Interactivity: Interactivity is one of the fundamental building blocks of VR environments. The ability of users to physically or symbolically relate to the content is at the center of the experience. Through interactivity, media participation takes place instead of media consumption. This engagement can have direct effects on the narrative. The level of interactivity is directly related to audience satisfaction and engagement time.

Narrative Forms: Virtual reality pushes the boundaries of traditional linear narrative structures, creating interactive and multifaceted stories. The viewer ceases to be a passive follower of the narrative and becomes an active participant. This requires a partial transfer of narrative control to the user. Non-linear, multi-perspective narratives create a new visual language. VR narratives therefore go beyond classical theories of cinema and television.

New Media Theories: New media theories provide a theoretical basis for analyzing the cultural, social and individual effects of technologies such as virtual reality. McLuhan's theory of "extensions", Manovich's analysis of digital narrative and Baudrillard's simulation approach are the cornerstones of this field. These theories help to understand VR's epistemological shift in media. They are also functional in explaining the ways in which VR reconstitutes media. The relationship between theory and practice is important in evaluating the impact of these technologies on visual media.

Virtual Reality: Virtual reality is a technology that isolates the user from the physical environment and fully immerses them in digitally created 3D environments. This technology creates a sense of reality by combining visual, auditory and tactile components. It radically transforms media production and experience. VR restructures narrative in cinema, gaming, education and cultural fields. In this context, it is considered not only as a technology but also as a new visual communication language.

Visual Communication Design: Visual communication design is a field of creative expression where messages, images and symbols are conveyed in a meaningful way. When combined with virtual reality, this field moves from two-dimensional surfaces to three-dimensional experience spaces. Aesthetics, user experience and interaction elements become more complex and multi-layered. Visual elements are no longer only seen, but also experienced. This transformation has brought about a shift in paradigms among design disciplines

Chapter 3
New Narrative Structures in Documentary Film in the Process of Digital Development and Transformation

Yusuf Yurdigül
https://orcid.org/0000-0002-9903-4176
Atatürk University, Turkey

Mustafa Gülsün
https://orcid.org/0000-0002-1202-6246
Muş Alparslan, Turkey

ABSTRACT

This study explores the profound impact of digital transformation on contemporary documentary cinema, particularly focusing on how digital technologies have reconfigured traditional narrative structures, production methods, and modes of audience engagement. In an era where media platforms are constantly evolving, documentary film has undergone significant conceptual and structural changes. Once rooted in linear storytelling and a singular authorial voice, the documentary genre is now increasingly characterized by interactivity, hybridity, and nonlinearity. These changes are not merely aesthetic or technological; they reflect a broader cultural and epistemological shift driven by the integration of digital tools into the filmmaking process.

DOI: 10.4018/979-8-3373-1967-4.ch003

INTRODUCTION

The emergence of digital technologies has profoundly shaped the face of documentary filmmaking, introducing new narrative structures that challenge traditional cinematographic narrative methods. Traditionally, documentary film narrative has been shaped by linear narrative structures, often with a narrator's voice and a fairly straightforward chronological progression of themes. However, digital developments have disrupted these narrative structures, causing filmmakers to move towards a non-linear, interactive and multi-perspective narrative. In addition, digital tools such as virtual reality (VR), augmented reality (AR) and interactive platforms offered by new developments not only change filmmakers, but also offer audiences the opportunity to engage with the film in more immersive and participatory ways. For example, VR documentaries allow viewers to navigate virtual environments, creating a sense of reality and action that goes beyond passive viewing. Similarly, interactive documentaries such as i-docs encourage a more personalized and dynamic viewing experience by allowing viewers to choose their individual preferences throughout the documentary narrative.

While developments and transformations have created new narrative structures, they also point to processes of democratization of the means of filmmaking. The accessibility of digital cameras, editing software, sound facilities and online distribution platforms minimizes the barriers in the production and narrative processes, enabling a wide range of factors to contribute to the documentary genre. Consequently, this democratization of digital tools forces documentary into hybrid forms that blend it with other genres such as fiction, graphics, animation and games. Hybrid genres, like social media content in general, adopt a fragmented and modular narrative structure that reflects the face of digital media consumption.

Another issue in shaping new documentary narratives is the role of data and algorithms. Digital developments have led filmmakers to incorporate innovations such as big data, machine learning and algorithmic processes into their work. This has enabled them to create data-driven narratives that respond instantly to real-time information or user comments. Compared to the traditional documentary narrative, this approach challenges the traditional narrative form that belongs solely to the documentary producer or director, and embraces a new narrative form that is fluid and more adaptable. For example, looking at some documentaries produced especially for social media platforms, we can see that algorithms are used to produce more personalized content according to audience preferences, blurring the line between the documentary creator and the audience.

As new narrative structures develop in documentary, professional and ethical issues arise. New narrative structures transforming from traditional ones raise important questions of authenticity, representation and ethics. For example, as digital developments allow for greater manipulation of images and sounds, the boundaries between reality and fiction are becoming increasingly permeable. Or filmmakers face the challenge of trying to maintain ethical standards while exploring the creative potential of digital technologies.

Setting a conceptual framework across documentary, digitalization and cinematographic narrative, the study attempts to demonstrate that digital developments and transformations in documentary filmmaking have led to the emergence of innovative narrative structures that reflect the complexity of the new media environment. In this new era, documentary filmmakers are redefining the possibilities of the genre by embracing non-linearity, interactivity and hybridity. Thus, they are creating new narrative forms that are more engaging, inclusive and responsive to the needs of different audiences. However, these developments and transformations challenge traditional notions of narrative, authenticity, and ethics, leading to ongoing debates about the future of documentary film narrative in the digital age.

Within these contexts, the study focuses on how digital development and transformation have redefined the ways in which documentaries are conceived, produced and consumed, and explores the reflection of innovative narrative forms reflecting the transforming media ecosystem on documentary.

DOCUMENTARY FILM AND ITS HISTORICAL DEVELOPMENT

The term "documentary" derives from the Latin word *docere*, meaning "to teach," and later evolved into the word "document." According to the *Oxford English Dictionary*, the original meaning of "documentary" was "a lesson; advice, or warning." In the early years of cinema, the French term *documentaire* was employed to distinguish travel films featuring landscape footage from other types of films. However, a significant shift in the meaning of the term occurred on February 8, 1926, when John Grierson, writing for *The New York Sun*, assigned a new interpretation to "documentary." Grierson noted the "documentary value" of Robert Flaherty's second film, *Moana*, thus marking a pivotal moment. From that point forward, "documentary" began to refer specifically to films that possessed educational or instructive qualities within the cinematic context (Ellis, 1989, p. 4).

The term "documentary" was first used in the history of cinema by the French to describe travel films. It was later adopted by British filmmaker and director John Grierson, who first mentioned the term in a 1926 article. In this piece, Grierson referred to the "documentary value" of American filmmaker Robert Joseph Flaherty's ethnographic film *Moana*. Grierson subsequently elaborated on the definition of the documentary film, describing it as a creative treatment of actuality, though he also criticized the term for being somewhat unwieldy. Paul Rotha further contributed to the discourse by defining documentary as an interpretation of reality that expands human knowledge and offers insights into various aspects of life, emphasizing its intellectual function. Similarly, Nijat Özön described documentary cinema as a form rooted in real events and nature, characterized by minimal or no fictional construction (Özgür & Bigiç, 2017, p. 21)

Documentary filmmaking can trace its origins back to early experiments in recording reality, beginning with work conducted at the Edison laboratory. Initial documentary attempts, such as *Fred Ott's Sneeze* and short films depicting Edison employees dancing, were primarily created for entertainment purposes. In 1895, the Lumière brothers presented their first documentaries in Paris. In contrast to Edison's large and cumbersome camera, the Lumières employed a lighter, more portable device that enabled them to film outdoors, capturing dynamic scenes of everyday life. Their screenings featured films such as *The Arrival of a Train at La Ciotat Station*, *Feeding the Baby*, and *Workers Leaving the Factory*. These early works documented scenes from daily life, circus performances, and brief staged sketches, while also incorporating elements of narrative and fantasy, as seen in the innovations of Georges Méliès. Together, these documentary experiments formed a crucial part of cinema's formative years (McLane, 2022, p. 9).

Documentary cinema is shaped by the filmmaker's expression of reality and their desire to serve society. It aims to contribute to social consciousness, examining reality and transferring cultural knowledge to future generations. Unlike entertainment cinema, documentaries appeal to a curious and questioning audience, presenting a unique form of expression without distorting reality. The purpose of making a documentary film is not just to pass time, but to bring together existing and desired creations, art and reality, and entertainment and thought. A successful documentary film informs the public about economic, social, cultural, and environmental issues, shedding light on the future (Aytaş & Yavuz, 2024, p. 60).

In the early days of cinema, two distinct forms of filmmaking emerged. While the Lumière brothers produced films that captured scenes from real life, Georges Méliès was among the first to experiment with narrative storytelling. By recording unmediated real-life events, the Lumière brothers unintentionally laid the groundwork for the development of documentary filmmaking. The first recognized documentary film is generally considered to be Robert Flaherty's *Nanook of the North* (1922), in which he consciously filmed the lives of the Inuit people in the Canadian Arctic. Initially approaching filmmaking as an amateur, Flaherty introduced a sense of coherence and narrative structure to his work with *Nanook of the North*. This film marked a significant shift toward a more systematic and deliberate approach to documentary production (Özgür & Bigiç, 2017, p. 23)

The term "documentary" originates from the act of recording filmed images and emerged as a category distinct from other types of cinema, such as newsreels and travelogues. This genre has evolved over more than a century, tracing its roots back to the earliest cinematic recordings by the Lumière brothers. Documentary films serve as powerful instruments that shape meaning through contemporary cultural narratives and play a critical role in the construction of culture. They possess the capacity to influence emotions, beliefs, and actions. Unlike entertainment genres that often detach from reality, documentaries engage directly with the complexities of real life. Although they draw upon real-world phenomena, documentaries do not simply represent reality; rather, they reconstruct it, offering filmmakers the freedom to interpret and shape the real. For historians, critics, and audiences alike, documentaries remain an essential form, valued for their perceived minimal intervention in capturing the fabric of reality (As & Parsa, 2023, p. 428).

DOCUMENTARY FILM TYPES

It is noted that theorists classify documentary films into different categories based on their form, content, and historical development. Among the identified types are news documentaries, travel documentaries, nature documentaries, social documentaries, investigative documentaries, scientific documentaries, historical documentaries, war documentaries, biographical documentaries, and propaganda documentaries. These categories facilitate a better understanding and analysis of the wide range of existing documentary films.

News Documentaries

News documentaries are an important genre of documentary film that report on events during a specific period of time. While they are no longer popular in theaters, they continue to attract attention on television. These documentaries, often presented in weekly or monthly episodes, use current and archival footage to provide a comprehensive view of the subject matter. They serve as a program type, with the functions of influencing public opinion and setting the agenda. News documentaries are known for their objective approach, providing enlightening information through good research and meticulousness. They use a gripping and comprehensible narrative to engage the audience (Yıldırım, 2019, p. 137).

Nature Documentaries

Nature documentaries focus on exploring the world we inhabit through both scientific and poetic approaches. The scientific perspective involves the use of various technologies to provide detailed views of different aspects of the natural world.

For instance, microscopic lenses allow us to observe the world of insects and gain insights into the behaviors of ants. Cameras offer glimpses into both the functioning of the human body and the behaviors of wild animals within their natural habitats. This visual information enhances our understanding of these subjects and fosters new perspectives on nature. Moreover, technological advancements have made it possible to capture images of the deep sea, revealing wonders that were previously inaccessible. Overall, these films offer a unique and insightful exploration of our world (Tağ, 2003, p. 56).

Project Documentaries

Research documentaries are a unique genre that combines artistic value with the exploration of social events. These documentaries aim to avoid the monotony of long narratives and instead focus on clear statements and concise durations to engage the audience. They often utilize technical tools, such as microscopes and telescopes, to delve into the minute details of the subject matter. The use of color shots and advanced shooting techniques also plays a significant role in these documentaries. Despite their seemingly benign nature, research documentaries often expose social corruption and highlight reality in its purest form. The documentarian operates within legal limits and employs the best research techniques available, patiently and meticulously capturing every stage of the process while risking personal dangers, all without flinching (Gündeş, 1998, p. 27).

War Documentaries

War documentary making, similar to guerrilla filmmaking, focuses on quickly documenting events and raising awareness. While the documentarian may lack experience and knowledge, they still strive to collect and present relevant information. This includes utilizing various sources like politicians' opinions, court records, news broadcasts, and even rumors. The key to interpreting the message lies in the individual's ability to perceive and analyze all the information, considering each new piece in relation to the old. Impartiality in war documentaries is challenging to achieve, but narration is a crucial narrative technique in presenting war. The filmmaker must balance the cinematic and dramatic aspects, as dramatization is often necessary. In recording a war document, the language used is often fragmented and haphazard, with the director often focusing on the parties they believe to have been wronged (Torun, 2019, p. 152).

During wartime, nations utilized documentary productions as a means of disseminating information about the progress of the conflict, highlighting military successes, and providing moral support to their citizens. German war documentaries were particularly prominent in this regard, as they relied on footage supplied directly by the German army, with military cameramen tasked with capturing the material. These films were subsequently shown to both domestic and international audiences, aiming to emphasize the accomplishments of the German military. In some instances, however, military operations were delayed due to the unavailability of cameramen to document the events. Overall, documentary productions emerged as a powerful instrument for influencing public opinion and sustaining morale throughout the war (Tağ, 2003, p. 73).

History Documentaries

The aim of historical documentaries is to uncover little-known historical events, present historical facts in a contemporary context and establish links between the past and the present. These documentaries are valuable in establishing the historical background and adding credibility and authenticity to the film. By shedding light on unknown aspects of history and relating them to modern times, history documentaries enable viewers to better understand the past and its impact on the present. By providing historical context, these documentaries act as a bridge between past eras and contemporary society. They play an important role in highlighting historical events that are less known to the general public and in turn contribute to a wider knowledge and appreciation of history (Cereci, 1992, p. 15).

Scientific Documentaries

Scientific documentaries serve as a means to present research findings in an easily understandable manner. These documentaries often incorporate elements of fiction and animation, while also having educational value. The aim of these documentaries is to create a learning environment and they are often used as teaching aids both in and out of schools. In order to be effective, these films must adhere to the principles of education and be tailored to the age and knowledge level of the audience. Additionally, scientific documentaries should be engaging and able to capture and maintain the viewer's attention. In these films, explanations are as crucial as the visuals in effectively conveying information. Overall, scientific documentaries play a significant role in making complex scientific topics accessible and interesting to a wider audience (Gündeş Öngören, 1991, p. 28).

Animation Documentaries

The animated documentary is a unique form of storytelling, combining animation and reality to create a persuasive and effective narrative. It can be used to convey facts or ideas, and is often used in a methodical and didactic way to teach lessons. While real images may have a greater sense of reality, animated documentaries are able to strengthen their narratives through different production methods and image characteristics. With the advancement of digital technology, animated documentaries are becoming more prevalent and powerful in their storytelling. By creating the illusion of movement through rapid succession of images, animated documentaries have the ability to captivate audiences and be successful in representing the truth. The use of fluent movement sequences, emphasis on reality, and musicality all contribute to the appeal of this genre (Cereci, 2022, p. 125).

Travel Documentaries

Travel documentaries are a form of film focused on documenting and promoting less well-known regions or current events. The key principle is to thoroughly research the topic and accurately depict the region through a narrative approach. Instead of just describing the content, it is important for the filmmakers to actually visit and experience the area at different times to capture its changes over time. The goal is to provide accurate information and highlight the most prominent features of the region. To make the documentaries more engaging, visual and audio elements are crucial, as well as the use of a compelling narrative. However, some recent travel documentaries have been criticized for being overly involved in the production process, suggesting they may not accurately represent the true nature of the region (Karadoğan, 2001, p. 18).

Biography Documentaries

Biographical and autobiographical storytelling is prevalent in modern society and is considered a cultural phenomenon. Life stories have the power to educate, inspire, guide, and stimulate individuals. The biography documentary genre aims to depict real lives and is a creative form of non-fiction interpretation. While it is based on real-life events, the content of a biography documentary also includes fictional elements to enhance the story of the subject. These documentaries serve as archival works, revealing important information about the subject's life, often through first-person accounts and perspectives from those who know them. By primarily relying on interviews and a wide range of supporting documents, biographical documentaries aim to portray the subject as an individual in a realistic and comprehensive manner. They are reminiscent of medical anamnesis, which uses interviews to diagnose and define a patient's condition, in that they use interviews and firsthand knowledge to tell the subject's story (Cereci, 2022, p. 109).

DIGITAL TRANSFORMATION AND THE EVOLUTION OF DOCUMENTARY CINEMA

The concepts of "digital" and "digitalisation" play a pivotal role in the realm of digital publishing and media platforms. The term "digital" derives from the Latin word *digitus*, meaning "finger," and is rooted in the use of integers ranging from 0 to 9, operating through a binary numerical system based on 0 and 1. Digitalisation, in contrast, refers to the process of converting analogue messages into binary signals. This transformation enables the seamless flow of information across diverse media platforms and facilitates easy access in multiple environments. In the context of television broadcasting, digital publishing replaces traditional analogue methods with new digital capabilities, reshaping the way audiences engage with content. This shift has led to the emergence of multimedia and interactive viewing experiences. Digital television broadcasting presents a more cost-effective alternative to conventional methods, offering streamed, predominantly one-way, and largely non-interactive programming. Ultimately, digitalisation carries profound social and cultural implications, influencing both media consumption patterns and broader communication practices (Kaynarca, 2024, p. 72).

Documentary filmmaking has experienced a digital transformation similar to that of fiction films, utilizing comparable production technologies. Nevertheless, notable differences exist between the two, particularly regarding digital conceptualization and the integration of interactive elements unique to documentary forms. Digitalization has profoundly influenced documentary culture, affecting not only production and distribution processes but also the ways in which content is constructed and manipulated. The rise of new media technologies has facilitated the shift from linear narratives to non-linear editing practices, making professional-quality documentary production more accessible than ever before. The rapid pace of technological advancement has further increased the appeal of documentary filmmaking among producers. This evolution has also given rise to new subgenres—including interactive documentaries, web-based documentaries, and multimedia documentaries—thus enabling forms of non-linear storytelling that were previously unattainable (Arda & Şentürk, 2021, p. 101).

Digital technology has profoundly transformed the landscape of documentary filmmaking by making equipment significantly lighter and more portable. High-resolution digital cameras now enable filmmakers to capture high-quality images at reduced costs, offering greater flexibility and efficiency throughout the production process. Editing software and visual effects tools have likewise become more widely accessible, allowing for quicker and more adaptable approaches to post-production work. This accessibility extends to sound editing, color correction, and other technical processes integral to documentary filmmaking. Notably, digital technology has played a critical role in lowering production costs, thereby enabling low-budget filmmakers to produce, distribute, and exhibit their work through online platforms. The affordability and availability of digital cameras, editing programs, and other production resources have democratized the field, empowering a broader and more diverse range of filmmakers to engage in documentary creation (Kaynarca, 2024, p. 73).

The rapid changes in the world and the integration of technology into our lives have influenced the development of new narrative forms in documentary production. The use of digital media has increased the speed at which recordings are created and consumed. However, this fast pace has led to a gradual decrease in the depth of storytelling, resulting in an absence of substance. Technological advancements have also brought about changes in the form and content of documentary films, leading to significant shifts in their narrative structure. As documentary filmmaking has transitioned from cinema to television and now digital media, it has adapted to the demands of the different mediums, altering its content and narrative language. This evolution highlights the ever-changing nature of documentary cinema and the importance of adapting to technological innovations (Aytaş & Yavuz, 2024, p. 62).

Documentary cinema is a branch of art that keeps social memory alive by capturing past periods. These films present their subjects to the audience by examining them in depth, encourage the audience to think and offer a different perspective on reality. Documentary films have evolved over time and changed in content and form from the Naturalistic Tradition to the True Cinema and Direct Cinema movements of the 1960s. Recent technological developments have led to the emergence of the Interactive documentary form, which allows the audience to participate in the production process (Yavuz, 2024, p. 113).

In Birchall's evaluation of the relationship between digital technology and documentary film, several key points are highlighted. Firstly, digital technology allows people from different parts of the world to collaborate on creating a common documentary film. This brings together diverse perspectives and creates new forms of community based on shared interests. Secondly, digital technology provides new avenues for political organizations to create and distribute documentaries, enabling them to influence public opinion and support election campaigns. Thirdly, the use of digital technology in filming daily events and debates allows for greater access to "bad facts," shedding light on important issues. Lastly, the use of video diaries and moving images gives us a closer look into people's private lives. Overall, digital technology has greatly impacted documentary filmmaking, allowing for increased collaboration, political mobilization, and access to information (Birchall, 2008, p. 278).

The technological advancements of the late 1990s and beyond have profoundly influenced the landscape of documentary cinema. The proliferation of diverse broadcasting channels, combined with the adoption of digital technologies, has transformed various aspects of film production. Decreases in production and post-production costs have enabled the creation of low-budget documentary projects, while digital tools have facilitated more accessible distribution and exhibition pathways. Consequently, there has been a notable increase in the volume of documentary films produced and circulated, marking an era of remarkable expansion for the genre. The rise in the number of television networks, broadband channels, and emerging mass media platforms has allowed documentary films to reach broader and more diverse audiences than ever before. In some instances, documentary productions have achieved significant recognition, both at international film festivals and in box office revenues. Furthermore, technological progress has reshaped the modes of film consumption, with documentaries now readily accessible in cinema theaters as well as through DVD, Blu-ray, and increasingly digital streaming formats (Erkılıç & Toprak, 2012, p. 13).

THE EFFECTS OF DIGITAL TRANSFORMATION ON DOCUMENTARY CINEMA

Digital transformation has profoundly reshaped narrative techniques and audience experiences within documentary cinema. The incorporation of drone footage and 360-degree cameras has introduced a new level of immersion, allowing viewers to engage with documentary narratives in increasingly dynamic and spatially expansive ways. Artificial intelligence-assisted editing programs have accelerated traditional post-production workflows, enabling faster and more efficient assembly of visual material. Simultaneously, data visualization techniques have enhanced the clarity and accessibility of complex subject matter, making intricate information more comprehensible to broader audiences. Moreover, digital platforms have revolutionized distribution models by facilitating the global dissemination of documentary films, thereby expanding access beyond traditional geographic and institutional boundaries.

The documentary film genre is undergoing significant changes in various dimensions such as technology, commerce, aesthetics, politics, and society. The advancements in technology have made it cheaper and more accessible, leading to debates on the transformation of documentary filmmaking. The use of digital technology and the internet has also impacted the arts by allowing for technology-based art productions and internet art. However, the internet has its own challenges, such as issues of copyright and unauthorized sharing of works. The rapid evolution of technology has become a crucial part of social life, affecting not only the art world but also various other fields. This technological intervention has facilitated the diversification of artistic elements in documentaries and other art forms (Arda, 2015, p. 4).

Digital communication and media technologies have significantly expanded the possibilities for production processes, content development, formal characteristics, and narrative structures within documentary filmmaking. These advancements have prompted critical debates concerning the definition of documentary, as traditional boundaries have become increasingly fluid. The emergence of formats such as the "interactive documentary," "multimedia documentary," and "web documentary" has further influenced the evolving conceptualization of documentary in the digital age. These new forms, alongside the application of storytelling techniques within digital environments, necessitate a fundamental reconsideration of documentary cinema's role and structure. Central to these innovations is the concept of interactivity, which enables audiences to actively engage in the communication process and contribute to the production and dissemination of media content. Furthermore, the influence of digital technologies is evident in evolving production models, where crowdfunding platforms and community-based participation now play a pivotal role in the creation and financing of documentary films (Toprak Ökmen, 2022, p. 193).

The advent of interactive digital media has profoundly influenced both the production processes and the viewing experiences of traditional audiovisual documentaries. Interactive documentaries, in particular, exhibit distinct formal and narrative characteristics. With the expansion of interactive and participatory cultural spaces, audiences now occupy an essential role within the narrative construction of documentary films. In certain instances, viewers are empowered to make decisions regarding the subject matter and even contribute their own narratives to the unfolding story. Users are given the opportunity to select among various options, thereby actively participating in and shaping the documentary experience, resulting in a more immersive and reality-based engagement. Rather than solely representing objective truth, interactive documentaries emphasize the negotiation and construction of truth. The focus in these emerging forms lies on the author's capacity to craft a specific version of reality and on the dynamic relationship between the user and the content. Notably, the degree of user participation and interaction varies across different projects (Toprak Ökmen, 2022, p. 194).

CONTEMPORARY MOVEMENTS IN DOCUMENTARY FILM

Contemporary movements in documentary film aim to explore reality in cinema, to encourage the audience to think more deeply and to explain social changes. Among these movements, Cinéma Vérité, Cinema Direct, Free Cinema and the Candid-Eye Movement come to the fore.

Cinéma Vérité

Cinéma Vérité, the French version of Dziga Vertov's KinoPravda, was revived in the late 1950s by directors who were inspired by it. This film movement emerged in the early 1960s. The French Cinéma Vérité movement was established in 1959 by Robert Flaherty's wife, Frances, at a seminar organized by Flaherty and his brother David. At the seminar, the prominent ethnographer-director Jean Rouch played a significant role. Filmmakers Michel Brault and Gilles Groulx, who later met Edgar Morin, were influenced by Jean Rouch's films, including "Chronique d'un été" (A Summer Diary). Cinéma Vérité aimed to capture reality without the use of scripted or staged elements, giving audiences a raw and authentic cinematic experience. The movement had a significant impact on the documentary genre, challenging traditional storytelling conventions and presenting a more honest and unfiltered portrayal of reality (Özdil, 2021, p. 57).

During the period when the term Cinéma Vérité was first coined in France, filmmakers Jean Rouch and Chris Marker were influential in this style of filmmaking. Jean Rouch, an ethnographer, made his initial films focusing on African tribes and their traditions, without concerns for aesthetics. One of Rouch's significant films that exemplifies his understanding of cinema is "The Mad Masters" (Les Maîtres Fous) from 1954, which portrays the Hauka religious sect. Rouch's approach involved living with the tribal people and utilizing participant observation to not only observe but also capture their reactions with his camera. He considered the camera as a means of shaping the reactions of the people being filmed. Overall, Cinéma Vérité, as demonstrated by Rouch and Marker, emphasized a raw and unfiltered documentation of reality (Armes, 2006, p. 81).

Cinema Verit Movement Documentary Film Example

Chronicle of a Summer" (1961), a collaborative work by Jean Rouch and Edgar Morin, represents a groundbreaking achievement in cinema history—not only within the documentary genre but also in its exploration of the representation of reality and the epistemological boundaries of film itself. Regarded as a foundational example of the ***Cinéma Vérité*** movement, which developed concurrently with the French New Wave and paralleled the principles of Direct Cinema, the film embodies a commitment to capturing reality with as much fidelity as possible. This aspiration significantly shapes both its formal structure and thematic content.

The film begins with a chain of experimental interviews aimed at capturing the social mood of the period. Centered around the deceptively simple question, "Are you happy?", these dialogues introduce viewers to individuals from diverse social, ethnic, and cultural backgrounds living in Paris. Jean Rouch's ethnographic approach, grounded in observational cinema, merges with Edgar Morin's sociological insight to provide the film with both empirical depth and theoretical substance. Thus, the film transcends a mere sociological snapshot, offering instead a self-reflective inquiry into issues of observation, representation, and intervention in cinema.

One of the film's most distinctive features is its alignment with the technical principles of *Cinéma Vérité*. The use of lightweight, portable camera technology enabled the filmmakers to cultivate more natural and spontaneous interactions with their subjects. Particularly notable is the concluding sequence—where the participants watch and critique the film themselves—serving as a rare meta-documentary moment in cinema history. These scenes not only introduce transparency into the production process but also encourage a critical engagement from the audience.

When evaluated in a comparative context, *Chronicle of a Summer* reveals both affinities and divergences with the Direct Cinema movement emerging in North America during the same period. For instance, Robert Drew's *Primary* (1960), which follows John F. Kennedy's election campaign, adopts a more observational stance that emphasizes the political actions of a public figure. In contrast, *Chronicle of a Summer* foregrounds personal narratives, subjective experiences, and self-awareness, while also questioning the filmmaking process itself.

Similarly, Frederick Wiseman's *Titicut Follies* (1967) stands as a paradigmatic example of American documentary filmmaking's adherence to non-interventionist principles. However, Wiseman focuses his lens on the operations of a state-run psychiatric institution, foregrounding structural critique. In contrast, Rouch and Morin's work prioritizes the emotional landscapes, social belonging, and subjective truths of individuals.

In this comparative framework, *Chronicle of a Summer* emerges not merely as a manifestation of the *Cinéma Vérité* ethos, but as a foundational text that interrogates the philosophical, ethical, and aesthetic boundaries of the movement. Here, the camera operates not just as a recording device, but as an instrument of social experimentation. The documentary invites the viewer to question not only the lives represented on screen but also the very process of representation itself. In doing so, it has carved out a distinctive place in the evolution of both documentary cinema and visual anthropology.

Cinema Direct

Direct Cinema is a contemporary documentary film movement that originated in America during the late 1950s and early 1960s. Key figures in this movement include Albert Maysles, David Maysles, Frederick Wiseman, Robert Drew, Richard Drew, D.A. Pennebaker, and Richard Leacock. These filmmakers aimed to capture unfiltered observations and placed importance on the use of television as a platform for screening their films. Albert Maysles, who studied sociology, added a social perspective to the movement, while Robert Drew brought in his experience in journalism. Richard Leacock's departure from a narrative style of filmmaking set him apart, and Pennebaker's background as a combat cameraman in World War II added to the technical expertise of the group. This collective's diverse backgrounds and skills contributed to their comprehensive approach to both technical and film-related issues (Özdil, 2021, p. 61).

The 'Direct/Indirect Cinema' approach, rooted in America, is a variation of the 'Cinema Real' approach, but with a greater emphasis on the director's involvement. It operates on the notion that everything should be documented as it occurs naturally in nature. In this approach, the director no longer controls the narrative, instead, the camera becomes almost invisible, capturing events organically in real locations with genuine witnesses. The focus shifts from the traditional 'subject-transmitter-viewer' dynamic to a 'subject-viewer' one, highlighting the power of the subject. Filmmakers in this movement aim to minimize their own presence and allow the camera to transparently record events without interference. This is made possible through the use of portable equipment for synchronized audio and video recording, granting the filmmakers more freedom and accessibility to social spaces and allowing the raw sounds and rhythms of the environment to be directly conveyed to the audience (Yavuz, 2024, p. 113).

Cinema Direct Movement Documentary Film Example

The 1960s marked a pivotal era in the history of documentary cinema, characterized by a heightened pursuit of "realistic" representation, facilitated by significant advancements in film technology. Within this context, two major documentary movements emerged: *Direct Cinema* in North America and *Cinéma Vérité* in France. Although these movements shared similar technological tools—such as lightweight cameras and synchronized sound—they diverged notably in their theoretical underpinnings and practical methodologies. In this regard, the documentary *Primary* (1960) serves as a key example for understanding the implementation of Direct Cinema and for drawing comparisons with Cinéma Vérité.

Directed by Robert Drew, and produced with the collaboration of pioneering figures such as Richard Leacock, D.A. Pennebaker, and Albert Maysles, *Primary* follows the Democratic primary campaign between John F. Kennedy and Hubert Humphrey. The film stands out for its immersive observation of events as they unfold, without inserting commentary or overt narrative framing. Here, the filmmaker assumes a detached, observational stance, refraining from direct intervention in the filmed reality.

At the core of Direct Cinema lies a commitment to minimal interference. Accordingly, *Primary* avoids voiceover narration, dramatic music, or formal interviews. The audience is invited to witness the backstage workings of a political campaign as though through an invisible lens. The camera, akin to a modern *flâneur*, roams through spaces, capturing candid expressions and spontaneous interactions without mediating their authenticity. This approach positions the documentary not as a constructed narrative, but as an experienced moment in time.

In contrast, *Chronicle of a Summer* (*Chronique d'un été*, 1961), directed by Jean Rouch and Edgar Morin, exemplifies the technical and philosophical premises of *Cinéma Vérité*. Although both movements utilize similar technological innovations, *Cinéma Vérité* diverges through its inherently interactive production model. Unlike Direct Cinema's observational mode, *Chronicle of a Summer* actively incorporates both the filmmaker and the participants into the creative process, generating a "dialogue within the film." Participants are not merely subjects of observation; they express opinions, engage in self-reflection, and ultimately shape the trajectory of the documentary itself.

In this context, Direct Cinema, as represented by *Primary*, aligns with the metaphor of covert observation, while *Cinéma Vérité*, as realized in *Chronicle of a Summer*, reflects a mode of participatory observation. The former documents what is present; the latter contributes to the construction and reflection of what is present.

In conclusion, both movements have made distinct contributions to the evolution of modern documentary film. *Primary* emphasizes the invisibility of the filmmaker and an ideal of objective representation, whereas *Chronicle of a Summer* foregrounds the reflexive nature of cinematic representation. Thus, Direct Cinema and Cinéma Vérité articulate two different epistemological frameworks: one concerned with witnessing reality, the other with engaging and transforming it.

Free Cinema

Free Cinema, a contemporary documentary cinema movement, emerged in England and aimed to liberate cinema from government propaganda, intellectual restrictions, and traditional film genre rules. It was influenced by television and transformed documentary filmmaking practices using technology within the socio-economic and political context of the time. This movement was characterized by a community of directors who sought intellectual and financial freedom in the filmmaking process. On February 3, 1956, the British Film Institute showcased three 16mm films at the National Film Theatre in London under the title Free Cinema. These films included Lindsay Anderson's O Dreamland, which documented the Margate Dreamland amusement park, Mamma Don't Allow by Karel Reisz and Tony Richardson, which focused on a jazz club in London, and Lorenza Mazzetti's Together, a feature-length documentary about two deaf dockers in London's East End (Özdil, 2021, p. 66).

The British Documentary School, led by John Grierson, produced social realist films and influenced many filmmakers to continue in this style. However, in the 1950s, the rise of technology and ideology led to serious ruptures in the principles of documentary film. The Free Cinema, a small but impactful organization, emerged as a strong example of this rupture and was the first significant reaction against Grierson's style. Although short-lived, it represented a new quest for documentary cinema and differed from Grierson's social realism by critically examining events and everyday life, seeking to capture the poetry of pure realism. However, when considering the documentaries of Özgür Cinema, which focused on the working class, it is challenging to categorize them as completely disconnected from Grierson's social realist films. In a sense, the foundations of Free Cinema can be traced back to the documentary realist tradition of British cinema (As & Parsa, 2023, p. 429).

Free Cinema Movement Documentary Film Example

One of the alternative currents in the history of documentary cinema, *Free Cinema*, emerged in Britain during the 1950s as a socially conscious and formally independent mode of expression. Initiated by filmmakers such as Lindsay Anderson, Karel Reisz, Tony Richardson, and Lorenza Mazzetti, the movement reflected both an aesthetic and ideological stance. Centered on portraying the everyday experiences of the British working class beyond the confines of mainstream media, Free Cinema advanced the belief that reality is not only to be observed but also to be emotionally and sensorially experienced. In this regard, *We Are the Lambeth Boys* (1959), directed by Karel Reisz, stands as a key example that encapsulates the movement's core principles.

The film focuses on the daily lives, social environment, and collective identity of a group of young people in London's Lambeth district. Shaped by the Free Cinema belief in the value of "the unshown ordinary," the documentary depicts seemingly mundane scenes—such as conversations in cafés, dancing, or playing football—without dramatization, yet with a strong sense of poetic construction. This stylistic approach positions Free Cinema within a tradition of *poetic realism* in documentary filmmaking.

In contrast, *Cinéma Vérité* represents a French documentary tradition that took shape during the 1960s, characterized by its deliberate aesthetic embrace of interaction with reality. Jean Rouch and Edgar Morin's *Chronique d'un été* (1961) exemplifies this method. The primary distinction between Free Cinema and Cinéma Vérité lies in the filmmaker's mode of engagement with representation. While Free Cinema maintains an observational perspective, it also employs narration, artistic editing, and stylized framing to guide the viewer's experience. Cinéma Vérité, on the other hand, emphasizes direct engagement: the camera's presence becomes an active agent, influencing participants and fostering an open dialogue between filmmaker and subject.

The treatment of sound in *We Are the Lambeth Boys* further highlights this contrast. The use of voice-over narration in the film provides socio-cultural context and guides interpretation, aligning with classical documentary conventions while also striving to give visibility to marginalized voices. Conversely, *Chronique d'un été* eschews narration entirely, building its structure around unscripted dialogue and personal reflection. This shift away from authoritative commentary allows the film to reveal subjective truths through participatory discourse.

From an epistemological standpoint, the two movements offer distinct approaches to representing reality. Free Cinema seeks to aesthetically reconstruct observed reality through lyrical observation, whereas Cinéma Vérité aims to reconstruct lived experience by immersing both filmmaker and subject in a shared cinematic process. While Free Cinema gently articulates working-class sensibilities through contemplative observation, Cinéma Vérité presents a more intrinsic critique of social structures through the subject's active engagement and self-representation in front of the camera.

Candid-Eye Movement

The Bee-Eye cinema movement, influenced by the work of Henri Cartier-Bresson and the idea of capturing the decisive moment in photography, produced films in Canada in 1958 at the National Film Board of Canada. This movement, also known as Candid Eye, had its own unique approach to cinema. The National Film Board of Canada was established in 1939 and operated under the belief that film production should be the responsibility of the state. John Grierson, the head of the organization, was influential in shaping the Bee-Eye Cinema Movement. In an interview with Wolf Koenig, a key figure in the movement, he explained that it was inspired by the combination of Cartier-Bresson's photographs and Free Cinema films. The movement aimed to create a cinema-truth that captured the essence of reality (Özdil, 2021, p. 69).

Candid-Eye Movement Movement Documentary Film Example

The transformation of documentary cinema throughout the 20th century was marked not only by technical innovations but also by a profound paradigm shift in approaches to reality. Representing the North American trajectory of this evolution, the *Candid-Eye Movement*, which developed under the auspices of the National

Film Board of Canada (NFB) in the late 1950s, advocated for an unobtrusive yet attentive cinematic gaze—one that observed reality without interference. Among the early and foundational examples of this movement is ***Les Raquetteurs*** (1958), co-directed by Michel Brault and Gilles Groulx, a film that exemplifies the aesthetic and ethical principles of the Candid-Eye approach.

Set during a traditional snowshoe festival in Quebec, *Les Raquetteurs* offers a portrait of ordinary people engaged in everyday activities and communal rituals. Eschewing voice-over narration, dramatized editing, or overt authorial guidance, the film aligns itself with the ethos of the "observational camera," a core tenet of Candid-Eye cinema. The camera, operating almost invisibly, follows events as they unfold organically, adopting the perspective of a spectator whose presence is unacknowledged by the subjects. In doing so, the film resists visual embellishment and seeks to render daily life in its unadorned state.

While Candid-Eye shares certain formal similarities with *Cinéma Vérité*—notably, the use of lightweight equipment and synchronized sound recording that enabled filming in uncontrolled, public environments—the two movements diverge significantly in philosophical orientation. Cinéma Vérité, as exemplified in *Chronique d'un été* (1961), does not merely observe reality but actively engages in its construction through dialogic encounters. Here, the camera becomes a medium of interaction, even confrontation, rather than remaining a passive recording device.

In contrast, the Candid-Eye method prioritizes the filmmaker's invisibility and minimizes the filmmaker's intervention in the filmed reality. *Les Raquetteurs* maintains a deliberate distance between the camera and its subjects, aiming to preserve their natural behavior and unfiltered responses. Editing in the film is rhythmic rather than narrative, with the intent not to dramatize but to aestheticize observation itself.

Although Candid-Eye is formally akin to *Direct Cinema*, it developed within Canada's bilingual and bicultural context, cultivating a unique ethics of observation. The participation of French-Canadian filmmakers fostered an implicit dialogue with European traditions of philosophical documentary, particularly the ethnographic sensibilities of Jean Rouch.

Ultimately, the Candid-Eye Movement provides a quiet yet incisive framework for documenting the visual rhythms of everyday life. By refusing to alter the flow of reality and maintaining fidelity to its pace and structure, it distinguishes itself from the reflexive mode of *Cinéma Vérité*. Remaining loyal to an observational-directorial paradigm, Candid-Eye not only expands but also critically examines the ethical and political boundaries of documentary filmmaking.

PURPOSE OF THE STUDY

The primary aim of this study is to explore the ways in which digital transformation has reconfigured the narrative structures, production processes, and audience interactions within contemporary documentary cinema. As digital technologies continue to evolve, they bring about a series of profound changes that challenge the conventional definitions of documentary filmmaking. This research seeks to examine how emerging tools and platforms—such as virtual reality, augmented reality, interactive media, and algorithm-driven personalization—have not only expanded the aesthetic possibilities of the genre but also reshaped the epistemological foundations upon which documentary narratives are constructed.

In this context, the study investigates the shift from linear and author-centric storytelling to more dynamic, participatory, and multi-perspective forms of narration. The integration of user interaction, modular narratives, and real-time data has led to new forms of engagement, allowing viewers to navigate content based on their preferences and experiences. This evolution demands a rethinking of authorship, narrative authority, and the traditional relationship between the filmmaker and the audience.

Additionally, the research aims to situate these developments within a broader socio-cultural and technological framework, analyzing how digital democratization—through accessible editing software, affordable production tools, and global distribution networks—has opened up new spaces for underrepresented voices and alternative storytelling practices. The convergence of these factors calls for a nuanced understanding of how digital innovation interacts with artistic intention, ethical considerations, and the documentary's foundational commitment to representing reality.

Ultimately, the purpose of this study is not only to document the transformations occurring in documentary film but also to critically engage with the implications of these changes for the future of the genre. By focusing on the intersection of technology, aesthetics, and audience participation, the study contributes to ongoing discussions about the evolving nature of documentary cinema in the digital age and offers insight into how these emerging practices reflect and reshape contemporary media ecologies.

METHODOLOGY

This study adopts a qualitative research approach, grounded in interpretive and analytical methods, to investigate the evolving narrative structures of documentary film in the context of digital transformation. Recognizing that documentary cinema is both an artistic and socio-cultural practice, the research is designed to capture the complexities, nuances, and contextual dynamics that underlie the transition from traditional documentary forms to emerging digital narratives.

To achieve this aim, the study employs a multi-dimensional textual analysis of selected documentary films and projects that exemplify key developments in the use of digital technologies. These include, but are not limited to, virtual reality documentaries, interactive documentary platforms (i-docs), and algorithmically personalized viewing experiences. The analysis focuses on how these films and formats reconstruct narrative flow, challenge representational conventions, and facilitate user participation. Particular attention is paid to narrative structure, authorship, interactivity, viewer agency, and the aesthetic integration of digital tools.

In parallel, the research engages in a critical literature review, drawing from contemporary scholarship in media studies, film theory, digital culture, and communication studies. The theoretical framework incorporates concepts such as remediation, hybridity, participatory culture, and post-cinema aesthetics, enabling a holistic examination of both the technical and ideological dimensions of digital documentary practices. By situating contemporary examples within this broader academic discourse, the study aims to contextualize the emergence of new narrative modes within the historical evolution of the documentary genre.

Additionally, comparative analysis is used to contrast digital-era documentary forms with those rooted in earlier traditions such as Direct Cinema, Cinéma Vérité, Free Cinema, and the Candid-Eye Movement. This comparative lens allows for a deeper understanding of how shifts in technology and media ecosystems influence the epistemological assumptions of documentary storytelling and redefine its communicative goals.

Rather than relying on statistical generalizations, this methodology privileges depth over breadth, emphasizing interpretation and critical reflection. Through the close reading of case studies and the synthesis of theoretical perspectives, the research seeks to elucidate the ways in which digitalization not only alters the documentary as a medium but also reconfigures the cultural and ethical landscapes in which it operates.

POPULATION, SAMPLE, AND LIMITATIONS

The scope of this study is situated within the domain of contemporary documentary cinema, particularly focusing on works that have embraced digital technologies in their narrative construction and production processes. The population of interest includes a broad array of digital-era documentary films produced across various cultural and geographical contexts from the early 2000s to the present. This time frame was chosen due to the significant acceleration of digital integration in both professional and independent filmmaking during this period.

Given the expansive and diverse nature of the digital documentary landscape, a purposive sampling method has been employed to select specific case studies that are representative of key narrative and technological innovations. The sample comprises a set of internationally recognized documentary projects that utilize virtual reality (VR), interactive media platforms, algorithmic personalization, or other forms of digital augmentation. These works were selected based on their visibility in academic and critical discourses, their adoption of emerging technologies, and their engagement with narrative forms that challenge traditional linear storytelling. By focusing on these particular examples, the study is able to explore the theoretical and aesthetic implications of digital transformation in depth.

However, this focus on selected case studies also introduces certain limitations. Firstly, the study does not attempt to provide a comprehensive survey of all digital documentaries produced in the specified timeframe. Instead, it seeks to generate analytical insights through detailed examination of exemplary works. Secondly, the research primarily engages with English-language and Western-produced documentaries, which may inadvertently exclude relevant contributions from non-Western or regional documentary traditions that have also been shaped by digital media, but which remain underrepresented in global discourse.

Additionally, the interpretive nature of the study, which emphasizes close textual analysis and theoretical synthesis, entails a certain degree of subjectivity. While this methodological orientation enables nuanced readings of form, structure, and meaning, it also relies on the researcher's academic perspective, which may not encompass all potential interpretations or audience responses. Furthermore, the rapid pace of technological change presents a moving target; thus, some of the findings and conclusions may become temporally bounded as newer innovations and practices continue to emerge.

In acknowledging these limitations, the study remains firmly grounded in its aim to offer a reflective, theoretically informed exploration of how digital tools are reshaping the documentary form. Rather than seeking universality, the research aspires to contribute to ongoing scholarly conversations about the aesthetics, ethics, and cultural implications of documentary filmmaking in the digital age.

CONCLUSION AND RECOMMENDATIONS

In light of the findings presented, it is evident that the digital transformation has not only altered the technical processes of documentary filmmaking but has also redefined its narrative structures, production paradigms, and audience engagement models. The transition from linear, narrator-driven formats to nonlinear, interactive, and algorithmically shaped narratives reflects a profound epistemological and aesthetic shift in the documentary tradition. These developments challenge the boundaries between fiction and reality, authorial control and audience agency, and passive reception versus participatory interpretation.

Digital technologies—ranging from virtual and augmented reality to interactive interfaces and algorithm-driven personalization—have expanded the expressive toolkit available to documentary filmmakers. This expansion enables the creation of hybrid forms that blend traditional documentary with elements of animation, data visualization, and gamification, generating multisensory, immersive experiences that go beyond observation toward affective and cognitive engagement. Furthermore, the democratization of production tools and distribution platforms has lowered barriers to entry, giving rise to a broader spectrum of voices and stories that were previously marginalized or excluded from mainstream discourse.

However, this proliferation of new narrative forms also brings to the surface critical ethical and professional concerns. As the line between authenticity and construction becomes increasingly blurred, the responsibility of the documentary filmmaker intensifies. Issues of manipulation, selective editing, algorithmic bias, and representational fairness must be continually interrogated. The challenge lies not only in navigating technological affordances but also in upholding ethical integrity in storytelling. Filmmakers are thus compelled to strike a balance between creative freedom and the duty to truthfully represent lived realities.

Moreover, the integration of real-time data, audience feedback, and participatory mechanisms into narrative architectures suggests a redefinition of authorship. In this new framework, the viewer is no longer a mere spectator but a co-creator whose decisions, preferences, and interactions shape the narrative trajectory. This shift necessitates a reassessment of narrative authority and questions long-standing conventions of objectivity and authorship in documentary practice.

In conclusion, the digital era has initiated a paradigmatic transformation in the field of documentary cinema—one that is as rich in creative possibility as it is fraught with philosophical and ethical complexities. The future of the genre will depend on the capacity of practitioners, theorists, and audiences alike to critically engage with these evolving dynamics, to innovate responsibly, and to preserve the documentary's enduring commitment to reality, however constructed it may become. The digital documentary, therefore, should not be seen merely as a technological novelty but as a reflective and reflexive form—one that interrogates its own mechanisms and continually renegotiates the relationship between form, content, and viewer in an ever-changing media landscape.

REFERENCES

Arda, Ö. (2015). Belgesel Filmde Yeni Yaklaşımların Idfa Doclab 2010-2015 Belgesel Film Örneklemi Üzerinden Yapısal Ve Kullanım Özelliklerine Göre İçerik Analizi. İstanbul Üniversitesi İletişim Fakültesi Dergisil Istanbul University Faculty of Communicatio*n Journal*, (48), 1-23.

Arda, Ö., & Şentürk, O. (2021). Dijital Çağda Belgesel Film: Dijital Dönüşüme Türk Belgesel Sinemacılarının Yaklaşımları Üzerine Bir İnceleme. *Bildiri Kitabı Proceedings Book*, 100.

Armes, R. (2006). *Sinema ve Gerçeklik Tarihsel Bir İnceleme, Çev*. ZeynepÖzen Barkot, Doruk Yayımcılık.

As, E. ve Parsa, A. F. (2023). Teknolojik Gelişmeler Bağlamında Belgesel Film Yaklaşımları. *Pamukkale Üniversitesi Sosyal Bilimler Enstitüsü Dergisi*, (59), 423–444.

Aytaş, M., & Yavuz, G. (2024). Netflix Çağı: Dijital Seyir Platformları Ve Belgesel Sinemada Anlatının Dönüşümü. *Idil: Journal Of Art & Language, 13*(113).

Birchall, D. (2008). Online Documentary. In Austin, T., & de Jong, W. (Eds.), *Rethinking Documentary New Perspectives, New Practices* (pp. 278–283). Open University Press.

Cereci, S. (1992). *Belgesel Filmde Senaryo. İstanbul Üniversitesi Sosyal Bilimler Enstitüsü Radyo Televizyon Bölümü.* Doktora Tezi.

Cereci, S. (2022). *Belgesel Yapımı*. İksad Yayınevi.

Erkılıç, H., & Toprak, A. G. (2012). Belgesel Sinemanin Alternatif Dağitim Ve Gösterim Olanaği Olarak İnternet. The Turkish Online Journal of Design, Art and Communication – TOJDAC, 2(2), 10-16.

Gündeş, S. (1998). *Belgesel Filmin Yapısal Gelişimi Türkiye'ye Yansıması*. Alfa Basım Yayım Dağıtım.

Gündeş Öngören, S. (1991). *Belgesel Filmin Yapısal Gelişimi ve Türkiye'ye Yansıması*. Der Yayıncılık.

Karadoğan, S. (2001). *Belgesel Film Ve Propaganda. Marmara Üniversitesi İletişim Fakültesi Radyo Tv Ve Sinema Bölümü*. Araştırma Projesi.

Kaynarca, F. (2024). Dijitalleşme İle Değişen Belgesel Filmlerdeki Anlatı Estetiğinin Dijital Yayın Platformları Üzerinden İncelenmesi. *Journal Of Media & Cultural Studies*, *6*(2).

McLane, B. A. (2022). *A new history of documentary film*. Bloomsbury Publishing USA.

Özdil, N. M. (2021). Dijital Platformlarda Belgesel Film Yapim ve Gösterim Pratikleri: Blutv Örneği. İstanbul Üniversitesi Sosyal Bilimler Enstitüsü Radyo, Televizyon Ve Sinema Anabilim Dali, (Yüksek lisan tezi), İstanbul, Türkiye.

Özgür, F., & Bigiç, N. Ç. (2017). *Belgesel/Kısa Film/Video Sanatı*. Doruk Yayımcılık.

Tağ, Ş. (2003). *Belgesel Sinema ve Türleri* (Master's thesis, Anadolu University (Turkey)).

Toprak Ökmen, A. G. (2022). Dijital dönemde belgesel film yapımındaki yenilikler üzerine bir değerlendirme. *Medya ve Kültür.*, *2*(2), 184–207.

Torun, H. (2019). Belgesel Sinema Dili Ve Savaş. *Karadeniz Uluslararası Bilimsel Dergisi*, (41), 144–158.

Yavuz, S. (2024). Belge Filmden İnteraktif Belgesele: Teknolojik Gelişmeler Bağlamında Belgeselin Değişen Estetiği. *İnönü Üniversitesi İletişim Fakültesi Elektronik Dergisi (İnif E-Dergi), 9*(1), 111-133.

Yıldırım, E. (2019). Televizyon Belgesel Filmleri Anlatı Yapısının İzler Kitle Beklentilerine Etkileri Ve Örnek Belgesel Film Çözümlemesi. *Dördüncü Kuvvet.*, *2*(2), 134–149.

ADDITONAL READING

Akbulut, H. (2010). Bellek olarak belgesel sinema: Son dönem Türkiye belgesel sinemasına bir bakış. *Sinecine: Sinema Araştırmaları Dergisi*, *1*(2), 119–124.

Aytekin, H. (2013). *Türkiye'de toplumsal değişme ve belgesel sinema* (Doctoral dissertation, Maltepe University (Turkey))

Budiana, A. A., Sutopo, D., & Rukmini, D. (2017). The use of translation techniques in subtitling the Dhaup Ageng documentary movie. *English Education Journal*, *7*(1), 01-05.

Gider, N. (2014). Yapısal özellikleri açısından belgesel sinema. *Marmara İletişim Dergisi*, *12*(12), 134–140.

Kavanagh, A. S. (1983). *Instructıonal And Persuasıve Effects Of Cınema Verıte And Narrated Documentarıes*. Michigan State University.

Lefebvre, M. (2011). On landscape in narrative cinema. *Canadian Journal of Film Studies, 20*(1), 61–78. DOI: 10.3138/cjfs.20.1.61

Mulvey, L. (2006). Visual pleasure and narrative cinema. *Media and cultural studies: Keyworks*, 342-352.

Mulvey, L. (2013). Afterthoughts on "visual pleasure and narrative cinema" inspired by Duel in the Sun. In *Feminism and film theory* (pp. 69-79). Routledge.

Safford, T., & Triplett, W. (1983). Haile gerima: Radical departures to a New Black Cinema. *Journal of the University Film and Video Association, 35*(2), 59–65.

Stoddard, J. (2013). Hillary: The movie, the History Channel, and the challenge of the documentary for democratic education. *Teachers College Record, 115*(3), 1–32. DOI: 10.1177/016146811311500302

KEY TERMS AND DEFINITIONS

Algorithmic-Personalization:: This term refers to the use of algorithms to tailor documentary content to individual viewers based on data such as preferences, behavior, or demographic information. Common in web-based or platform-distributed documentaries, this approach enables more personalized and dynamic user experiences but also raises concerns about filter bubbles and editorial transparency.

Data-Visualization:: Data visualization refers to the graphical representation of complex datasets within documentary narratives to enhance audience comprehension and engagement. When used effectively, it transforms abstract information—such as statistics, geolocation, or behavioral patterns—into compelling visual arguments, thereby contributing to both narrative clarity and persuasive impact.

Digital-Transformation:: Digital transformation refers to the integration of digital technologies into all aspects of documentary filmmaking, from production and editing to distribution and audience interaction. This process has not only revolutionized technical workflows but also catalyzed new narrative structures and viewing experiences. It challenges traditional cinematic paradigms by introducing interactive, nonlinear, and immersive forms of storytelling.

Documantary-Film:: A documentary film is a nonfictional cinematic form that seeks to depict reality, people, or events with the intention of informing, educating, or provoking reflection. Unlike fictional cinema, documentaries are grounded in factual representation, although the extent of objectivity may vary depending on stylistic and ideological choices. In the digital age, the boundaries between representation and construction have become increasingly fluid, prompting critical debates about authenticity and mediation.

Ethics-of-Representation: The ethics of representation concerns the moral responsibilities filmmakers hold in portraying individuals, communities, and events. This includes considerations of consent, accuracy, manipulation, and cultural sensitivity. In the digital era, where editing tools and AI technologies blur the line between real and constructed, ethical questions surrounding authenticity and accountability are more critical than ever.

Hybrid-Narratives:: Hybrid narratives combine elements from various genres—such as fiction, animation, data visualization, and documentary—to create multifaceted storytelling structures. These narratives challenge genre purity and embrace the complexity of modern media consumption, often reflecting fragmented, modular, or multi-perspective approaches. Hybridity allows for a more experimental and adaptive form of representation.

Interactivity:: Interactivity in the context of documentary refers to the capacity of the viewer to influence the narrative experience through active engagement. This can include choosing narrative paths, contributing content, or navigating through layered information. Interactive documentaries, such as i-docs, redefine the audience's role from passive observer to co-author, thereby reshaping the power dynamics between filmmaker and viewer.

Nonlinear-Structure:: A nonlinear structure in documentary filmmaking disrupts the traditional chronological flow of information, allowing for stories to unfold in a modular, associative, or user-directed manner. This structure aligns with digital platforms that support segmented or hyperlinked content, and it invites viewers to construct meaning through their own navigation of the material.

Participatory-Culture:: Participatory culture denotes a media environment in which users are no longer passive consumers but active participants in content creation and circulation. In the documentary context, this culture enables collaborative storytelling, crowdsourced narratives, and community-based filmmaking, expanding the democratic potential of the genre while introducing new ethical complexities.

Virtual-Reality-(VR):: Virtual Reality in documentary cinema enables audiences to experience immersive environments, often through headsets, where they can explore spatial narratives. VR documentaries simulate a presence within the filmed or constructed world, allowing for heightened emotional engagement and a sense of embodied observation. This medium expands the sensory and experiential dimensions of nonfiction storytelling.

Chapter 4
Fictional Space in the Context of Space-Time and Spatial Relationships

Yelda Özkoçak
https://orcid.org/0000-0002-9617-3787
Doğuş Üniversitesi, Turkey

ABSTRACT

With the increasing influence of technology, narratives are gradually transforming into grander stories. Undoubtedly, one of the most critical stages in a film's storytelling and creation process is the construction of space. The concept of space-time and spatiality, empowered by technological possibilities, gains a multidimensional significance, enabling creators in cinema to produce "surreal" spaces. In cinema, surreal creations emerge through the linkage of narrative time and narrative space via editing. This study will focus on space creation and the concept of constructed space in cinema. Based on Kuleshov's notion of "Creative Geography," it will also explore Eisenstein's theory of montage and J. Arnheim's concepts of aesthetics and cinematic reality, aiming to articulate the process of fictional space construction.

INTRODUCTION

In his 1989 study, H. Avni Öztopçu defines the human being as the only living creature capable of objectifying everything and characterizes fictional space as a form of reality created by the artist through the arrangement, condition, and sequencing of elements such as line, color, and plane within an abstract intellectual world.

Considering that the creation of space in cinema is a multidimensional phenomenon, it is important to address this subject from theoretical and practical perspec-

DOI: 10.4018/979-8-3373-1967-4.ch004

tives. The creation of space in cinema transcends the mere depiction of a physical environment, integrating symbolic, emotional, and psychological dimensions.

In her doctoral dissertation, *The Relationship between Fictional Characters and Fictional Space in Fantasy and Horror Narratives* (2015), Elif Özdoğlar emphasizes that fictional spaces must be constructed by composing elements that, in the context of their contemporary conditions, could not entirely or largely exist in reality.

The visual language of cinema is a powerful narrative tool that shapes the audience's perception by manipulating time and space. In filmmaking, space functions not merely as a backdrop for characters' movements but also as a fundamental element of dramatic structure and aesthetic experience. In cinema, space holds significance beyond physical reality, carrying aesthetic and emotional meanings. These layers of meaning constitute some of the most critical tools that enhance the visual storytelling power of cinema.

With technological advancements, filmmakers have gained unprecedented capabilities to create spaces. Moving beyond traditional notions of "real space," they have begun crafting surreal spaces and abstract realities within cinema. In this context, the concept of fictional space emerges as a mode of narration that develops the film's narrative, enabling the audience to establish emotional and psychological connections with what they see on screen.

The creation of space in cinema, by bringing together time and space fictionally, enriches the aesthetic and psychological layers of the narrative.

As Bob Foss (2012) notes, the audience is unaware of anything beyond what is shown or suggested within the frame. The real-world space must be recreated to provide the viewer with a continuous and uninterrupted impression of a physical environment; the three dimensions of reality must be translated into the two-dimensional spatial relations of the screen through a defined set of rules.

Editing and the flow of spaces must evoke a range of emotions in the audience—such as excitement, astonishment, fear, joy, and tension—to ensure perceptual coherence in form, reinforce the subject matter being conveyed, and define the film's concept through the use of spaces. According to Jane Barnwell (2011), "The concept is the complementary element that ensures the visual identity of the film remains consistent."

In this context, the section titled *Fictional Space in the Context of Space-Time and Spatial Relationships* will address the creation of space and the concept of fictional space in cinema through the theories of montage by Sergei Eisenstein, Rudolf Arnheim's understanding of cinematic reality, and Lev Kuleshov's notion of "Creative Geography."

In his works, Eisenstein explains how editing techniques manipulate the audience's perception and how space and time are transformed into a fictional reality. On the other hand, Arnheim argues that cinema does not merely reflect reality but constructs a new visual reality based on the audience's perception. Kuleshov's "Creative Geography" concept emphasizes how cinematic space is integrated with visual language and meaning, asserting that space is not merely a backdrop but a component that carries dramatic significance.

Space affects the viewer on both a physical and emotional level, visually expressing the characters' inner worlds and the dramatic structure of the narrative. From this perspective, the study demonstrates that the creation of space in cinema goes beyond the representation of a physical environment; it serves as a component that deepens the thematic and emotional structure of the film. It highlights that space's meaning and fictional construction lie at the heart of the audience's connection with the film.

THEORETICAL BACKGROUND and LITERATURE REVIEW

Kuleshov and The Concept of Creative Geograph

Lev Vladimirovich Kuleshov (1899–1970) was a Soviet filmmaker and theorist who played a pivotal role in founding the Gerasimov Institute of Cinematography, the world's first film school. He is particularly renowned for his work in cinema and film editing. Kuleshov was one of the first figures to shape the visual and narrative language of cinema, and his approach to editing laid the groundwork for significant ideas regarding the creation of space in film.

Kuleshov argued that constructing space and atmosphere in cinema is a mental process. His famous "Kuleshov Effect" (1924) helps us to understand the relationship between film and fictional space. In the experiment, Kuleshov combined a single shot of a neutral facial expression with three different shots, altering the audience's perception of emotion and space. In this context, space functions as a physical setting and a mental construct.

According to Kuleshov, the creation and meaning of space within the cinematic world are directly linked to the editing process. For him, space is not merely a backdrop; it is a structural element that guides audience perception, reflects characters' psychological states, and produces dramatic effects. He argues that space must acquire meaning within the film through its relationship with time.

Core Principles of Kuleshov's Concept of "Creative Geography"

- **The Meaning-Making Function of Space**: According to Kuleshov, space in cinema is not merely a physical environment but a structure that carries emotional and psychological meaning. Cinematic spaces should be constructed fictionally, and this construction must be connected to the film's themes and the characters' emotional states. Space should become an element that reinforces the characters' inner worlds and the film's dramatic structure.

- **Fictional Manipulation**: Kuleshov asserts that editing can shape the meaning of space. Filmmakers should not treat space solely as a location but rather as a construct that generates meaning and emotional resonance. Kuleshov argues that editing can alter the meaning of space and guide the audience's perception.
- **The Psychological Dimension of Space**: Kuleshov's concept of "Creative Geography" emphasizes that space is not merely a physical entity but one that reflects characters' emotional and psychological states. In cinema, space serves as a tool that conveys the mental conditions of the characters and influences the viewer's emotional response while watching the film.

Kuleshov's concept of "Creative Geography" demonstrates that space in cinema derives its meaning through editing and transcends physical reality to become a tool that influences audience perception. According to Kuleshov, the sense of space is constructed through montage, which generates the film's dramatic impact. In cinema, space is a psychological and aesthetic element that guides the viewer's perception, symbolizes characters' inner worlds and social structures, and contributes to the film's overall narrative.

Kuleshov emphasized that the boundary between reality and imagination becomes blurred in creating cinematic space. This is particularly evident in how editing can alter the meaning of space. As previously noted, Kuleshov regarded cinematic spaces as physical realities and constructs that can be transformed into psychological, symbolic, and dramatic elements through the director's aesthetic choices and the relationship established with the audience. His "Creative Geography" concept offers a theoretical framework for understanding the construction of space in cinema, thus illustrating how filmmakers can utilize space in innovative and expressive ways.

Eisenstein and Montage: The Manipulation of Tıme and Space

Sergei Mikhailovich Eisenstein (1898–1948) was one of the most prominent directors of the Soviet era, known for his contributions to film theory and practice. Through the theories he developed on montage, Eisenstein significantly shaped the visual language of cinema, offering a profound examination of how time and space can be manipulated through editing. In this way, his concept of "montage" became a stylistic approach and a method for demonstrating cinema's influence over the perception of time and space.

Eisenstein is a pioneering figure in manipulating space and time in cinema. In his montage theory, space and time are not merely technical components but also powerful emotional, ideological, and psychological tools. His montage theory illustrates how space and time in cinema can be transformed into meaningful elements and how they can be employed to guide the audience. According to Eisenstein, fictional time and space convey intense dramatic tension and ideological messages. This perspective reveals that space and time in cinema are narrative elements and instruments for constructing meaning.

Furthermore, Eisenstein argued that the fictional manipulation of space was a deliberate strategy to influence the psychological responses of the viewer. In this context, Eisenstein's montage theory allows us to understand how space is constructed and altered within the viewer's perception. His conceptualization of montage emphasizes that space and time in cinema are not merely informational elements but powerful tools for emotional and ideological manipulation, evoking diverse emotional and intellectual responses from the audience.

Eisenstein's foundation of montage theory on the manipulation of cinematic space and time is a key element that distinguishes his cinematic expression. According to Eisenstein, montage should not merely produce an intellectual effect but evoke emotional and ideological impact. Space and time generate broader dramatic, ideological, and symbolic meanings in this regard. This understanding of montage is evident in Eisenstein's films and demonstrates how space can be utilized aesthetically in cinema.

The key points of Eisenstein's understanding of montage are as follows:

The Aesthetic Power of Montage: According to Eisenstein, montage in cinema is a powerful tool for manipulating time and space. The relationship between two independent images generates a new meaning in the viewer's mind. Eisenstein's theory of montage, also known as "dialectical montage," emphasizes that the conflict between two images creates an intellectual context. This allows space and time to function as a backdrop and as elements that carry intrinsic meaning.

The Manipulation of Time: In Eisenstein's theory, the manipulation of time is significant. By employing techniques such as temporal discontinuities, ellipses, and parallel editing, Eisenstein aimed to guide the viewer's psychological perception. This manipulation compels the viewer to engage with the rhythm and psychological progression of events in the film. Techniques like continuous temporal shifts and slow-motion sequences disrupt the linear flow of time, offering psychological depth to the cinematic experience.

The Manipulation of Space: For Eisenstein, space is not just a background but a meaningful element. Through fictional manipulation, space becomes a tool that reflects the characters' inner worlds. Altering spatial representations can influence the audience's emotional connection with the characters and direct their emotional responses. Additionally, space reinforces the characters' psychological states and the film's ideological messages.

Emotional and Ideological Manipulation: Eisenstein emphasized that montage techniques should produce visual, emotional, and ideological effects. In this context, space and time are used to convey ideological messages. Spatial and temporal elements in the film are manipulated to dramatize characters' internal conflicts, social structures, and individual experiences.

J. Arnheim and Cinematic Reality

Rudolf Arnheim (1904–2007), a German-born American author, art and film theorist, and perceptual psychologist, gained recognition as an art critic and psychologist through his seminal work *Film as Art*, published in 1932. A Gestalt School of Psychology graduate, Arnheim approached cinema as a visual art form and explored its effects on visual perception. According to Arnheim, cinema does not merely reflect reality; it recreates it, and space plays a crucial role in this recreation process.

Arnheim emphasized that space in cinema becomes symbolic, distinguishing between visual reality and aesthetically meaningful space. Cinema presents space as a physical setting and an emotional and aesthetic experience. His work focused on how cinematic space is shaped and manipulated. Cinema employs space not only as a backdrop but as a tool to reflect characters' inner worlds. Arnheim examined cinema through the lenses of visual perception, aesthetic meaning, and as a specific mode of artistic expression, arguing that cinema offers a reflection of reality and a creative reality.

Arnheim's work on cinema plays a crucial role in our understanding of *cinematic reality*. This concept posits that cinema possesses the capacity not merely to reflect the real world but to create a new reality altogether. Cinema presents not a reality obtained through direct observation but one shaped through visual perception. Arnheim's perspective helps us understand how cinema constructs a psychological and aesthetic reality.

According to Arnheim, although cinema may claim to reflect the real world, this reflection is not an accurate reproduction of reality. Cinematic reality is a creation based on the viewer's perception, shaped by visual language and editing techniques. The way cinema reflects reality not only replicates it but also reconstructs it aesthetically meaningfully.

Arnheim asserts that cinema's visual world is not a superficial replication of reality but a construct that shapes the viewer's perception and cognitive processes. This suggests that cinema's visuals go beyond mere representations of the real world, instead forming a distinct reality governed by cinema's unique conventions and psychological perception.

Cinema does not merely present a physical reality; it also shapes how the viewer perceives and interprets that reality. The power of visual representation in cinema transforms reality in psychological and aesthetic ways. In his analysis of the visual elements of cinema, Arnheim highlights the camera's ability to manipulate visual perception. He associates this capacity with elements such as lighting, color, composition, camera movement, and editing, arguing that cinema presents reality to the viewer emotionally and perceptually.

According to Arnheim, cinema does not represent a "distancing from reality" but rather a reconstruction. A film appeals to the viewer's inner world, and cinematic reality carries an internal meaning rather than simply reflecting the external world. While shaping the viewer's sensory and cognitive processes, cinema aesthetically reflects the meaning of reality. The techniques used in cinema do not seek to imitate reality exactly but instead incorporate aesthetic and dramatic elements that appeal to the viewer's perception.

At this point, cinematic reality involves abstraction and symbolic representation. Every image in cinema does not merely depict reality; it also creates an aesthetic reflection of that reality and introduces a new perception. For example, the use of light, color, and shadow in cinema does more than mirror the real world—it symbolizes emotional tones and internal states. Arnheim emphasizes that through such abstraction, cinema can create an inner reality. The arrangement of lighting and color in a film does not simply replicate the visuals of the real world; it guides the viewer's emotional perception, thereby establishing cinematic reality as an aesthetic creation.

Key Characteristics of Cinematic Reality According to Arnheim

The fundamental characteristics of cinematic reality can be summarized as follows:

Aesthetic Manipulation: Cinema offers a physical representation of reality and an emotional and aesthetic one. Visual elements enable the abstraction and stylized presentation of reality. Lighting, color, and composition used in the film are not direct reflections of reality but representations imbued with meaning.

Symbolic Representation: Cinema conveys more than reality; it also creates deeper meanings through visual symbols and metaphors. Space, characters, and events are represented through symbolic language that influences the viewer's perception.

Cognitive and Emotional Impact: Cinematic reality appeals to the viewer's mental and emotional processes. Cinema does not merely reflect reality through observation but constructs it through visual perception and psychological influence.

Manipulation of Time: Cinema also manipulates time. Time within a film may function differently from the linear progression of real-time. Fictional time and space, shaped by cinema's abstracting power, alter the viewer's perception, making cinematic reality a perceptual construct.

Explaining Fictional Space Through Cinematic Examples

Cinema frequently observes concrete examples of the creation of fictional space. In such instances, space transcends its role as a mere physical setting and offers the viewer diverse meanings, emotions, and perceptions.

"Battleship Potemkin" (1925) – Sergei Eisenstein

In Sergei Eisenstein's *Battleship Potemkin* (1925), space is fictionally manipulated to convey social unrest and the impact of war within an urban setting. The film's iconic "Odessa Steps" sequence is a prime example of how space can be dramatically manipulated. Through montage, space becomes an emotionally charged backdrop used to depict the tragic human condition.

https://www.imdb.com/title/tt0015648/mediaviewer/rm3902838274/?ref_=tt_ov_i

The film can also be seen as an example that concretizes Kuleshov's concept of fictional space. Eisenstein employs montage to create the psychological meaning of space. Through the Odessa Steps sequence, he constructs a fictional space that allows the audience to grasp the magnitude of violence and the uprising of the revolutionary people. In this context, space functions not merely as a physical location but as a symbol of the collective emotional state and the spirit of resistance.

A wide shot of the massacre on the Odessa Steps https://en.wikipedia.org/wiki/Battleship_Potemkin#/media/File:Potemkinmarch.jpg

With this presentation, the Odessa Steps sequence is one of the most renowned examples of montage and spatial manipulation. The steps are not merely a physical location but a symbol of social tension and ideological struggle. This scene exemplifies how the manipulation of time intensifies dramatic tension and transforms space into a socially meaningful element.

The boots of the soldiers shown marching down the Odessa Steps

https://en.wikipedia.org/wiki/Battleship_Potemkin#/media/File:Odessastepsboots.jpg

A baby in a carriage falling down the Odessa Steps

https://en.wikipedia.org/wiki/Battleship_Potemkin#/media/File:Odessastepsbaby.jpg

In this scene, Eisenstein utilizes the power of montage to deliver a profound emotional impact to the viewer. The shots taken from various angles of the steps simultaneously symbolize transformation. Here, time is fragmented so that events unfold rapidly and dramatically, while the space—the steps—acquires symbolic meaning as an emblem of the people's revolutionary movement. In doing so, Eisenstein employs space as a psychological domain, conveying a powerful ideological message to the audience.

"Strike" (1925) – Sergei Eisenstein

Strike is a 1925 Soviet silent propaganda film directed and edited by Sergei Eisenstein. In *Strike*, the conflict between the working class and the bourgeoisie is visualized by manipulating space and time. The stark contrast between the confined, cramped spaces inhabited by the workers and the expansive, luxurious offices of the factory owners dramatizes the ideological divide and class struggle.

Poster for *Strike* (1925)

https://en.wikipedia.org/wiki/Strike_(1925_film)#/media/File:Strike_(1925)_original_poster.jpg

The film, which is arranged in six parts, depicts a strike in 1903 by the workers of a factory in pre-revolutionary Russia and its subsequent suppression. It is best known for a sequence towards the climax, in which the violent suppression of the Strike is cross-cut with footage of cattle being slaughtered. Similar animal metaphors are used throughout the film to describe various individuals.

https://www.imdb.com/title/tt0015361/mediaviewer/rm1947657473/

Through the manipulation of time, Eisenstein intensifies the contrast between scenes depicting the forced labor of workers and the leisurely indulgence of the factory owners. Eisenstein employs space and time to present physical reality and convey an ideological tension to the audience.

https://www.imdb.com/title/tt0015361/mediaviewer/rm2014766337/

Like Kuleshov's "Creative Geography" concept, Eisenstein utilizes space as a fictional device to construct a visual language that conveys more dramatic meaning than mere reality. Like Kuleshov, Eisenstein emphasizes that in cinema, space is not simply something to be shown, but rather an element that carries and generates meaning.

"October" (1927) – Sergei Eisenstein

As Lev Kuleshov had before him, Eisenstein declared that editing was the soul of cinema. He argued that the juxtaposition of shots—not the shots themselves—gave moving images their meaning and that actors were props used to serve that meaning (Kizirian,2025).

https://en.wikipedia.org/wiki/October:_Ten_Days_That_Shook_the_World#/media/File:Octyabr_poster.jpg

In Eisenstein's film *October*, the manipulation of space and time and the representation of revolutionary movements and opposing forces guide the audience emotionally and ideologically. In scenes featuring the "enemy" characters, the use of temporal pauses and the depiction of lavishly designed spaces instill a sense of the enemy as a significant threat. This approach aligns with Eisenstein's montage theory centered on the struggle against the enemy, demonstrating that space and time are employed to create a psychological impact.

https://www.imdb.com/title/tt0018217/mediaviewer/rm3468381184/

The film can be defined through the manipulation of enemy spaces and time. Eschewing traditional protagonists, Eisenstein further advanced the "mass-epical" approach—where the collective masses, rather than a singular heroic figure, propel the narrative forward. *October* was only his third feature film, yet Eisenstein had already established himself as a master among Soviet montage filmmakers (Kizirian, 2025).

"Citizen Kane" (1941) – Orson Welles

In Orson Welles' film *Citizen Kane*, the manipulations of space and time are excellent examples of Arnheim's understanding of cinematic reality. The film's use of deep focus transforms space into not merely a visual element but also one that carries psychological depth.

https://www.imdb.com/title/tt0033467/mediaviewer/rm684416000/?ref_=tt_ov_i

Released in 1941, the film made numerous innovative contributions to the language of cinema, particularly in montage, the deep focus technique, and makeup artistry. One of its most significant contributions was the effective use of low-angle shots. Prior to this, low-angle perspectives were rarely employed, resulting in ceilings being left out of frame and lighting equipment often being placed above. However, following the innovations introduced by this film, side lighting began to be used effectively, allowing ceilings to be included naturally within the mise-en-scène.

https://www.imdb.com/title/tt0033467/mediaviewer/rm3287822593/

The characters' emotional states and personal histories are conveyed through the cinematic reality of space and time. Welles manipulates the audience's perception of reality through visual language and montage.

https://www.imdb.com/title/tt0033467/mediaviewer/rm2319133953/

"2001: A Space Odyssey" (1968) – Stanley Kubrick

2001: A Space Odyssey, one of the films in which Kubrick pushes the cinematic possibilities of space to their limits, symbolizes the transformation of space from a realistic domain into a philosophical and existential dimension through its scenes set in outer space. Kubrick utilizes cinematic space as a symbolic element that carries meaning parallel to time.

In the film's opening sequence, "The Dawn of Man," prehistoric apes discover a mysterious stone monolith that appears from an unknown origin shortly before they use tools for the first time. The scene then makes a temporal leap into the future to the year 2001.

https://www.imdb.com/title/tt0062622/mediaviewer/rm2918849793/?ref_=tt_ov_i

The film is renowned for its scientifically accurate portrayal of spaceflight, groundbreaking special effects, and exploration of ambiguous, philosophical themes. Kubrick deliberately eschewed conventional cinematic and narrative techniques to create a unique, immersive audiovisual experience.

https://www.imdb.com/title/tt0062622/mediaviewer/rm3806714369/

"A Clockwork Orange" (1971) – Stanley Kubrick

A Clockwork Orange demonstrates Eisenstein's montage theory's influence on the representation of space. Kubrick constructs the film's settings in a highly dramatic manner to reinforce his critique of violence and societal decay.

https://www.imdb.com/title/tt0066921/mediaviewer/rm3313613825/

The Korova Milkbar is used as a symbolic backdrop representing the characters' relationship with violence and society's moral decay. In this context, the fictional manipulation of space parallels the characters' psychological states. Like Eisenstein's understanding of montage, Kubrick utilizes space as a narrative device to convey the characters' internal deterioration to the audience.

https://www.imdb.com/title/tt0066921/mediaviewer/rm3079518464/

A Clockwork Orange is a film that illustrates the intertwining of fictional space and violence. The film's settings symbolize society's decayed structure and individuals' internal turmoil. Highly stylized and distorted spaces, such as the Korova Milkbar, reflect the characters' relationship with violence and their disconnection from society. Kubrick uses space not merely as a physical environment but as a medium that mirrors the psychological states of his characters.

https://www.imdb.com/title/tt0066921/mediaviewer/rm3828316416/
https://www.imdb.com/title/tt0066921/mediaviewer/rm2347058688/

"Eraserhead" (1977) – David Lynch

David Lynch's *Eraserhead* presents a striking example of how fictional space can be constructed to reflect the viewer's psychological state. Filmed at various AFI locations in California, including the Greystone Mansion and an abandoned stable where Lynch resided, the film tells the story of a man (played by Jack Nance) who is forced to care for his grotesquely deformed child in a desolate industrial environment.

https://www.imdb.com/title/tt0074486/mediaviewer/rm1938112000/?ref_=tt_ov_i

In the film, an industrial cityscape and bleak, monotonous settings reflect the characters' bizarre, terrifying, and disoriented states of mind. Lynch employs space to materialize the fear and alienation within the characters' inner worlds. In this context, space is not merely a physical reality but also an element that defines the protagonist's psychological condition.

https://www.imdb.com/title/tt0074486/mediaviewer/rm464539648/

"The Shining" (1980) – Stanley Kubrick

Stanley Kubrick's psychological horror film *The Shining* is a compelling example of how space can function as a physical setting and a psychological tool. The film follows a recovering alcoholic and aspiring novelist (Nicholson) who takes a job as the winter caretaker of a haunted mountain resort hotel, accompanied by his wife (Duvall) and clairvoyant son (Lloyd).

Much of the film was shot on sets based on real locations. One of the film's central settings, the Overlook Hotel, is not merely a physical space but is designed as a psychological environment that reflects the characters' mental states. Kubrick utilizes the hotel's vast and labyrinthine architecture to evoke the audience's sense of the characters' inner collapse. In this context, space reinforces themes of isolation, paranoia, and the descent into madness. Furthermore, Kubrick's use of symmetrical camera compositions and long corridors contributes to a sensation of claustrophobia and disorientation within the space.

https://www.imdb.com/title/tt0081505/mediaviewer/rm2518225920/

In alignment with Kuleshov's propositions, Kubrick designs the hotel not merely as a physical setting but as a geographical reflection of Jack Torrance's psychological deterioration. The hotel's expansive and isolated structure is closely tied to Jack's descent into loneliness and madness, allowing the audience to deeply feel his unraveling. In this way, Kubrick uses space as a mirror of the character's inner worlds.

https://www.imdb.com/title/tt0081505/mediaviewer/rm1805287938/

According to Kuleshov, space is shaped through montage, and this shaping serves to convey the character's emotional state and the film's themes to the audience. Kubrick employs space not merely as a backdrop but as a structural element that deepens the character's psychological collapse and sense of isolation. This approach represents a cinematic reflection of Kuleshov's concept of "Creative Geography."

https://www.imdb.com/title/tt0081505/mediaviewer/rm3109967105/

“Blade Runner” (1982) – Ridley Scott

According to Kuleshov, space is not merely a backdrop; it is a tool that carries fictional meaning and elicits specific emotional responses. In *Blade Runner* (1982), the city of Los Angeles functions not only as a physical setting but also as a symbol of dystopia and the tension between humanity and machines. In line with Kuleshov's concept of “Creative Geography,” the film's spaces become reflections of the characters' inner worlds and sources of dramatic power. The gloomy, rain-soaked streets construct a psychological space of a dystopian society and reinforce the characters' emotional states.

https://www.imdb.com/title/tt0083658/mediaviewer/rm3676975360/?ref_=tt_ov_i

Set in the dystopian Los Angeles of 2019, the film portrays synthetic humans known as replicants, who are bioengineered by the powerful Tyrell Corporation to work in off-world colonies. Interrogating the boundaries between humanity and machines within a dystopian future, *Blade Runner* demonstrates how space can carry both dramatic and symbolic significance.

https://www.imdb.com/title/tt0083658/mediaviewer/rm3447881984/

In the film, the city of Los Angeles is depicted with a constantly rain-soaked, bleak atmosphere, transforming space into a tool that guides the audience’s emotional responses. As one of the leading examples of the cyberpunk genre, the film employs the city and its surroundings as symbolic representations of the conflict between humanity and artificial intelligence. In *Blade Runner*, space functions not merely as a backdrop but as an element that reflects the characters' inner worlds and their environment's moral and societal decay.

https://www.imdb.com/title/tt0083658/mediaviewer/rm2854916354/

Ridley Scott’s *Blade Runner* emphasizes the symbolic significance of space as it explores a dystopian future and the complex relationship between humanity and machines. The city of Los Angeles, with its gloomy atmosphere and rain-soaked streets, carries a deeper meaning. In parallel with Kuleshov's concept of “Creative Geography,” space in the film functions not merely as a backdrop but as a visual element that reflects the decaying structure of society and the tension between humans and machines. Space shapes the viewer's social, cultural, and psychological perception within this fictional context.

https://www.imdb.com/title/tt0083658/mediaviewer/rm3431104768/

“Pulp Fiction” (1994) – Quentin Tarantino

In Quentin Tarantino’s *Pulp Fiction*, space plays a significant role parallel with the film's nonlinear structure and the intertwining narratives of its characters. Set in Los Angeles, this crime—and violence-laden film features "Jackrabbit Slim's" restaurant as a location that holds symbolic significance.

https://www.imdb.com/title/tt0110912/mediaviewer/rm1959546112/?ref_=tt_ov_i

The film opens with a diner robbery staged by a couple, and the spaces throughout are not merely physical locations but areas intrinsically connected to the characters' lives and shaped by their choices. Tarantino transforms space into an integral part of both time and narrative structure, elevating cinematic storytelling to a deeper level.

https://www.imdb.com/title/tt0110912/mediaviewer/rm3022720513/

"The Matrix" (1999) – Lana Wachowski, Lilly Wachowski

In *The Matrix*, the fictional manipulation of space closely aligns with Eisenstein's understanding of montage. The "Matrix" world transcends conventional reality and transforms the viewer's mental perception. In this context, space becomes a construct that reflects the characters' subconscious and perception. As in Eisenstein's montage theory, manipulating space in *The Matrix* guides the audience's emotional responses and generates a dramatic impact.

https://www.imdb.com/title/tt0133093/mediaviewer/rm177362689/

The film's events occur in a distant future where technology has advanced to the point that machines with artificial intelligence have been created. In this world, a war has erupted between humans and machines. As a result of the conflict, the human race, in an attempt to defeat the machines, blocked sunlight—their primary energy source—by covering the sky with dark clouds using chemical weapons.

https://www.imdb.com/title/tt0133093/mediaviewer/rm2182645504/

Depicting an undesirable dystopian future, the film is an example of cinematic space used as a simulated reality. In *The Matrix*, space is shaped by the characters' physical and mental states. The film allows the audience to perceive space from a perspective distinct from the real world.

https://www.imdb.com/title/tt0133093/mediaviewer/rm416093953/

https://www.imdb.com/title/tt0133093/mediaviewer/rm2712652545/

"Enter the Void" (2009) – Gaspar Noé

Gaspar Noé's *Enter the Void* completely alters the perception of space and time. Set in the neon-lit nightclub environments of Tokyo, this fantastical psychological drama is described by Noé himself as a "psychedelic melodrama." Locations were used in Kabukichō and other parts of Shinjuku. Since much of the film was shot in neighborhoods known for gambling and prostitution, the producers secured agreements with Yakuza crime syndicates before filming certain on-location scenes (Adams, 2010).

https://www.imdb.com/title/tt1191111/mediaviewer/rm3216645632/?ref_=tt_ov_i

The film presents the nightlife and streets of Tokyo as a hallucinogenic experience. Space functions not only as a physical space but also as a symbolic reflection of experiences of death and beyond. By offering the viewer an outward perspective from the "present moment," Noé shows that space is not always fixed but is shaped by a state of mind.

https://www.imdb.com/title/tt1191111/mediaviewer/rm1208048640/

"Inception" (2010)- Christopher Nolan

In the film, Christopher Nolan shows how space is flexibly created and transformed in the dream world. The fictional shaping of space in the film creates different meanings in the minds of the audience.

https://www.imdb.com/title/tt1375666/mediaviewer/rm733803009/?ref_=tt_ov_i

Nolan's Inception explores reality and perception, inviting viewers to question the nature of reality and the boundaries between the conscious and subconscious mind. As the film navigates through multiple dream levels, it challenges the viewer to distinguish the fine line between reality and illusion and makes us question what is real and what is a construct of the mind.

https://www.imdb.com/title/tt1375666/mediaviewer/rm1069515520/
https://www.imdb.com/title/tt1375666/mediaviewer/rm1035961088/

"The Tree of Life" (2011) - Terrence Malick

In The Tree of Life, the creation of space is based on Arnheim's aesthetic approach. Malick associates the spaces that symbolize the vastness of the universe and the power of nature with the spiritual and existential themes of the film. Here, space is both a visual reality and a reflection of the characters' universal experiences. In Arnheim's view, the aesthetic creation of space directs the viewer's emotional and mental responses and transforms cinematic reality.

Nolan's Inception explores reality and perception, inviting viewers to question the nature of reality and the boundaries between the conscious and subconscious mind. As the film navigates through multiple dream levels, it challenges the viewer to distinguish the fine line between reality and illusion and makes us question what is real and what is a construct of the mind.

https://www.imdb.com/title/tt0478304/mediaviewer/rm2065465857/

In the film, Terrence Malick shows how space can have both universal and personal meaning. The spaces symbolize the meaning of life and the human connection with nature. Using space as a metaphor, Malick presents the viewer with the deeper meanings of the universe and human life.

https://www.imdb.com/title/tt0478304/mediaviewer/rm1451280385/

The Tree of Life exemplifies cinematic reality constructed through abstractions and visual images. The nature scenes and the narration of family dynamics in the film offer the viewer an inner experience through the manipulation of time and space. By using visual images of nature and time jumps in the film, Malick shapes the viewer's perception on an aesthetic and spiritual plane. Terrence Malick shows how space can be used as a space with both universal and personal meaning. The spaces in the film symbolize the meaning of life and the human connection with nature. Using space as a metaphor, Malick presents the viewer with the deeper meanings of the universe and human life.

https://www.imdb.com/title/tt0478304/mediaviewer/rm663668737/

"Her" (2013) - Spike Jonze

In romantic drama and science fiction, Spike Jonze questions the effects of technology and human relationships on space. The film was shot in Los Angeles and Shanghai. It presents a monotonous life in a city in the future with minimal visuals. The space symbolizes the character's loneliness and technology's effects on personal life.

https://www.imdb.com/title/tt1798709/mediaviewer/rm1539299328/?ref_=tt_ov_i

In the midst of technological developments, the individual's mental state is shaped by the environment in which he or she lives, and the space becomes an element that reflects the character's emotional journey. Here, the environment in which the character lives carries meaning in both a physical and emotional context.

https://www.imdb.com/title/tt1798709/mediaviewer/rm3430365440/

https://www.imdb.com/title/tt1798709/mediaviewer/rm2088188160/

"The Grand Budapest Hotel" (2014) - Wes Anderson

Wes Anderson's "The Grand Budapest Hotel" is characterized by its colorful and stylized location design. The space is used as an aesthetic element and also sets the characters' emotional states and the story's tone. Ralph Fiennes leads a seventeen-actor ensemble cast as Monsieur Gustave H., the famed concierge of a twentieth-century mountainside resort in the fictional Eastern European country of Zubrowka.

https://www.imdb.com/title/tt2278388/mediaviewer/rm1834474496/?ref_=tt_ov_i

Principal photography took place at Babelsberg Studio in the suburbs of Berlin and in Görlitz, a medium-sized border town in the Lusatian Neisse on Germany's eastern border. Anderson uses locations to express the characters' worlds and relationships multi-layered.

The Grand Budapest Hotel universe is depicted with a nostalgic longing, in which the characters maintain "the illusion of a time to which they do not belong" (Marshall, 2014).

https://www.imdb.com/title/tt2278388/mediaviewer/rm2714124801/

The interior structures of the hotel, its exterior surroundings and the layout of the rooms become symbols, each with a different aesthetic meaning. Anderson creates both a real place and a symbolic space, so each place tells a story.

https://www.imdb.com/title/tt2278388/mediaviewer/rm3106971137/

"The Revenant" (2015) - Alejandro González Iñárritu

The biographical western thriller The Revenant shows how nature can be used dramatically as part of the setting. It becomes not just a backdrop but a symbol of the characters' struggle for survival and inner strength. The wild and ruthless face of nature conveys to the audience the characters' psychological and physical struggles. The wilderness in the film functions as a space that symbolizes a character's quest for survival and revenge.

https://www.imdb.com/title/tt1663202/mediaviewer/rm1354918144/?ref_=tt_ov_i

The movie was shot in 12 locations in 3 countries, including Canada, the USA and Argentina.

(https://www.imdb.com/title/tt1663202/locations/)

The movie tells the story of fur trapper Hugh Glass, who, in 1823, is brutally attacked and wounded by bears while on an expedition in what would later become Dakota Territory, and his quest to find his betrayer and avenge his son's death.

https://www.imdb.com/title/tt1663202/mediaviewer/rm1042145280/

The above-mentioned films show that space is not only physical but is also used in cinema to manifest emotions, symbols, character development and social structures. The creation of space in cinema has a fictional, aesthetic and symbolic function, and in each film, these spaces become an element that shapes the viewer's experience.

CONCLUSION

Space creation and fictional space in cinema refers to a physical space and a structure that carries meaning, emotion and symbolism. The creation of space in cinema, as a combination of fictional processes and aesthetic choices, is a tool that guides the viewer's perception, reflects the inner worlds of characters and carries social, psychological or cultural meanings. In this context, Kleşov's "Creative Geography", Eisenstein's fiction, and Arnheim's concepts of aesthetic and cinematic reality provide theoretical frameworks for reflection on the creation of space in cinema.

In creating space, cinema expresses meaning, emotion and mental state. Kleşov's concept of "Creative Geography", Eisenstein's montage theory and Arnheim's aesthetic perspective reveal that space in the cinema is not just a physical place but an experience shaped by the emotional and psychological perceptions of the viewer. While these theories lay the theoretical foundations for creating space in cinema, they also show how these creations are embodied in exemplary films.

According to Kleşov, space is a mental construction, shaping the viewer's perception as part of the fictional processes. By manipulating space through fiction, Eisenstein creates a dramatic effect and uses space as a psychological tool. On the other hand, Arnheim explores the aesthetic dimension of cinematic reality by making space an aesthetic experience. In light of these theories, the creation of space in cinema is the construction of a physical space and the presentation of an emotional, psychological and symbolic experience to the audience. This reveals the complexity and depth of the creation of space in cinema.

REFERENCES

Adams, S. (24 September 2010). "Interview: Gaspar Noé". https://web.archive.org/web/20110731142928/http://www.avclub.com/articles/gaspar-noe,45554/ (Access 24.04.2025)

Arnheım, R. (1985). *Film ve Gerçeklik-Sinema Kuramları, Derleyenler: Seçil Büker Oğuz Onaran*. Dost Kitabevi.

Bergson, H. (2017). *Yaratıcı Tekamül. (Çev. M. Tunç)*. Dergah Yayınları.

Clarke, J. (2012). *Sinema Akımları: Sinema Dünyasını Değiştiren Filmler. Şahan Yatarkalkmaz (Çev.)*. Kalkedon.

Comolli, J. L. (1974), Teknik ve İdeoloji, Çev. Yakup Barokas Çağdaş Sinema Dergisi, Sayı:1,İstanbul

Doane, M. A. (2002). *The Emergence of Cinematic Time*. Harvard University Press.

Eisenstein, S. (1977). *Film Form-Essays in Film Theory* (Leda, J., Ed. & Trans.). Harcourt Brace&Company.

Foss, B. (2012). *Sinema ve Televizyonda Anlatım Teknikleri ve Dramaturji*. Hayalperest Kitap.

Gültekin, G. (2019). Cınematıc Use Of Tıme And Space Through Example Fılms, *V Human and Civilization Congress From Past To Future Full Text Book*, Antalya.

Kizirian, S. (2025). 1927: *October* (Sergei Eisenstein) https://www.sensesofcinema.com/2017/soviet-cinema/1927-october-sergei- eisenstein/

Kuleshov, L. (1971). *Kuleshov on Film: Writings by Lev Kuleshov*. University of California Press.

Marshall, L. (Summer, 2014). "The Budapest Hotel: Wes Anderson's Fabulous Fancy". *Queen's Quarterly*. Vol. 121. Kingston, Ontario. pp. 242–251.

Özdoğlar, E. (2015). Fantezi ve Korku Kurguda Kurgusal Karakter ve Kurgusal Mekân İlişkisi Yayınlanamış sanata yeterlik tezi.

Özkoçak, Y. (2014). *Ses'li Sinema*. Der Yayınları.

Öztopçu, H. A. (1989) Kurgusal Mekân, https://www.avnioztopcu.com/avni/kurgusalmekan.htm (Erişim, 20.04.2025)

Shaviro, S. (1993). *"The Cinematic Body" (1993)*. University of Minnesota Press.

Stam, R. (2004). *Film Theory: An Introduction*. Wiley-Blackwell.

ADDITIONAL READING

Andrew, J. D. (1976). *The Major Film Theories*. Cambridge Univ. Press.

Armes, R. (2011). *Sinema ve Gerçek, Çev*. Zeynep Özen Barkot Doruk Yayımcılık.

Balazs, B. (1995). *Theory of The Film*. Dennıs Dobson Ltd.

Bazin, A. (2007). *Sinema Nedir?* Çev. İbrahim Şener, İzdüşüm Yayınları.

Büker, S. (1989). *Film ve Gerçek*. A. Ü. Eğitim Teknolojisi Ve Yaygın Eğitim Vakfı Yayınları.

Kracauer, S. (1960). *Theory of Film*. Oxford University Press.

Parkinson, D. (2002). *History of Film*. Thames & Hudson Ltd.

Pezzella, M. (2006). *Sinemada Estetik*. Dost Kitapevi.

Pudovkin (1966), Sinemanın Temel İlkeleri Çev. Nijat Özön, Bilgi Yayınevi, İstanbul

Tarkovski, A. (1992). *Mühürlenmiş Zaman, Çev*. Füsün Ant, Afa Yayıncılık İstanbul.

KEY TERMS AND DEFINITIONS

Cinema: Cinema, which has emerged and developed as a result of both ideological and economic demands (Comolli, 1974, p. 15), can be defined as a product created to express creativity and imagination. Although the word cinema is used in the sense of a place where products, i.e., films, are shown, it also refers to the process of detecting pictures by dividing any movement into parts at regular intervals and then recreating the movement by projecting these pictures on a wall/screen in a dark room with the help of a projector. Cinema is an art that has the power to depict the technique and its effects on human perception, and its main feature is the result of efforts to copy reality, as mentioned before (Özkoçak, 2014).

Creative Geography: A film technique invented in the early 1920s by Russian film director and theorist Lev Kleşhov. He emphasized that creating space and time in cinema should go beyond the physical space to construct meaning in the viewer's mind. Kleşov's concept of Creative Geography suggests that cinematic space can be fictionally shaped and will affect the viewer's emotional, psychological and intellectual perceptions. In one of his works, Kleşov fictionalized shots that took place at different times and places as if they were shot in a single location. In the experiment, a man and a woman are shown walking in different neighborhoods of Moscow. They greet each other before a statue and look out of the frame, where the White House appears. Then they start walking on the street. Another shot shows them climbing the stairs of the cathedral. When the shots are combined, it is shown as if the couple is walking up the stairs of the White House. This scene, shot without using camera tricks, proves the power of montage. It has been seen that all kinds of perceptions can be made and recreated with montage (Clarke, 2012).

Dramatic Space: In the filmic sense, this space is used to identify and surround the psychological states of the film characters. For example, the dark streets used in films intensely evoke death, immorality, crime and violence.

Fictional Space: This is space created through fiction and with the help of various technologies. Thus, it aims to give the audience the impression that a space that does not exist in reality exists.

Geographical Space: It is the real, non-artificially created space where the movie takes place and can be considered a background that accompanies the atmosphere of the movie. A geographical location can also be characterized as a dramatic location as long as a specific meaning is attributed to it. Many films that follow the realism movement use geographical space as it is without changing its structure.

Time in Cinema: The manipulation of time in cinema is carried out off-screen, that is, in invisible areas or the intervals between shots. Bergson's understanding of time puts the concept of time in cinema on a philosophical basis. In the Bergsonian approach, the basic criticism of the relationship between cinema and human beings is made. Accordingly, cinema works the same way as human perception as a tool that allows fixed plans to come together and re-motion (Bergson, 2017; Doane, 2002). In movies, time can be extended, frozen or shortened. These three situations show that time is used differently in films through appropriateness, intensity, tension, continuity, temporality, storytime, discourse time, flashback, subtraction and time transitions (Gültekin, 2019). Cinematic spaces are not only a platform where cinematic action takes place but also contain the economic, political, cultural, social, ideological, class, gendered and similar elements experienced by the cinematic subject. According to Metz (2012: 58), cinematic space is not abstract but perceived and prioritized over time. This priority does not harm cinematic time; on the contrary, it creates a fulcrum for it. Therefore, time necessitates space and moves in space. Gültekin (2019) states in his study that spaces in films are generally designed in three different ways: geographical, dramatic and fictional.

APPENDIX

Sample Movies

"Battleship Potemkin" (1925) – Sergei Eisenstein
"Strike" (1925) – Sergei Eisenstein
"October" (1927) – Sergei Eisenstein
"Citizen Kane" (1941) – Orson Welles
"2001: A Space Odyssey" (1968) - Stanley Kubrick
"A Clockwork Orange" (1971) – Stanley Kubrick
"Eraserhead" (1977) – David Lynch
"The Shining" (1980) – Stanley Kubrick
"Blade Runner" (1982) – Ridley Scott
"Pulp Fiction" (1994) – Quentin Tarantino
"The Matrix" (1999) -Lana Wachowski, Lilly Wachowski
"Enter the Void" (2009) – Gaspar Noé
"Inception" (2010)- Christopher Nolan
"The Tree of Life" (2011) – Terrence Malick
"Her" (2013) – Spike Jonze
"The Grand Budapest Hotel" (2014) – Wes Anderson
The Revenant" (2015) – Alejandro González Iñárritu

Chapter 5
Audience Information and Ethical Marking System in Cinema

Elif Çanğa Bayer
https://orcid.org/0000-0002-3305-3016
Independent Researcher, Turkey

Gökhan Toptepe
https://orcid.org/0000-0003-0825-1237
Independent Researcher, Turkey

ABSTRACT

The integration of artificial intelligence into cinema has brought profound changes to creative processes and audience perception. This chapter examines how AI becomes part of the cinematic message and its cultural implications through the lenses of McLuhan's "the medium is the message" and Gerbner's theories of resonance and mainstreaming. The ethical implications of AI technologies used in scriptwriting, visual production, sound design, and editing are critically discussed. The chapter proposes an "Intelligent Marking Model" to transparently indicate the extent of AI involvement in film production. This model aims to inform both audiences and industry stakeholders, promoting ethical and responsible use. Ultimately, transparency and ethical approaches are essential for the sustainable integration of AI in cinema.

INTRODUCTION

The integration of AI into the cinematic realm has precipitated a paradigm shift in the creative process, giving rise to significant ethical and engagement-related

DOI: 10.4018/979-8-3373-1967-4.ch005

dilemmas. The implications of AI in cinema extend beyond the realm of storytelling, giving rise to debates concerning content authenticity and the potential manipulation of audience perceptions. This development consequently initiates a novel discourse domain concerning the manner in which filmmakers configure the audience experience and the criteria that are employed to delineate ethical standards. This raises questions regarding the sensitivity of audiences to AI-generated content and the manner in which the film industry can employ this technology in a responsible manner. Concurrent with the transformation of audience expectations and experiences engendered by emergent technologies, there is an imperative for filmmakers to adopt innovative creative approaches. This change has the potential to represent a significant turning point in the evolution of cinema as an art form, encouraging the development of innovative methods to enhance audience engagement alongside ethical debates. A strategy that prioritises transparency in the use of AI is essential to ensure that audiences are aware of the impact of these technologies on storytelling and content creation. In addition, there is a necessity to establish ethical guidelines for navigating the complexities of AI's role in cinema, whilst preserving artistic integrity and audience trust.

The ethical implications of AI's use in cinema necessitate a thorough examination of the boundaries between creativity and technology. Filmmakers must judiciously balance innovation and responsibility in their work. In addition to the possibilities offered by AI, it is imperative to consider its potential dangers and long-term effects on the audience. It is therefore imperative to develop a comprehensive strategy on how to integrate AI for the future of the film industry. This strategy should aim to balance human contribution and the role of AI in creative processes. Furthermore, it is incumbent upon filmmakers to continuously assess the impact of AI on content production, taking into account audience feedback (Vives, 2024).

As the film industry navigates this transformative environment, it is imperative to consider not only the ethical implications of AI-generated content, but also its potential to enhance storytelling techniques. For instance, the utilisation of artificial intelligence (AI) in the cinematic domain entails the capacity to analyse audience preferences and adapt narratives in a manner that is more likely to resonate profoundly with viewers. This, in turn, engenders a more personalised cinematic experience. However, this raises concerns about the potential for algorithmic bias, where certain demographics may be underrepresented or misrepresented in AI-driven narratives, potentially serving to perpetuate stereotypes rather than challenge them. The delicate balance between leveraging AI's capabilities and maintaining the integrity of storytelling will ultimately shape the evolution of cinema in profound ways.

It is imperative to educate audiences on the extent of AI's contribution to the creation of the content they consume and to establish ethical frameworks within this process. This is a pivotal step in fostering audience trust. Moreover, the provision of transparency regarding the role of AI in filmmaking will facilitate audience comprehension of the impact of this technology on the creative process. As Taşçı and Çelebi (2020) observe, further research is required in order to comprehend the impact of AI-supported content on audiences. Moreover, the establishment of ethical guidelines is becoming an imperative to ensure that filmmakers utilise this novel technology in a conscientious manner. As posited by Tekbey and Alaybeyoğlu (2023), it is imperative to devise innovative methodologies with a view to enhancing audience engagement and fortifying ethical standards. While the integration of AI within the film production process presents a range of opportunities to enhance both the creative potential and the audience experience, it concomitantly brings with it a concomitant responsibility. This responsibility is manifested in a structure where the medium is messaging.

SECTION 1: THEORETICAL SUCCESSION

Theoretical Underpinnings of Gerbner's Concept of Resonance and Its Influence on the Mainstreaming

In the domain of media and communication studies, George Gerbner's notion of resonance stands as a pivotal element within his overarching theoretical framework. The term refers to the process by which media narratives align with the cultural, social, and psychological predispositions of audiences, thereby amplifying their impact and shaping public perceptions. This notion is firmly embedded in Gerbner's cultivation theory, which proposes that prolonged exposure to media content fosters the development of particular beliefs, attitudes, and values in individuals. The resonance concept extends this idea by emphasising the dynamic interplay between media messages and audience worldviews, leading to the mainstreaming of certain narratives over others.

Gerbner's research on resonance is founded on several key theoretical traditions, including cultivation theory, framing theory, and critical cultural studies. As expounded in his seminal work, the cultivation theory posits that repeated exposure to media content gradually shapes individuals' perceptions of reality. In this particular context, resonance is defined as the phenomenon in which media narratives are found to align with pre-existing cultural beliefs and values, thereby reinforcing and amplifying their impact (Morgan, 2011; Gerbner & Morgan, 2002).

Framing theory, another influential tradition, highlights the role of media in shaping public perceptions through the selection and emphasis of particular aspects of a story. Gerbner's concept of resonance can be regarded as an extension of framing theory, insofar as it elucidates the manner in which specific frames are more efficacious in their resonance with audiences, a phenomenon attributable to their congruence with cultural narratives and shared meanings (Giorgi, 2017; Sheets et al., 2022). Resonance has been identified as playing a crucial role in the mainstreaming of media narratives by amplifying their reach and impact. When media narratives resonate with audiences, they are more likely to be accepted, shared, and internalised. The process is driven by several factors, including cultural alignment, emotional appeal, and social identity. The influence of resonance on media narratives is profound and far-reaching. It has been demonstrated that advertising exerts a significant influence on the content of media messages, the perceptions of the audience, and the reinforcement of prevailing cultural and social norms.

The phenomenon of resonance has been demonstrated to exert a significant influence on the content of media narratives, primarily through the guidance of selection and framing of stories. Media producers frequently devise narratives that they believe will resonate with their target audience, taking into account cultural, social, and psychological factors. This process has the potential to result in the homogenisation of media content, as producers may place a premium on narratives that align with prevailing cultural values and beliefs (Gerbner & Morgan, 2002; Ettema, 2005). Gerbner's concept of resonance provides a critical framework for a deeper understanding of audiences' interactions with media content and the impact of these interactions on social perceptions (Shrum, 2017). Resonance emphasises how specific media narratives integrate with cultural and social contexts and shape audience perceptions in the process (Busselle & Bulck, 2019).

Although Gerbner's field of study was media content and its role on the audience, he claims that media texts reach the audience as messages, and the impact of these messages deepens with the resonance effect when they interact strongly with the cultural and social contexts of the audience. From this standpoint, screenings and performances produced with artificial intelligence, and in parallel with the general world acceptance of the audience, have the capacity to engender a comprehensive acceptance among the audience. This is predicated on the premise that the audience will tend to accept the medium (artificial intelligence) that produces the message rather than the message itself. Consequently, it has the potential to engender a novel perception and experience for the audience in a resonance relationship with artificial intelligence as a medium. As posited by Gerbner in his mainstreaming formula, a form of intelligence designated by the term "artificial" may, through the resonative cinematic content it engenders, effectively mainstream the audience, instilling within them an acceptance of the artificial as a tool, a phenomenon that is perceived as both natural and commonplace. This may further exacerbate the impact of AI-generated content on the audience, thereby enhancing audience participation.

"Medium is the Message"

The integration of artificial intelligence (AI) into the film industry has precipitated a transformative shift, not only altering creative processes but also raising significant questions regarding audience engagement and ethical considerations. The impact of AI extends beyond mere storytelling, prompting discussions about the potential for manipulation of content reality and viewer perceptions. This evolution necessitates a re-evaluation of the manner in which filmmakers shape audience experiences and establish ethical standards in an era where AI-generated content is increasingly prevalent. As the audience's sensitivity towards AI-produced materials grows, critical inquiries arise about the responsible application of this technology within the film industry.

The advent of new technologies has compelled filmmakers to adopt innovative approaches in their creative processes, thereby reshaping audience expectations and experiences. This transformation has the potential to be a pivotal moment in the evolution of cinematic art, fostering not only ethical discussions but also the development of innovative methods to enhance viewer interaction. A strategic focus on transparency regarding AI's role in storytelling and content creation is essential, ensuring that audiences are aware of how these technologies influence narrative construction. Moreover, it is imperative to establish ethical guidelines to navigate the complexities surrounding AI's role in cinema, preserving artistic integrity while maintaining audience trust. Consequently, as the film industry navigates this evolving landscape, it becomes imperative to strike a balance between the opportunities presented by AI and the potential hazards it poses, ensuring that the future of cinema remains rooted in ethical responsibility and audience engagement.

The phrase "the medium is the message", coined by Marshall McLuhan, posits that the medium through which content is disseminated exerts a pivotal role in shaping the message itself. This concept underscores the notion that the characteristics of the medium wield a greater influence on how the message is perceived and understood in comparison to the content itself. This notion has been thoroughly investigated and implemented across diverse disciplines, ranging from communication theory to marketing and data analysis. It has been demonstrated that media exerts a pervasive influence on human interaction and perception.

McLuhan's seminal concept underscores the notion that media such as electric light function as pure information, thereby reshaping human work and association by eliminating time and space constraints. This decentralization and involvement in depth are intrinsic to the medium itself, independent of its content (McLuhan, 2005). The notion that the medium itself carries a message is further explored in the context of data communication networks, where channel descriptors in packet data networks can be used to convey data, thereby enhancing transmission capacity significantly (Fitzek, 2006).

In this context, a preconditioned communication process will be realized between the audience and artificial intelligence productions, with artificial intelligence-based cinema content and the notification of these contents to the audience prior to experience through the utilization of smart signs. When viewers with a high level of readiness in terms of cinematic experience encounter content from an unfamiliar source that they already know exists in a familiar form, they will have the opportunity to make a more informed choice about the content by evaluating the smart beacon message they receive before starting their viewing experience. This kind of communication process will allow viewers to develop a more open and critical approach towards content produced with artificial intelligence (Kuipers, 2012). From this perspective, intelligent signage that underscores the role of artificial intelligence in the production process can function as a novel medium for communication between artificial intelligence and human beings. The signage could convey a message that is complementary to the cinematic content. Consequently, it is proposed that from this point onwards, the pre-informative intelligent signs themselves, and by extension artificial intelligence itself, should be regarded as messages.

The concept of resonance, as proposed by Gerbner, and the notion that "the medium is the message", as espoused by McLuhan, provide a critical framework for understanding the impact of AI-supported cinematic content on the audience. The resonance of AI-generated content with cultural and social contexts has been demonstrated to facilitate more seamless acceptance by audiences and to deepen the mainstreaming effect. In this process, the AI itself becomes a message, and the preliminary information provided to the audience through smart signs creates a new context between the media content and the generative tool. It is evident that the audience is engaged in the experience, cognisant of the AI's contribution, and is presented with the opportunity to cultivate a more critical relationship with the content. This has resulted in a re-evaluation of the role of AI in cinematic production processes and the audience experience, thereby paving the way for the mainstreaming of AI not only as a tool but also as a message and media element.

In this context, it is possible for a mainstreaming process to begin when the medium itself is the message, as McLuhan states, and when this message resonates with the audience, as Gerbner points out. From this standpoint, it is evident that the influence of AI-supported content on the audience will extend beyond the content itself. When approached from this framework, the phenomenon of cinema reveals a new area of responsibility for both the producing mechanisms and cinematic designers.

SECTION 2: CURRENT SITUATION

The Relationship between Artificial Intelligence and Cinema

The relationship between cinema and artificial intelligence is twofold: firstly, it is transforming audiences' experiences, and secondly, it is having a profound impact on the creative processes of the film industry. This interaction has the potential to redefine the audience's engagement with film and pave the way for the emergence of new narrative forms and experiences. This situation gives rise to debates on the use of artificial intelligence as a cinematic tool and the manner in which filmmakers can integrate this technology. The advent of these new narrative forms has precipitated an evolution in audience expectations and experiences, thereby generating a plethora of creative opportunities for the film industry.

In this context, the role of artificial intelligence in cinema enables the re-evaluation of traditional narrative structures and the development of innovative methods to increase audience engagement. Consequently, filmmakers can utilise this technology more effectively to engage audiences and enhance their experiences. As asserted by Çalişkan (2024), this process is of significance in terms of the manner in which artificial intelligence shapes audience perception and gives rise to ethical concerns, in addition to the convenience it provides in content production. Consequently, the cinema industry must embrace this technology in a responsible manner and cultivate a stronger relationship with its audience.

The relationship between artificial intelligence and cinema should be evaluated separately in terms of production and consumption. The production phase involves the utilisation of artificial intelligence (AI) technologies and the aesthetic and ethical aspects created by these technologies. Conversely, the consumption phase shapes the manner in which audiences interact with these technologies and their perceptions. It is imperative to achieve a balanced perspective between these two stages in order to comprehensively grasp the repercussions of AI on the realm of cinema. As posited by Çelikel and Akin (2017), the manner in which artificial intelligence (AI) is employed during the production stage, and the subsequent impact on audience perception, constitutes a pivotal factor in the formulation of strategies by filmmakers. Furthermore, achieving this balance has been demonstrated to enhance the quality of content created with AI and reinforce audience satisfaction (Sun, 2024).

The Most Used Artificial Intelligence Tools in Cinema

Script Writing

AI tools for scripting enhance the authoring process by generating text, suggesting plot ideas and improving dialog. Important tools include:

OpenAI's GPT-3: Generates human-like text, assisting writers in creating scripts and dialog.

Final Draft: Offers AI-driven features to efficiently format and structure scripts.

Celtx: Provides collaborative scripting tools with AI suggestions for character development and story arcs.

Image Production

Artificial Performer

AI tools in games, often referred to as artificial players, simulate human behaviour to enhance gaming experiences. Key tools include:

Unity ML-Agents: Allows developers to train intelligent agents using machine learning to create dynamic non-player characters (NPCs).

IBM Watson: Used in game development to create responsive and adaptive characters that learn from player interactions.

OpenAI's Gym: It is often used as a toolkit to develop and compare reinforcement learning algorithms used to train AI agents in various gaming environments.

Deepfake

Deepfake technology uses artificial intelligence to create realistic synthetic media, primarily video and audio. Featured tools include:

DeepFaceLab: A popular open-source tool for creating deepfake videos by manipulating faces in images.

Zao: An app that allows users to add their faces to movie scenes using deepfake technology.

Reface: A mobile app that uses deepfake algorithms to allow users to swap faces with celebrities in videos and GIFs.

Poster Design

Artificial intelligence tools accelerate the creative processes in poster design and help filmmakers create more impressive and original visuals. In this context, taking audience feedback into account in the design of AI-supported content can help filmmakers improve their creative processes. Some of the most commonly used AI tools in the design of movie posters are as follows:

Canva: An intuitive graphic design platform that offers AI-driven features to help users create visually appealing movie posters with customizable templates, images and text options.

Adobe Spark: A web-based design tool that leverages AI to help users create stunning graphics, including movie posters, by providing design suggestions and easy-to-use templates.

DeePart: This tool uses AI algorithms to transform images into artistic styles, allowing designers to create unique and visually stunning movie posters by applying different artistic effects to their designs.

Sound & Music

Artificial intelligence is also revolutionizing sound and music production. AI-powered music creation tools accelerate the creation of soundtracks, enabling filmmakers to reflect the emotional atmosphere more effectively. In this context, AI technologies offer the possibility to analyse music genres and create original music tracks that better target the emotional reactions of the audience. Through AI, film dubbing has been enhanced with innovative methods such as voice cloning and voice synthesis. These technologies allow filmmakers to create characters' voices more effectively and enrich the audience experience. Practices such as recreating the voice of a deceased person or the voice of a character can have a profound impact on the viewer. Such innovative sound technologies have the potential to transform the creative process for filmmakers, providing audiences with more immersive and memorable experiences. Informing audiences creates a new area of responsibility.

Some of the most frequently used tools are:

AIVA (Artificial Intelligence Virtual Artist): AIVA is specially designed to compose music. It allows filmmakers to create unique soundtracks tailored to their projects by entering parameters such as mood, style and instrumentation.

OpenAI's MuseNet: MuseNet can generate compositions of various genres and styles. It can create complex arrangements that blend different musical elements, making it a useful tool for film composers looking for inspiration or unique soundtracks.

Amper Music: Amper Music is an AI-powered music composition tool that enables users to create and customize original music tracks. Filmmakers can choose the desired mood, style and length, allowing for personalized soundtracks that fit their cinematic vision.

Jukedeck (now part of TikTok): Jukedeck was an AI music creation platform that allowed users to create custom music tracks for videos. Although integrated into TikTok, its technology laid the foundation for AI-created soundtracks in film.

Soundraw: Soundraw is an AI music generator that allows users to create and edit music tracks. It provides filmmakers with the ability to create soundtracks that can be tailored to their own scenes or themes.

These examples illustrate the growing role of AI in sound and music production in cinema and how technology can enhance the creative process in filmmaking.

Automated Editing

Automated editing tools streamline the video editing process using AI to increase productivity. Key tools include:

Adobe Premiere Pro's Sensei: AI features that automate tedious tasks like colour correction and audio editing.

Magisto: An AI-powered video editing tool that analyses footage and automatically creates polished videos.

Filmora: AI-driven features for automatic scene detection and editing suggestions, making the editing process more efficient.

In conclusion, the integration of AI into the film industry is not only transforming the creative process, but also playing an important role in reshaping the audience experience and ethical standards. From science fiction films to real-life applications of AI, this transformation is an important factor that will determine the future of cinema. While creating a significant cultural shift, it is also causing audiences to become more sensitive to AI-enhanced content. This further increases the ethical responsibilities of filmmakers and requires transparency to ensure audience trust.

Audience Information Needs

It is imperative to acknowledge the significance of viewer information needs in facilitating enhanced connections between filmmakers and their intended audiences. In this context, it is imperative to understand the expectations of viewers and create content that will appeal to them if the film industry is to thrive. In order to respond to these demands, filmmakers are moving towards the provision of more personalised experiences by means of artificial intelligence tools such as data analytics and machine learning. This process has been shown to increase viewer interest in film content, while also enabling creative teams to develop innovative ideas.

It has thus become imperative for filmmakers to disclose the involvement of artificial intelligence (AI) in the filmmaking process, whether it pertains to entirely AI-generated content, and partially AI-generated content, or collaborative efforts with AI. In the context of ethical filmmaking, transparency is of paramount importance. By elucidating the extent of AI's involvement, filmmakers can foster a greater degree of audience confidence and enable viewers to make informed decisions regarding the content they consume. This ethical approach not only respects the viewer's autonomy but also contributes to ongoing discussions about the impact of AI in creative industries. It is imperative for the film industry to establish explicit regulations and standards regarding the utilisation of AI in film production, with the objective of safeguarding both the creative integrity of the film and the audience's right to information. Transparency and ethical principles have been identified as fundamental elements that will shape the future direction of the film industry. It is therefore incumbent upon producers to fulfil their responsibilities to their audience. In this context, it is imperative for producers to adopt proactive approaches to informing viewers and to implement the principle of transparency, thereby fostering trust within the industry. In order to engender greater levels of trust amongst viewers in AI-generated content, it is imperative that the film industry develop ethical oversight mechanisms for such projects.

These mechanisms have been designed to assist in preventing viewers from questioning the quality and reliability of AI-generated content, while also emphasising the importance of the human factor in the creative process. In this process, it is also of great importance that all stakeholders in the industry collaborate to establish common standards and implement them. The establishment of standards will enhance viewer trust and enable producers to develop their creative visions and produce innovative content.

On the other hand, advanced artificial intelligence tools enable viewers to become potential content producers, and as this production becomes increasingly normalised, the need for knowledge about how content is produced is increasing.

SECTION 3 A MODEL SUGGESTION FOR INFORMING THE AUDIENCE

Methods of Informing the Audience

As the film industry navigates the complexities of integrating artificial intelligence, it is imperative to consider the implications of audience engagement strategies that leverage this technology. One promising avenue is the development of interactive viewing experiences, where AI can tailor content in real-time based on viewer preferences and reactions, creating a dynamic narrative environment. Such innovations have been shown to enhance viewer satisfaction, but also to raise questions about the authenticity of the cinematic experience and the potential loss of the filmmaker's original intent. Furthermore, it is imperative to establish ethical guidelines for interactive systems to ensure that audience manipulation does not compromise artistic integrity or lead to algorithmic biases that could misrepresent diverse perspectives in storytelling (Shropshire et al., 2016). It is therefore vital to foster a collaborative relationship between AI and human creativity in order to facilitate the evolution of cinema. Such a relationship will enable filmmakers to explore new storytelling frontiers while maintaining accountability to their audience. This collaboration has the potential to enhance the cinematic experience by respecting audience autonomy and simultaneously challenging the boundaries of creativity.

As filmmakers adopt these technologies, it is imperative that they prioritise ethical considerations and transparency to cultivate trust and foster meaningful connections with viewers. As the film industry continues to explore the integration of artificial intelligence, it becomes increasingly vital to address the ethical landscape surrounding the usage of audience data. The aggregation and examination of viewer preferences, whilst concomitantly enhancing personalisation, engender risks of privacy violations and algorithmic bias. This has the potential to misrepresent diverse demographics and perpetuate stereotypes (Dove et al., 2020). Furthermore, filmmakers must consider the implications of creating content that is excessively tailored to audience desires, a practice which has the potential to compromise the artistic vision and authenticity of storytelling. This tension between commercial viability and creative integrity necessitates a robust framework of ethical guidelines that not only safeguard audience interests but also promote a more inclusive representation of narratives. By cultivating an environment in which ethical considerations are of the utmost importance, the film industry can ensure that the evolution of cinema through AI remains a collaborative endeavour that respects both the artistry of filmmaking and the diverse voices of its audience.

Percentage Information System

The Percentage Information System is a crucial tool for effectively communicating the extent of artificial intelligence's involvement in film production. By providing clear metrics and data regarding AI contributions, filmmakers can enhance audience understanding and trust, ultimately fostering a more engaged viewer experience. As posited by Cincinatus and Sheffet (2016), this system has the capacity to incorporate visual representations, such as graphs or infographics, with a view to elucidating the proportion of AI-generated content in comparison to human-created elements. This approach engenders a dual effect: it both informs the audience and encourages discussions surrounding the ethical implications of AI in storytelling.

Moreover, the implementation of the Percentage Information System can serve as a benchmark for assessing the evolving relationship between technology and creativity in cinema. By establishing clear guidelines for AI contributions, filmmakers can navigate the ethical landscape while promoting a more informed and participatory audience experience.

This approach has been demonstrated to enhance audience awareness and to equip them with the capacity to critically evaluate the narratives presented. In conclusion, it is imperative to cultivate a culture of transparency and ethical consideration in the utilisation of AI in cinema, thereby ensuring the continued relevance and appeal of the art form to a diverse audience.

This transparent approach not only provides viewers with information regarding the extent of AI involvement, but also encourages them to engage in critical discussions about the ethical implications of such technologies in filmmaking. By clearly communicating the percentage of AI contributions, filmmakers can foster a deeper understanding of the creative processes behind their films, thereby enhancing audience trust and engagement.

An Intelligent Marking Model for Determining the Contribution of Artificial Intelligence in Film Production

The increasing influence of artificial intelligence (AI) in creative production processes necessitates a re-evaluation of conventional methodologies and conceptual frameworks within the domain of cinema. The direct involvement of generative AI technologies in areas such as screenwriting, visual design, sound, and music demonstrates that this contribution goes beyond mere technical support, taking on a dimension that affects the integrity of the work.

In the context of contemporary film production, the question of the relative contributions of human labour and artificial intelligence (AI) has become increasingly salient from ethical, aesthetic, and academic perspectives. The Smart Tagging Model proposed in this study is designed as an academic tool for measuring and classifying AI's contribution to film production. The model incorporates a range of components associated with film production and utilises a scoring system that evaluates the AI contribution in percentage terms. It also assigns meaningful categories (smart tags) to films based on this contribution.

Purpose of the Model

The main objective of the model is to provide a set of metrics that can quantitatively evaluate AI contribution, allow for cross-domain comparisons, and support the principle of transparency in the production process. In this context, the model aims to:

- Make AI contribution in film production measurable,
- Analyse the ratio of human and AI collaboration on a field-by-field basis,

- Provide transparent and understandable information to audiences and industry stakeholders,
- Serve as a decision-support tool for issues related to copyright, ethical responsibility, and creative expression.

The purpose of this model is not to propose legal regulations; rather, it offers an academically validated measurement proposal for researchers, producers and creative industry actors.

Method: Field-Based Scoring System

The model takes into account six key components that play a decisive role in the film production process:

- Screenplay
- Visuals (visual design/cinematography)
- Acting
- Editing (montage)
- Sound design
- Music

The AI contribution for each domain is estimated as a percentage based on expert evaluation or analysis of records kept during the production process. The values assigned to each domain range from 0 to 2 points, with the percentage value determining the point allocation.

Table 1. AI Contribution Rate

AI Contribution Rate (%)	Score	Description
0–19	0	Human production
20–60	1	Human–AI collaboration
61–100	2	AI-heavy production

The total score given for these six areas determines the film's overall "smart sign":

Table 2. Total Score

Total Score	Smart Sign	Description
0–2	Human-made	Very low AI contribution, human production dominant
3–7	Hybrid Construction	Balanced contribution from humans and AI
8–12	AI Construction	Film largely produced using AI

Scorecard (Application Template)

Below is a standard scorecard recommended to increase the applicability of the model:

Table 3. Scorecard

Production Area	AI Technology Used	AI Contribution (%)*	Score (0–2)	Description
Screenplay	ChatGPT, Sudowrite	%		Example: Written entirely by AI
Cinematography	Midjourney, Runway	%		Example: Scene and atmosphere creation
Acting	Synthesia, Deepfake	%		Example: Virtual character or face
Editing	Adobe Sensei, Magisto	%		Automatic editing tools
Sound Design	ElevenLabs, Descript	%		AI-generated sound effects
Music	AIVA, Soundraw, Amper	%		Artificial music composition

Visual Representation of Smart Signs

In order to facilitate the implementation of the model and ensure its acceptance in the industry, it is recommended that each "smart sign" category be visualised using simple and universal symbols. These labels facilitate expeditious information retrieval for viewers and enhance transparency of information on digital platforms and in festival catalogues.

Table 4. Example of Smart

Example of Smart Sign	Recommended Symbol	Description
Human-made		Figure 1. No or very low AI contribution
Hybrid Construction		Figure 2. Human and AI contribution together
AI Construction		Figure 3. Content where AI makes a dominant contribution to production

Uses of labels:

- Next to film titles in festival programs
- In digital platform content information boxes
- In production files or funding applications
- In promotional materials such as posters and trailers
- Verification and Verifiability

To increase the applicability and reliability of the model, various supporting documents and methods that enable the measurement of AI contribution are recommended:

- Production Log: Historical records documenting which tools were used at which stages.
- Prompt and File Archive: Documentation of AI inputs used in script, visual, and audio production.
- Version History: Revision records of content production processes (e.g., Google Docs, Figma, etc.).

- AI Usage Reports: Usage analyses automatically generated by software.
- Expert Panel Evaluation: Sample analysis by a panel of independent experts.
- Open Data Sharing: Transparent data presentation in cases such as festival submissions and funding applications.
- These documents ensure that academic analyses are conducted in a sound manner and contribute to the development of an ethical and transparent production culture in the sector.

Limitations

While the model has many advantages, it also has some limitations:

- Risk of Subjective Estimation: Contribution percentages may not always be measured objectively.
- Qualitative Contributions Are Not Considered: Creative value at the aesthetic, artistic, or narrative level is not measured.
- Differences Based on Genre: The meaning of AI contribution may vary in genres such as documentary, animation, and experimental film.
- Timeliness Issue: AI technologies are rapidly evolving, so the model needs to be revised periodically.

Evaluation

This study introduces the Intelligent Marking Model, which has been developed for the purpose of evaluating the contribution of artificial intelligence in the domain of film production. The impact of artificial intelligence on the production process is measured using a domain-based scoring approach and classified using a three-level marking system. The model can be regarded as a practical instrument that seeks to engender transparency and objectivity in academic and industrial evaluation processes.

The role of artificial intelligence in the art of cinema is still in its early stages of evolution. The overarching objective of this model is twofold: firstly, to comprehensively grasp the prevailing production forms, and secondly, to establish a foundation for fruitful discourse concerning the nexus between creative labour, ethical responsibility, and technological transformation.

CONCLUSION

The Importance of the Ethical Marking System

Furthermore, the implementation of an Ethical Marking System has the potential to serve as a pivotal mechanism for addressing audience concerns regarding the authenticity and integrity of AI-generated content. By categorising films according to the degree of AI involvement, filmmakers can enhance transparency and empower viewers to make informed choices about the media they consume. It is argued that this system may be able to draw parallels with existing ethical frameworks in the field of advertising. Research conducted by Shahaby et al. (2024) demonstrates a significant correlation between adherence to guidelines and audience trust and engagement. As the film industry grapples with the complexities of AI integration, the establishment of such a system could facilitate a dialogue about the balance between technological innovation and artistic expression, ultimately fostering a more responsible and inclusive cinematic landscape. By prioritising ethical considerations in the marking process, filmmakers can navigate the delicate interplay of creativity and technology while ensuring that diverse narratives remain at the forefront of storytelling. It is evident that this system will serve to illuminate the role of artificial intelligence in content production, while concomitantly empowering filmmakers to fulfil their ethical responsibilities. This will empower viewers to make informed choices, thereby laying the groundwork for the evolution of cinema and the strengthening of ethical standards.

In addition to the Ethical Marking System, the film industry must consider the implications of audience feedback mechanisms that utilise artificial intelligence to refine content. By employing AI-driven analytics, filmmakers can gain insights into viewer preferences and emotional responses, allowing for a more responsive storytelling approach that aligns with audience expectations. However, this raises critical questions about the potential for homogenising narratives, as creators might prioritize popular trends over innovative storytelling, risking the dilution of artistic expression. Moreover, as AI tools become increasingly incorporated into the creative process, the challenge lies in maintaining a balance between data-driven decisions and the inherent unpredictability of human creativity, thereby ensuring that the essence of storytelling remains vibrant and diverse (Mohamed et al., 2024). In conclusion, it is imperative to cultivate an environment in which ethical considerations guide the utilisation of AI in audience engagement. This approach is pivotal in preserving the integrity of cinematic art while simultaneously enhancing viewer satisfaction.

The ethical labelling system will facilitate viewer understanding of the role of artificial intelligence in film production, whilst also encouraging filmmakers to act responsibly. It is anticipated that this system will emerge as a critical tool in shaping the future direction of the film industry.

Artificial Intelligence and the Future of Cinema

The integration of artificial intelligence into the film industry is effecting a paradigm shift in the manner in which films are created and experienced. This technology has been demonstrated to simplify production processes and enhance storytelling techniques. As filmmakers employ AI tools, they are able to analyse audience preferences with greater efficacy. The advent of machine learning algorithms has precipitated a paradigm shift towards personalised viewing experiences. Furthermore, the utilisation of artificial intelligence in the editing suite has led to the development of innovative visual effects and seamless transitions. As this trend continues, the future of cinema is poised for a remarkable evolution.

Impacts on the Audience Experience

The impact of artificial intelligence on cinema is twofold. Firstly, it is profoundly shaping the audience experience. Secondly, it is transforming viewers' expectations and engagement. In the future, the increased use of AI-powered content that is more personalised and interactive is likely to increase audience loyalty and drive the evolution of cinema. As posited by Joomuck (2024), the influence of artificial intelligence on the cinematic realm is twofold. Firstly, it has been demonstrated to engender an augmentation in audience participation. Secondly, it has been demonstrated to engender a transformation in the creative processes of filmmakers. Consequently, the film industry will embrace innovative narrative forms while also adapting to better meet the expectations of its audience.

REFERENCES

Busselle, R., & Van den Bulck, J. (2019). Cultivation theory, media, stories, processes, and reality. *Media effects*, 69-82.

Çalışkan, Ö. (2024). Yapay Zekâ ile Retorik Farklılıklarla Üretilen Türkçe ve İngilizce Film Diyalogları. *Yedi*, (Sanatta Dijitalizm Özel Sayısı), 357-374.

Çelikel, A., & Akın, M. B. (2017). Yenilebilir filmler ve peynir teknolojisinde kullanımı. *Batman Üniversitesi Yaşam Bilimleri Dergisi, 7*(2/2), 50-58.

Cincinatus, R. B., & Sheffet, M. (2016). "With Percentages the 100 Is Always in the Denominator": From the Field to Pre-Service Teachers. *International Journal of Research in Education and Science*, *2*(1), 143–155. DOI: 10.21890/ijres.75271

Dove, M., Balasubramanian, A., & Narayanan, B. G. (2020). Transparency as a way of attaining quality, safety and optimal food purchases.

Ettema, J. S. (2005). Crafting cultural resonance: Imaginative power in everyday journalism. *Journalism*, *6*(2), 131–152. DOI: 10.1177/1464884905051005

Fitzek, F. H. (2006, June). The medium is the message. In *2006* [). IEEE.]. *IEEE International Conference on Communications : [proceedings]. IEEE International Conference on Communications*, *11*, 5016–5021.

Gerbner, G., & Morgan, M. (2002). *Against the mainstream: The selected works of George Gerbner*. No Title.

Giorgi, S. (2017). The mind and heart of resonance: The role of cognition and emotions in frame effectiveness. *Journal of Management Studies*, *54*(5), 711–738. DOI: 10.1111/joms.12278

Joomuck, N. (2024). Exploring AI in Movies: Present, Future, and Implications. [IJSR]. *International Journal of Scientific Research*, *13*, 1066–1069.

Kuipers, G. (2012). *Medium is the Message*. The Wiley-Blackwell Encyclopedia of Globalization. DOI: 10.1002/9780470670590.wbeog381

McLuhan, M. (2005). The medium is the message.

Mohamed, E., Abbasher, M. H., & Mohamed, M. A. (2024). Navigating the Ethical Horizon: Artificial Intelligence-Generated Content and the Imperative for Transparency and Ethics.

Morgan, M. (2011).George Gerbner: A Critical Introduction to Media and Communication Theory.

Shahaby, A., Setiawan, S. A., & Handoko, D. (2024). *Dampak Penerapan Etika Periklanan Terhadap Efektivitas Iklan Layanan Masyarakat*. Harmoni.

Sheets, P., Rowling, C. M., Gilmore, J., & Melcher, N. (2022). Us and Them: The Role of Group Identity in Explaining Cultural Resonance and Framing Effects. *Mass Communication & Society*.

Shropshire, J., Gowan, A., & Guo, C. (2016). Defining Audience Awareness for Information Systems Research.*ACM SIGMIS Conference on Computers and People Research*. DOI: 10.1145/2890602.2890614

Shrum, L. J. (2017). Cultivation Theory: Effects and Underlying Processes. In Rössler, P., Hoffner, C. A., & Zoonen, L. (Eds.), *The International Encyclopedia of Media Effects*. DOI: 10.1002/9781118783764.wbieme0040

Sun, P. (2024). A Study of Artificial Intelligence in the Production of Film.SHS Web of Conferences. DOI: 10.1051/shsconf/202418303004

Taşçı, G., & Çelebi, M. (2020).Eğitimde Yeni Bir Paradigma: "Yükseköğretimde Yapay Zekâ."

Tekbey, O., & Alaybeyoğlu, A. (2023). Yapay Zekanın Kullanım Alanları ve Etik Sorunlar. *International Journal of Advanced Natural Sciences and Engineering Researches*, *7*(9).

Vives, V. V. (2024). *El presente de la IA en la actividad cinematográfica*. Revista Policía y Seguridad Pública. DOI: 10.26439/vent.indiscreta2024.n032.7474

ADDITIONAL READING

Carey, J. W. (1967). Harold Adams Innis and Marshall McLuhan. *The Antioch Review*, *27*(1), 5–39. DOI: 10.2307/4610816

Ergen, M. (2019). What is artificial intelligence? Technical considerations and future perception. *The Anatolian Journal of Cardiology*, *22*(2), 5–7. DOI: 10.14744/AnatolJCardiol.2019.79091 PMID: 31670719

Fisher, R. B. (2001). AI and cinema–does artificial insanity rule? In *Twelfth Irish Conference on Artificial Intelligence and Cognitive Science* (pp. 1-11).

Frey, M. (2013). The ethics of extreme cinema. In *Cine-Ethics* (pp. 145–162). Routledge.

Gerbner, G. (1998). Cultivation analysis: An overview. *Mass Communication & Society, 1*(3-4), 175–194. DOI: 10.1080/15205436.1998.9677855

Gerbner, G., Gross, L., Morgan, M., & Signorielli, N. (1987). Charting the mainstream: Television's contributions to political orientations. American media and mass culture: Left perspectives, 441-464.

McLuhan, E., & Zingrone, F. (1997). *Essential McLuhan.* Routledge. DOI: 10.4324/9780203992968

Siau, K., & Wang, W. (2020). Artificial intelligence (AI) ethics: Ethics of AI and ethical AI. [JDM]. *Journal of Database Management, 31*(2), 74–87. DOI: 10.4018/JDM.2020040105

Singer, J. B., Báez, J. C., & Rios, J. A. (2023). AI creates the message: Integrating AI language learning models into social work education and practice. *Journal of Social Work Education, 59*(2), 294–302. DOI: 10.1080/10437797.2023.2189878

Song, J., Wang, B., Wang, Z., & Yip, D. K. M. (2023, September). From expanded cinema to extended reality: How AI can expand and extend cinematic experiences. In *Proceedings of the 16th International Symposium on Visual Information Communication and Interaction* (pp. 1-5). DOI: 10.1145/3615522.3615556

KEY TERMS AND DEFINTIONS

AI and Ethics: Focuses on the moral challenges and frameworks associated with the integration of artificial intelligence into creative and cultural production.

AI-Generated Content: Refers to film or media content created wholly or partially using artificial intelligence, raising issues of authenticity, authorship, and ethical responsibility.

Artificial Intelligence in Cinema: Examines the applications and implications of AI technologies in film production, narrative design, editing, and audience interaction.

Audience Studies: Explores the analysis of viewer behaviour, preferences, and the evolving relationship between audiences and cinematic content, particularly in the context of AI-generated media.

Digital Media Regulation: Addresses the legal and institutional frameworks that oversee digital content creation, distribution, and ethics, particularly with the rise of AI and automated storytelling.

Ethical Labelling in Cinema: Involves the implementation of transparent markers or guidelines that inform audiences about the ethical dimensions of film content, especially when AI technologies are used.

Film and Digital Communication: Analyses the intersection of traditional cinematic forms and new digital communication technologies, particularly in how narratives are shaped and delivered.

Film Classification: Refers to systems that rate films based on content, age suitability, and societal norms, now challenged by the complexities of AI-generated narratives.

Information Systems in Media: Covers digital infrastructures and platforms used for distributing, tagging, and regulating media content, including those enabling AI integration in cinema.

Media Studies: An interdisciplinary field that investigates the content, history, and effects of various media, including emerging technologies in audio-visual storytelling.

Viewer Guidance Systems: Includes tools, labels, and standards designed to help viewers make informed choices about content, adapted to the presence of algorithmically generated media.

ENDNOTE

* The percentage fields in this scorecard are intentionally left blank, as they are designed to be filled out by filmmakers or production teams based on their own AI usage data during the film production process. These values will vary across projects and are essential for accurate self-assessment within the proposed model.

Chapter 6
Narrative Flows in Digital Movies:
The Effect of Digital Technologies on Non-Linear Storytelling

Semra Kotan
https://orcid.org/0000-0002-1583-3772
Atatürk Üniversitesi, Turkey

ABSTRACT

This study examines the effects of digital technologies on non-linear narrative structures in cinema. With the rise of digital cinema, the flexibility provided in areas such as editing, camera use, visual effects and sound design in filmmaking processes has transformed traditional narrative patterns. Non-linear narrative is characterized by flashbacks, parallel plots and structures where events are not presented in chronological order, and where time is broken. Digital technologies play a decisive role in the creation of such narratives; they support the development of new narrative forms that question the perception of time and space and direct the audience through memory, perception and emotions. The study aims to present a theoretical framework through post-2000 digital film examples and academic sources with a descriptive literature review method.

INTRODUCTION

The rise of digital cinema has created a radical change in filmmaking processes and reshaped the narrative forms of cinema. Especially in recent decades, the opportunities offered by digital technologies have provided filmmakers with great flexibility in many areas such as editing, camera use, visual effects and sound design.

DOI: 10.4018/979-8-3373-1967-4.ch006

These technological developments have allowed new narrative forms to emerge by transcending traditional cinematic structures. Non-linear narrative refers to a structure in which time does not progress linearly and events are presented with breaks, flashbacks and parallel narratives (Bordwell, 2008). In digital cinema, such narrative techniques detach the audience from the flow of time and direct them more through abstract concepts such as memory, perception and emotional transitions (Mulvey, 2006). The role of digital technologies in this transformation has become evident with innovations in filmmaking, especially in the use of editing techniques and visual effects.

The use of digital cameras, speed and flexibility in editing, elements such as computer-aided visual effects and sound design allow themes such as the bending of time and emotional inner journeys to be processed more effectively (Elsaesser, 2015). However, the non-linear narrative form of digital cinema challenges the narrative boundaries of cinema by offering both a mental and emotional journey to the audience (Gunning, 2004). The continuous questioning of the perception of time and space in digital cinema has become possible, especially with technological developments. Thanks to Digital Editing, the breaking of time, transitions between past and present, and the interaction between parallel narratives have become easier. While digital technologies provide filmmakers with the opportunity to break the structure of the narrative and detach the audience from traditional perceptions of time and space, they also transform non-linear narrative forms.

This study is based on the descriptive literature review method, which is one of the qualitative research approaches. Descriptive literature review; It aims to systematically compile the existing knowledge on a specific subject and to reveal the basic trends, theoretical approaches and research gaps (Snyder, 2019, p. 335). In this context, in order to evaluate the effects of digital cinema on non-linear narrative structures, current academic literature, theoretical texts and post-2000 digital film examples were scanned. During the literature review, studies on what kind of transformations digitalization in cinema has created at technical, aesthetic and narrative levels were systematically analyzed. In addition, theoretical approaches and sample films that examine the role of digital technologies in restructuring the narrative in time and space were focused on. Through this method, a structured framework is presented on the existing knowledge, and the effect of digital technologies on the narrative structure is discussed both at the conceptual level and through applied examples. The study aims to compile the existing knowledge in the light of the information already in the literature and restructure it in a meaningful context.

The impact of digital cinema on non-linear narrative structures is a highly interdisciplinary topic, and different researchers have addressed this issue from various perspectives. Kinder (1999) argues that digital media enables multiple narrative layers, which questions the traditional linear story structure. In his work "Rewiring the Spectator," he draws attention to the fact that digital technologies transform the interaction with the audience. In his work "The Language of New Media," Manovich (2001) analyzes the effects of digital media on cinema in detail. He argues that cinema has moved away from linear narrative and gained structures based on database logic with the digitalization process. Thompson (2003) examines how the aesthetic innovations provided by digital cinema are reflected in the narrative. He states that techniques such as editing, camera movements, and time jumps have been redefined by digital technologies. Bordwell (2006), in his work "The Way Hollywood Tells It," accepts the effects of digital technologies on cinema narrative, but argues that the classical narrative structure is still valid. However, he states that in some "puzzle films" (e.g. Memento, Mulholland Drive), the time-space structure is intentionally broken. Pisters (2012) argues in his study "The Neuro-Image: A Deleuzian Film-Philosophy of Digital Screen Culture" that digital narrative has gained a neurological dimension through Deleuze's concepts of "time-image" and "movement-image". He examines how non-linear structures in digital films relate to audience memory and perception. Brown (2013) focuses on the philosophical dimension of digital cinema in his study "Supercinema: Film-Philosophy for the Digital Age". He discusses how digital images transform cinematic reality and what effects this has on the narrative plane. On the other hand, Suner (2006) examines Turkish cinema after 2000, particularly in his study "Ghost House: Belonging, Identity and Memory in New Turkish Cinema", and focuses on narrative structures and transformations in the time-space relationship. It discusses how non-linear narratives open up to the individual's inner world and social traumas. Öztürk (2024) stated in his study titled "Multi-Layered Narrative in Cinema: A Study on the Movie Inception" that the layered narrative structure is a powerful narrative tool in cinema that involves the audience in the multi-dimensional structure of the story and mentally activates them, and that this technique is shaped by elements such as the non-linear use of time, intertwined narratives, and the consciously directed sense of uncertainty of the audience. Erzurum (2019), in his article titled "Did Transmedia Narrative Begin with Movies?", stated that non-linear narratives allow for the active participation of the audience, and stated that Star Wars movies are suitable examples for transmedia storytelling. These studies reveal that digitalization creates not only a technical change in cinema, but also an aesthetic, perceptual, and theoretical transformation. Each author has addressed the subject from a different dimension and contributed to our understanding of how digital cinema reshapes the narrative structure.

The literature review and the film examples analyzed within the scope of the study reveal that digital technologies encourage non-linear structures in cinematic narratives and that these structures have allowed the traditional boundaries of time-space perception to be exceeded. In particular, digital editing software and advanced visual effects technologies have enabled the reconstruction of the plot in non-linear ways; thus, narratives have been shaped more around psychological depth, subjective perception, memory, dreams and parallel realities. In addition, the speed and flexibility offered by digitalization in the editing process has provided filmmakers with the opportunity to make dynamic transitions between temporal leaps, flashbacks, alternative timelines and subjective narratives. The findings show that in digital age cinema, narrative has become not only a storytelling but also an area of experience that increases the cognitive participation of the audience.

TRADITIONAL-LINEAR NARRATIVE STRUCTURE IN CINEMA

Narrative is a form of expression that has been going on since the existence of man, beyond being just a term or technical concept. The individual's effort to understand and make sense of both himself and his environment is shaped through narratives. In this respect, narrative is a cultural and intellectual practice that permeates every area of life. Jerome Bruner (1991, p. 4) argues that narrative is a cognitive structure and states that people make sense of the world by making stories. According to him, narrative is one of the basic ways of organizing our mental lives. This view shows that narrative is a basic tool of thought and expression not only in literature but also in disciplines such as psychology, history and sociology.

Narrative comes from the verb "to tell". "Narratives are languages (langues) conveyed through the words (paroles) of concrete verbal communication or other means of communication." (Chatman, 2009, p. 21). Throughout history, humans have tried to understand themselves, their surroundings, and their lives by telling them; thus, narrative has become a deep-rooted tradition encompassing every aspect of life. Therefore, narrative is the telling of an event, situation, or story. According to Randall, narrative is the narration of events, experiences, or similar situations in an orderly manner (1999, p. 92). In other words, it is the expression of what a person or character experiences in a specific order and in a meaningful manner. Narratives are usually constructed on a story or series of events and have a purpose; this purpose may be to explain a thought, to describe a situation, or to convey an emotion. In parallel with this, Paul Ricoeur draws attention to the relationship between narrative and the experience of time. According to him, narrative provides structure to time and thus establishes meaningful connections between the past, present, and future. Narrative does not only order events; it also establishes causality, purpose, and value relationships between these events, thus providing integrity to human experience (Ricoeur, 1984, p. 52). In addition, in the sociological context, Arthur Frank particularly examines narratives of illness and reveals how individuals reconstruct their identities through such narratives. According to Frank, individuals reconstruct their lives through narratives in times of crisis such as illness (Frank, 1995, p. 55).

Different media environments (oral culture, written texts, digital platforms) shape the narrative in line with their own technical and aesthetic codes. In this diversity, cinema has become one of the most effective tools of modern narrative forms thanks to its audiovisual narrative power. This perspective shows that narrative is not only a form of expression, but also a healing and transformative tool. Narrative in cinema is not only the ordering of events, but also the presentation of these events within a certain framework of order and meaning. When defining the classical narrative structure, David Bordwell and Kristin Thompson state that narrative in cinema generally includes "a structure based on cause-effect relationships, continuity in time-space, and character-oriented" (Bordwell & Thompson, 2013, p. 81). This structure is functional in terms of the audience being able to easily follow the story and identify with the characters. On the other hand, narrative in cinema is not limited to content alone; the language of cinema also plays a decisive role in the construction of the narrative. According to Christian Metz's structuralist film theory, cinema is a "sign system" and the narrative is established through visual and auditory codes within this system (Metz, 1974, p. 69).

Arguing that cinema functions like language, Metz emphasizes that meaning arises not only from events but also from how these events are shown. The basic elements that determine the types of narrative in cinema are the structure of the film, the use of time and space, the treatment of characters and the way information is presented to the audience. The classical narrative structure, which is one of the most common forms, is referred to by names such as traditional, Hollywood-style, popular and dramatic narrative, which is especially dominant in Hollywood cinema. The traditional narrative structure is a classical narrative model in which events are presented in a certain order and mostly consists of introduction, development and conclusion sections. This structure makes it easier for the reader or viewer to understand the story and empathize with the characters. According to Aristoteles, fiction is a structure that is constantly reproduced in line with the creative horizon of the artist. According to him, what is essential in narrative is the consistency and continuity of the cause-effect relationships between events. The result of one event can turn into the cause of another event; Thus, the narrative gains a dynamic and mobile structure. For this reason, events should progress by being connected to each other in a logical chain rather than by coincidences (Aristoteles, 1987, p. 11-12).

The basic elements of the traditional narrative structure are as follows:

- The traditional narrative structure is based on the classical dramatic structure understanding, which ensures that the narrative is organized in a logical and temporal order. This structure usually consists of three basic sections: introduction (beginning), development (middle section) and conclusion (resolution). In the introduction section, the basic components of the story are introduced: the main character(s), the time and place frame in which the narrative takes place are determined, and the audience is informed about the basic situation or order of the narrative. In this stage, there is often a "disruption of balance" that will trigger the events, in other words, a development that changes the character's routine life. In this way, the audience both begins to identify with the character and becomes involved in the universe of the narrative. The development section is the longest part of the narrative and is the section where the conflict blossoms and the character's struggle with the obstacles he encounters is discussed in detail. These conflicts can be internal (dilemmas in the character's own inner world) or external (struggle with factors such as other characters, society, nature). In this section, the character experiences various failures while trying to achieve his goals, goes through turning points, and both the events and the character himself undergo transformation in this process. In this section, the tension increases, the complexity of the events deepens and the audience begins to wait in curiosity for the solution. The conclusion section is the section where the conflict is resolved, it becomes clear whether the character has achieved his goals or not and the narrative is connected to an ending that gives a sense of completion. Here, the open ends in the narrative are closed, the relationships between the characters are clarified and a new state of balance is established, offering the audience a satisfying closure. It is seen that this structure overlaps with the classical narrative model, which is especially identified with Hollywood cinema. According to the three-act structure model developed by the famous screenwriter Syd Field, these sections determine not only the content but also the mathematics of the script in terms of duration and structure (Field, 2005, p. 23–27). This structure allows the audience to easily follow the cause-effect relationships in the narrative; thanks to the linear flow of time, character development can be clearly observed and the events in the narrative progress in a logical integrity. Thus, the traditional narrative structure stands out as a functional model that facilitates the interpretation of the narrative and allows a strong identity to be established between the audience and the character.

In traditional narratives, the progression of events is based on a logical cause-effect relationship rather than randomness. Each event occurs as a result of a previous event and is structured in a way that prepares the next event. This linear and causal structure ensures the consistency of the narrative and the viewer/reader's orientation within the narrative. The actions of the characters in the narrative produce certain results in line with their goals; this contributes to the establishment of a logical structure in the progression of the story. This principle of causality is considered one of the most fundamental elements of the classical narrative form. This linear time order, in which time is generally presented in a linear manner, that is, by adhering to the chronological order of events, simplifies the flow of the narrative and makes it easier for the viewer or reader to establish a logical relationship between events. In his study on the organization of narrative time, Gérard Genette states that in classical narratives, instead of complex structures such as (time jumps), a straight time flow is generally preferred (1980, p. 33–34). Similarly, Mieke Bal emphasizes that the sequential presentation of events in narratives stabilizes the narrative plane and facilitates the mental follow-up of the receiver (2009, p. 83).

- In the traditional narrative structure, conflict creates the driving force of the narrative through the obstacles the character encounters. These conflicts may be dilemmas experienced in the character's inner world, or they may arise from environmental conditions or relationships with other characters. Certain turning points in the progression of the narrative give direction to the story and represent critical moments that change the character's fate. These structures increase the dramatic tension of the narrative while also facilitating the audience's emotional connection. McKee emphasizes that turning points are structures that determine the rhythm of the narrative and that these moments provide "motion" to the narrative (1997, p. 196–205).
- In the traditional narrative structure, the plot is mostly shaped around a single main character. This character is the carrier element of the narrative and all developments are shaped around his efforts to achieve a certain goal. The difficulties and obstacles the character encounters complicate the process of achieving this goal and feed the dramatic structure of the narrative. The progression of the narrative is directed by the decisions the character makes during this process and the consequences of these decisions. This fictional structure makes the narrative more impressive by allowing the reader or audience to establish an emotional bond with the character. Özdemir states that in the classical narrative form, a character's struggle towards a goal is placed at the center of the events and the tension of the narrative is built around whether this goal is successfully achieved or not (2005, p. 87).

-In classical narrative structures, events are resolved at the end of the narrative and the audience or reader is brought to a clear conclusion without leaving any ambiguity. Such structures are based on a principle of closure where all the basic questions that arise in the narrative are answered, the fates of the characters are determined and dramatic conflicts are resolved. In this context, a satisfying and completed narrative is presented to the audience.

-Hutcheon emphasizes that traditional narratives design the audience as a "passive observer" (2006, p. 27). In classical narratives, the audience generally cannot change the outcome of events and does not have the chance to intervene in the obstacles that the characters encounter. In classical narrative structures, according to the principle of mimesis (imitation) based on Aristotle's understanding of drama, the audience is expected to identify with the character and achieve catharsis (emotional purification) from this process. While watching the representation seen on the stage, the audience identifies with the character and experiences his feelings, thoughts and behaviors as their own. This identification allows the audience to empathize with the character. The events and emotional processes experienced by the character also affect the audience, so the audience can sometimes even accept a wrong or negative behavior because they are in the character's world at that moment and look from his perspective (Kotan, 2022, p. 50).

- In linear narratives, space is the concrete or abstract physical environment in which the narrative evolves over time and events take place. In this form of narrative, space is not just a backdrop or background, but a structural element that supports the development of events in a logical order. Like time, space also contributes to the orderly flow of the narrative; events occur within a certain spatial order, and this order helps the viewer make sense of the movement in the narrative. The continuity and consistency of space makes it easier for the viewer to grasp the story over time. For example, in scenes where a character leaves home and goes to work, the transitions between spaces are connected by a logical causal chain. In this context, space mediates the visualization of time: as time progresses in the story, the spaces the characters are in also transform or change places. Thus, the viewer intuitively perceives the passage of time and the evolution of the character's journey through spatial movements. In addition, in linear narratives, space also reflects the character's mood, social position, and the tone of the narrative. For example, in a classical dramatic structure, the story usually progresses through a "regular starting place" (e.g., a warm home environment), followed by "crisis places" (dangerous streets, foreign cities, battlefields, etc.), and finally a "resolution or return place" (a return to the starting point or a new order). This type of spatial evolution supports both the development of character and the dramatic structure of the narrative.

EVOLUTION OF NARRATIVE FLOW: NON-LINEAR STORYTELLING

Classical Hollywood narrative has dominated theatrical texts, novels, and cinema for many years. Frequently used in Hollywood cinema of the 1940s and 1950s, this structure follows the obstacles and solutions that characters encounter as they attempt to achieve a goal. Traditional linear structure is also widely used in adventure, drama, and romantic films. Films present the development and struggles of a main character to the audience in a specific order. However, with the development of digital technologies, especially in digital films and new media narratives, this linear structure has evolved significantly. With the influence of digital technologies and postmodern narrative techniques, non-linear narratives have become more prevalent in cinema, allowing for greater flexibility in narrative forms. In contrast to the linear and centralized nature of traditional narrative, digital narratives are often diversified with elements such as multiple perspectives, time jumps (non-linear storytelling), open-ended structures, and participatory audience experiences. Digital narratives enable the viewer to be not only a passive follower but also an active participant. The non-linear narrative form can be considered a structure more suitable for the nature of the human mind. However, considering the globalization process and the defining characteristics of new media environments, it is no longer possible to fix any phenomenon in the context of time and space (Thakur, 2008, p. 262). In this context, it is clear that cinematic narrative is constructed not only at the content level but also at the formal and structural level. While the classical narrative model has been dominant throughout most of the history of cinema, alternative narrative forms that transcend the boundaries of this structure have become more visible, especially with digitalization. The temporal, spatial and structural diversification of narrative shows that the classical model is not the only valid form, on the contrary, the language of cinema is in a dynamic evolution.

The characteristics of Non-Linear Storytelling are as follows:

-In non-linear narrative structures, events are presented without adhering to a chronological order; forward and backward jumps can be made within the timeline. In this type of narration, interruptions occur in the narrative time and information about the events before or after is given without adhering to traditional causal relationships. Thus, the flow of time within the narrative is intentionally interrupted and the viewer or reader is expected to actively establish the relationship between the events. According to McKee (2011, p. 46), this method is a technique that disrupts temporal continuity, but thus allows the narrative to gain multi-layered depth. This structure consciously plays with the perception of time of both the narrator and the receiver, allowing the narrative to become more impressive and thought-provoking. In this type of narration, events can be presented to the audience with various techniques such as flashbacks, jumps to the future, parallel time periods or momentary interruptions. While returning to a character's past experiences or events is interpreted as a flashback, this technique can usually reveal the character's past or bring a different perspective to the story. The concept of jumping to the future can be defined as jumping to a future point in events. This situation offers the audience a clue or prediction about what will happen in the future. -Postmodern narratives[1], time does not function in a linear integrity; past, present and future are intertwined. The narrative is based on a fragmented understanding of time and this fragmentation forms the basic structure of the narrative. According to İspir and Kaya, "According to the evaluation and perspective of the postmodern understanding, there is no definite boundary between the past and the present. In addition to this state of limitlessness, it is not possible to talk about a holistic concept of time and space" (2011, p. 856). Non-linear narrative is a form of narrative that deliberately disrupts the order of events without adhering to the classical chronological flow of time and thus creates a different dynamic in the narrative structure. In this type of narrative, the beginning, development and conclusion of the story can be presented as independent sections, separated from the traditional structure. On the other hand, the story can progress simultaneously or in parallel in different time periods within the film. Such structures also offer the opportunity to create narratives in which more than one character or event develops independently of each other but comes together in the end.

- The viewer may have to think carefully from time to time to bring the events together and create a meaningful whole. This allows the viewer to interact more deeply with the film. In this context, non-linear narratives require the viewer or reader to establish the cause-effect relationships within the narrative themselves, creating a more active reception process. In this context, non-linear narratives offer a structure that stimulates the mental activities of the receiver and differ from traditional narratives in this respect. In other words, the ultimate meaning of the narrative emerges not from the narrator, but from the efforts of the viewer or reader. Thus, the meaning that each individual derives from the narrative may differ according to their own experiences and interpretation style.
- Such narratives apply certain techniques to prevent viewers from identifying with the characters or the fictional. These techniques are related to the narrative clearly showing that it is aware of its own fictional nature. The story does not only tell a plot, but also deals with the telling itself.

In this way, the narrative provides content on the one hand and conveys the message “I am a story” and “Don't get carried away” on the other. This situation constantly reminds the reader or viewer that the narrative is not real but fictional. The narrative plays with its own structure, questioning the classical rules of narrative[2]. Some of the basic examples of postmodern narrative techniques are a character talking to the author or reader, a story being told within a story, the narrator intervening directly in events or making explanations by interrupting the flow of the narrative, or breaking reality in the middle of the narrative with sentences like “this is fiction.” In this context, the following table has been prepared to more systematically compare the differences between classical linear narratives and nonlinear narrative structures that have evolved in the digital age. The table presents a comparative overview of key components such as the time-space framework, viewer position, and causal structure. This comparison concretely demonstrates how digital technologies have transformed classical narrative structures (see Table 1). For example, while linear narratives typically position the viewer as a passive observer, nonlinear narratives transform the viewer into an active interpreter. Similarly, unlike the strong causal principle of classical narratives, digital narratives exhibit a fragmented, often causal, structure.

Table 1. Structural and Experiential Differences Between Linear and Non-Linear Storytelling in Cinema

Linear - Characteristics of Traditional Narrative	Characteristics of Non-Linear Storytelling
Logical Chronology (Linear Progression of Time)	Disruption of Chronology (Disorder of Time)
Holistic Structure	Fragmented Structure
Strong Cause-and-Effect Relationship	Weakened Causality
Fixed, Limited, Measurable Time and Space	Relative Time and Space
Pacified Audience	Active Audience
Predictable, Definite, and Clear Endings	Ambiguous or Open-Ended Endings
Identification (Catharsis)	De-Identification Techniques (Distancing Effects)

Nonlinear storytelling in cinema is a powerful narrative form that pushes the boundaries of classical narrative, shatters the perception of time and space, and makes the audience more active and questioning. This structure, which can be seen as a part of the postmodern cinema approach, reflects the dynamic structure of contemporary cinema narratives in terms of the interaction it establishes with the audience.

Nonlinear narratives, which are further developed with digital editing techniques, expand the expressive possibilities of cinema. Quentin Tarantino's film "Pulp Fiction" (1994), which tells the intersecting lives of different characters, is considered an example of nonlinear storytelling as a production that does not conform to the classical timeline, where the episodes are not sequential and the end of the events is given at the beginning and the beginning in the middle. This structure disrupts the audience's perception of time and expectations about the events, while the audience is forced to establish mental connections between the events in the film. "Fight Club" (1999) is a film with a nonlinear narrative directed by David Fincher. The events are conveyed to the audience from the present, but most of them have happened in the past. Throughout the film, time frequently jumps back and forth. This also parallels the narrator's mental state. In the film, the events are conveyed to the audience from the present, but most of them have happened in the past. Throughout the film, time frequently jumps back and forth. This also parallels the narrator's mental state. The film deviates from the classical three-act structure. The final scene at the beginning (the moment when the narrator puts a gun in his mouth) is explained with later sections showing how it got there. The story progresses not according to a single timeline; it depends on the narrator's mental map. Christopher Nolan's film "Memento" (2000) also tells the story of Leonard, who has short-term memory loss, as he finds his wife's killer, and the narrative progresses in two separate timelines: black-and-white scenes are given linearly, while colored scenes are presented in reverse chronological order. This structure forces the audience to experience the main character's confusion firsthand. This active viewing process creates an experience that is opposite to the passivity of classical narrative.

"Mulholland Drive" (2001) is another example of breaking the classical narrative patterns directed by David Lynch. In the film, which tells the dream and awakening of a woman experiencing an identity crisis in Hollywood, reality and dreams are intertwined. Most of the events in the film are not connected to a clear temporal order; there is uncertainty between the characters and events. "Irreversible" (2002), directed by Gaspar Noé, tells the story of revenge after a rape. The film progresses completely in reverse chronological order: the last scene is at the beginning, the first scene is at the end. "Eternal Sunshine of the Spotless Mind" (2004), directed by Michel Gondry, is another film that tells the inner journey of a man who decides to erase his memories with his ex-lover from his mind. While the events in the film progress on a mental plane, in the order of remembrance, the memories are presented in mixed, dream-like transitions in terms of time. The story is presented in emotional order, not chronological. The timeline is broken.

EFFECTS OF DIGITAL TECHNOLOGIES ON NON-LINEAR STORYTELLING

Digital transformation refers to a multi-layered process of change experienced at social, cultural, economic and individual levels due to the rapid developments in information and communication technologies. In this process, the transition from analog systems to digital infrastructures is evaluated not only as a technical transformation, but also as a comprehensive paradigm shift that restructures ways of thinking, producing, representing and communicating (Castells, 2010). Digital transformation has radically changed all stages of content, from production to distribution, access to consumption, especially in the fields of media and cultural production. Manovich (2001) underlines elements such as digitization, automation, interchangeability, multiple connectivity and visualization while defining the characteristics of digital media. These characteristics provide a basic framework for understanding how digital technologies transform traditional narrative forms. Automation and editability, especially in media production, have made it possible to re-edit content and paved the way for the development of experimental narrative forms with non-linear structures. Jenkins (2006) states that digitalization is not only a technological development, but also a cultural transformation. According to him, the digital age is defined by the concept of "participatory culture"; the audience is no longer just a consumer of content, but also a producer and meaning-giver. This situation has led to a reshaping of the relationship between the audience and the narrative, especially in visual narrative arts such as cinema; the audience has begun to establish a much more interactive relationship with the text thanks to non-linear narratives.

The digital transformation that has been taking place since the beginning of the 21st century has radically transformed many areas, from cultural production methods to communication tools, from individual experiences to artistic expressions. One of the areas most affected by this transformation has been cinema. Digital technologies have fundamentally changed not only film production processes but also narrative structures, viewing styles and the relationship established with the audience. Digital technologies, which have replaced traditional analog production tools, have provided faster, more flexible and more affordable production opportunities for the art of cinema, while also paving the way for new forms of expression on a narrative level. In this context, digitalization is not only a technical innovation in cinema; it also means an aesthetic paradigm shift.

The development of digital technologies has brought about radical transformations in storytelling formats. Especially with the widespread use of the internet, interactive media and digital editing technologies, non-linear narrative structures have become more functional and creative in the digital environment. While events are presented in chronological order in traditional narratives, narratives developed with digital technologies are fragmented in terms of time and space and reach a multi-layered structure (Ryan, 2006, p. 104). The development of digital editing programs has facilitated the manipulation of time in film and video production; it has allowed techniques such as flashback, flashforward and parallel editing to be used more freely. In this context, digital technology has offered directors not only technical but also structural freedom, making the breaking of the temporal line an aesthetic choice (Manovich, 2001, p. 245). This provides a narrative experience that requires the viewer to combine the story in their own mind, making them more active. In addition, hypertext structures and interactive stories developed thanks to digital technologies remove the viewer or user from the position of being a mere viewer and make them a part of the narrative. Such narratives do not have a single outcome; they have multiple possibilities, branching stories and structures open to user intervention (Aarseth, 1997, p. 79). Game-based narratives in particular, thanks to this flexibility offered by digital technology, break down the classical time-space integrity and challenge the boundaries of postmodern narrative.

Digitalization has also transformed the production and consumption forms of the story. Digital platforms such as Netflix and YouTube have individualized the viewing experience by allowing users to experience content at any time and in any order they want. This new narrative culture forces both the content producer and the consumer to think outside the linear structure and redraws the boundaries of linear narrative (Jenkins, 2006, p. 121).

The effects of digital technologies can be listed as follows:

- Digital editing technologies have made it technically possible to disrupt the linear flow of time and space and have also ensured its widespread adoption as a method in cinema aesthetically. Digital editing technologies have eliminated the necessity of linear time and space in cinema and have allowed directors to freely shape the narrative structure. Thanks to this technological opportunity, techniques such as flashback, flashforward, parallel editing and temporal fracture have become more effective and creative. To elaborate on an example film, Alejandro González Iñárritu's "21 Grams" (2003) presents events with a disrupted timeline, establishing relationships between the characters' past traumas and their current decisions. Such digital editing stands out not only as an aesthetic choice, but also as a structural strategy that enables the audience to actively participate in the process of making sense of the narrative.
- Digital environments have technically facilitated the processing of narrative from different perspectives, thus developing multi-layered narrative forms that go beyond the classical linear structure. This structure allows the audi-

ence to present the different experiences and interpretations of the characters regarding the same events in parallel (Ryan, 2006, p.108). In Doug Liman's film "Go" (1999), the events that occur during one night are presented from the perspective of three different characters, in a structure where time is rewound. The repetition of the same events but from different angles allows the audience to discover the full picture of the narrative in pieces. Since this technique can be applied more fluently and consistently thanks to digital editing, the time-space integrity in the narrative is replaced by a polyphonic and multi-perspective narrative structure.

-Digital narratives that allow the user to determine the flow of the story or choose alternative paths disrupt the traditional "beginning-middle-end" line and allow the narrative to evolve in multiple directions. Such narratives turn the viewer into a participant who directs the story, not a passive consumer. Interactive film and game narratives in particular are the pioneers of this structure (Jenkins 2006, p. 136). In this context, new media forms turn storytelling into an experience. Interactive cinema and VR technologies in particular transform the viewer into a subject who is not just a viewer but also an "existing" within the narrative universe. (Jenkins & Thorburn, 2003, p. 126) For example, Netflix's interactive film "Black Mirror: Bandersnatch" (2018) offers the viewer the chance to make the character's decisions and reach different endings with these decisions. While some choices in the film completely change the flow of scenes, others open up previously unseen narrative branches. These structures create a narrative experience specific to each user by exceeding the boundaries of classical linear narrative with the flexibility offered by digital technology. In this context, with the development of digital technologies, cinematic narratives have begun to take shape not only with visual and auditory elements, but also with big data and algorithms. In these new narrative forms, content can be dynamically shaped according to the viewer's preferences, interaction history or behavior on the platform.

- In digital narratives based on hypertext logic, the user can access different parts of the narrative out of order by clicking on multiple links (Landow, 1997, p. 56). Hypertextuality is a structure that allows the reader or viewer to choose their own reading/guidance path by establishing connections (links) between texts. This approach has become a fundamental paradigm for digital narratives because it breaks the traditional linear narrative structure and offers a multi-faceted and unordered reading/comprehension experience. In digital narratives, this practice allows the user to access different parts of the story in the order they want. For example, a user who wants to access information about a character's past can click on a certain link to go to that part of the story; then they can take another path again. Thanks to this structure, the narrative becomes modular, multi-layered and customizable instead of progressing in a single line.

- Simulations created with digital technologies blur fiction and reality, creating a postmodern narrative aesthetic. Jean Baudrillard's theory of simulation argues that the boundaries between reality and fiction have become increasingly blurred as representations replace reality in the postmodern era. According to him, simulation no longer represents reality but creates a hyperreality that replaces reality; in this case, "reality" becomes merely reproduced impressions (Baudrillard, 2003, p. 146). Similarly, in cinema narratives produced with digital technologies, virtual universes created through tools such as visual effects, computer-aided animations, augmented reality (AR) and virtual reality (VR) offer the viewer a hyperrealistic experience. Although this experience is visually convincing, it is fundamentally completely disconnected from reality and has a simulative structure. Therefore, digital cinema ceases to be a mere representation tool and offers a fictional universe that draws the viewer in, imitates the "aesthetics of reality" but is essentially artificial. In this context, narratives constructed with digital technologies, as described by Baudrillard, produce simulations powerful enough to erase the trace of reality and replace it, thus forming the basis of postmodern narrative aesthetics. Virtual worlds, visual effects, augmented reality (AR), virtual reality (VR) and hyperrealistic animations created through digital technologies offer the viewer experiences that "seem real" but are actually completely fictional.
- Transmedia storytelling, another element that shows that the boundaries of storytelling have expanded in the digital age, is a form of narration in which a story is told in parts in different digital media (film, game, social media, website, etc.) and each medium makes a unique contribution to the story. According to Jenkins, this process occurs when the basic elements of a fiction are systematically disseminated through multiple distribution channels, and each media platform makes its own unique contribution to the development of the narrative (Jenkins, 2006, p. 97-98). This form of narration is a product of media convergence, one of the defining characteristics of the digital age; because media convergence refers to a situation in which different media systems coexist and content circulates fluidly between these systems (Jenkins, 2016, p. 436). Thus, transmedia narratives offer the audience a multi-layered and interactive experience by using the transitivity between media platforms.

-Digital editing facilitates techniques such as stream of consciousness, flashback, fragmentation, allowing the narrative to be closer to the character's mental processes (Bordwell, 2008i p. 221). In this way, the narrative can get closer to the character's mental processes; thoughts, memories, hallucinations or emotional fluctuations can be directly represented with visual and auditory tools. For example, in the film "Memento" (2000), the forward-backward jumps allow us to experience the character's memory loss, while in the film "Mulholland Drive" (2001), the boundaries between reality and dreams are blurred through digital editing with the stream of consciousness technique. Fragmentation, on the other hand, reflects the chaos in the character's mind by replacing the classical linear structure with a fragmented, temporally dispersed narrative structure. These narrative forms integrate the narrative more deeply into the character's inner world thanks to the editing flexibility offered by digital technologies.

- Digital editing software has created a significant transformation in filmmaking processes, allowing filmmakers to edit scenes in much faster, more flexible and creative ways. In traditional analog editing techniques, when film strips were physically cut and pasted, this process was time-consuming, limited, and difficult to reverse. However, digital editing programs (such as Adobe Premiere Pro, Final Cut Pro, Avid Media Composer) make it possible to cut scenes from any point and put them in a different order, and to enrich them with visual effects, transitions, and time manipulations. This makes it especially easy to implement non-linear narrative structures. In this way, free transitions can be made between time periods within the story, scenes can be ordered according to thematic or psychological associations rather than sequential, and events can be edited without adhering to the classical cause-effect relationship. Thus, digital editing becomes not only a technical convenience, but also an aesthetic tool that shapes the narrative structure.
- The development of artificial intelligence technologies is leading to radical transformations in digital narrative structures, both formally and in terms of content. The linear time structure, fixed plot and character-centered progression that define traditional narrative structures are being replaced in AI-supported narrative systems by new narrative forms that are open to multiple possibilities, data-based and interactive with the user. Thanks to natural language processing (NLP), artificial learning and algorithmic editing systems in particular, narratives are no longer dependent solely on human creativity, but rather can be shaped by artificial intelligence and even restructured in real time. This eliminates the fixity of the narrative and transforms it into a structure that is living, changing and can evolve in different directions according to audience interaction. Thus, artificial intelligence removes digital narrative from being just a matter of technology and opens the door to a conceptual rethinking of the nature of narrative and its forms of reception. Artificial intelligence technologies have not only accelerated production processes in the cinema sector, but have also begun to play a decisive role in the evolu-

tion of narrative structures. Especially through automatic editing systems and script writing software, artificial intelligence can analyze cinematographic narratives by feeding on large data sets and reconstruct them in line with non-linear structures. Such systems can suggest different narrative paths by separating elements such as theme, symbol, character development and dramatic structures in a story; thus, it makes it possible to develop new narrative forms beyond classical structures, such as multi-layered, parallel progress or giving the audience a choice. For example, an AI-supported editing software can reconstruct the story from different perspectives or make it adaptable according to the audience's preferences by creating alternative timelines of a film. These developments mean not only a technical convenience in cinema, but also an expansion of creative expression possibilities. This situation eliminates the fixity of the narrative and transforms it into a structure that is living, changing and can evolve in different directions according to the audience interaction. Thus, artificial intelligence opens the door to a conceptual rethinking of the nature of the narrative and the forms of reception, removing the digital narrative from being a matter of technology alone (Ensslin & Goggin, 2020). The impact of digital technologies on nonlinear narratives is not merely a technical convenience, but also a dynamic that transforms the structural and conceptual framework of narrative. In this context, the table below (see Table 2) summarizes the innovations digital technologies bring to narrative structure under various headings. Each row represents a transformation not only technological but also narrative aesthetics. For example, AI-powered narrative formats have the potential to reshape the narrative based on the viewer's past interactions. Narratives structured with hypertext logic, on the other hand, offer the viewer not a single plot, but multiple paths filled with possibilities.

Table 2. Narrative Transformations Enabled by Digital Technologies in Non-Linear Storytelling

Audience Interaction and Selective Narratives
Personalized Narratives with Big Data and Artificial Intelligence
Disordered and Multiple Narrative Possibilities with Hypertext Structure
Bringing Narrative Closer to Mental Processes
Elasticity of Fiction
Dynamic Narrative Production with Artificial Intelligence

continued on following page

Table 2. Continued

Audience Interaction and Selective Narratives
Blurring Boundaries Between Reality and Fiction
Transmedia Narrative

Examples of non-linear narratives in digital cinema are productions that go beyond the classical narrative structure and offer multi-layered transitions between time, space and consciousness planes. In such films, events are mostly constructed in line with the mental processes, emotional states or thematic relationships of the characters, rather than following a chronological order. The editing possibilities and visual narrative tools offered by digital technology have enabled this form of narrative to gain a more flexible, dynamic and experimental structure. In this way, the audience is not only presented with a story, but also given an aesthetic and conceptual experience aimed at reconstructing the perception of time. The film "21 Grams" (2003) is an important example of non-linear narrative, directed by Alejandro González Iñárritu. The scenes in the film are not told in chronological order. While it is understood that a character has died in one scene, that character is still alive in the next scene. These transitions are made consciously and thus offer the audience an active process of meaning, not passive viewing.

Terrence Malick's "The Tree of Life" (2011) film includes a man's (Jack) thoughts on his childhood, his relationship with his family, his mother's spiritual guidance, and the meaning of the universe. The director presents this not in chronological order, but in the form of a mind flow. Jack's flashbacks to his past are presented intermittently, like fragmented memories. This presentation of time makes the audience experience the nature of remembering. The post-production phase and visual effects of the film (scenes of the formation of the universe, etc.) are adorned with digital technologies. In addition, light effects, color transitions, and abstract nature images are also important in strengthening the visual narrative.

Christopher Nolan's "Dunkirk" (2017) is one of the important examples of non-linear narration. In the film, the Dunkirk Evacuation is told from three different perspectives, but each takes place in a different time period. Although these three stories start at different times, they come together at the same time at the end of the film. In this way, the audience experiences the events from different angles and emotional tones. Again, the film does not progress chronologically. In one scene, while the pilot Farrier has not yet taken off, in the next scene we see the plane he shot down crash into the sea. Later, the same event is shown again from another time plane with its reasons. This narration allows time to intertwine: the audience does not feel the events exactly as they happened, but in pieces and with emotional intensity. Thanks to digital editing and effects, the transitions between these time planes are quite reasonable. On the other hand, the color palette, sound editing and music (Hans Zimmer's time sound effect) provide clues that connect the three planes. For example; the ticking sound effect maintains the tension on all planes and makes us feel that time is shrinking. While the transformative role of digital editing—such as temporal manipulation exemplified in *Dunkirk*—has been pivotal in reshaping narrative structures, recent advancements in artificial intelligence and algorithmic storytelling warrant greater attention. Platforms like Netflix are increasingly integrating data-driven algorithms to generate dynamic, interactive, and personalized content experiences. These emerging technologies challenge traditional concepts of authorship, linearity, and audience passivity, suggesting a paradigm shift in how narrative flows are constructed and perceived in the digital age.

Another Christopher Nolan film, "Tanet" (2020), is not only about past-future jumps, but also about inverted time travel. The protagonist enters an art warehouse in Oslo. The same scene is shown twice in the film. The first time is given in normal time flow, and then we witness the same scene in reverse time (the inverted character is actually the past version of himself). This structure goes beyond the classical non-linear narrative and includes the physical aspect of time in the narrative. Digital effects are important in the film to visually show the reversal of time. Fires flowing in reverse, explosions retracting, and cars turning upside down are meaningful in this context. Again, the sound design in the film works in reverse; reverse conversations, reverse gunshots can be given as examples of this situation. The fictional cuts that create conscious confusion are presented in a way that makes it easier for the audience to understand whether they are going forward or backward. In addition, "Palindrome editing" is used in the film. In other words, it offers an example of a narrative form in which the narrative is structurally constructed like a palindrome (can be read the same way from beginning to end and from beginning to end).

The Green Knight (2021) is considered as an example of a non-linear narrative by David Lowery that breaks the classical narrative in both time and meaning. Throughout the journey in the film, the events experienced by Gawain are sometimes ambiguous between reality and fantasy, as if they were a dream. On the other hand, the character's journey in the film is presented in an immersive and symbolic way, deviating from the linear flow of time and space.

Daniel Kwan and Daniel Scheinert's Everything Everywhere All at Once (2022) is another example of a non-linear narrative that stands out with its transitions to different universes, times, and identities. Time and space are fragmented in the film, where fast cuts, color transitions, and harsh cuts between universes are used.

CONCLUSION

This study examines the transformative effects of digital technologies on cinema narrative structures, especially in the context of non-linear narrative forms, and demonstrates that digitalization has brought not only technical innovations but also profound narrative transformations to the art of cinema. Digital technologies have broken the cause-effect relationship-based, time-and-space continuity of classical narrative, paving the way for new narrative forms that are multi-layered, temporally discontinuous, and semantically pluralistic. This transformation represents a fundamental paradigm shift not only at the level of cinematic language but also in terms of audience experience. In contrast to the linear time flow of classical narrative, digital cinema offers a narrative regime in which time is run in reverse, parallel universes are intertwined, or structures in which events are presented in a fragmented and non-chronological manner have become widespread.

Within this framework, Christopher Nolan's film "Memento" (2000) overturns the linear time perception of classical narrative by placing the audience at the center of the narrative with its structure that constructs time backwards. This structure is not only a narrative innovation; it also becomes a narrative strategy that represents the character's mental state (short-term memory loss). The use of time as a structural narrative element takes the viewer out of the passive position they are accustomed to in classical cinema and turns them into an active participant who undertakes the task of establishing a connection between events. Similarly, Alejandro González Iñárritu's film "21 Grams" (2003) fragments narrative time and presents themes such as trauma, death and identity in a non-chronological structure. The film's plot is woven with a puzzle logic that reflects the psychological states of the characters; the conscious disruption of the causal order of events necessitates the viewer's constant repositioning throughout the narrative and their active participation in the process of producing meaning. Here, the structure of the narrative presents a fragmented perception of reality parallel to the scattered mental states of the characters.

Terrence Malick's film "The Tree of Life" (2011), on the other hand, completely suspends the linear flow of time and establishes a transition between personal memory and universal time. Malick's narrative preference is based on a Bergsonian understanding of duration; The narrative violates all the codes of the classical dramatic structure while establishing a bridge between the flow of individual memory and cosmic existence. Instead of the causal relationship between events, the film adopts a narrative strategy that progresses through emotions, images and associations. Nolan's "Dunkirk" (2017) questions the classical understanding of fictional time by simultaneously processing time on three different axes (one hour, one day, one week). This structure manages to present individual and collective experiences of war in different continuities, offering a multi-layered approach to the ways time is experienced. On the other hand, in Nolan's "Tenet" (2020), time is included in the fiction not only as a structural element but also as a physical force. Through characters moving forward and backward in time, the film opens the viewer an intellectual space beyond the linear perception of time. Tenet analyzes the basic pillars of the classical narrative, the beginning, development and conclusion structure, and establishes an alternative narrative order with a cyclical time structure. Finally, "Everything Everywhere All at Once" (2022), directed by the Daniels duo, places the multiverse theory at the center of the narrative with the possibilities offered by digital effects technologies.

The film destroys the singular perception of reality in the classical narrative and puts the audience in a constant transition between multiple universes. This structure not only creates temporal and spatial diversity; it also creates a postmodern plurality by dispersing the central authority of the narrative. In the light of these examples, it can be said that the non-linear structures that digital cinema narrative puts forward as an alternative to the classical structure are not only formal preferences, but also take shape around fundamental epistemological problems such as human memory, perception of time and perception of reality. These narrative forms transform the relationship between the audience and the narrative; they move the process of meaning production from a central and closed structure to a participatory, open to interpretation and pluralistic plane. As a result, digital technologies redefine the technical, aesthetic and conceptual boundaries of cinema narrative; It replaces the one-way, linear and centralized structure of the classical narrative with open-ended and audience-centered narrative regimes based on multiple time-space sequences. This transformation requires evaluating digital cinema not only as a set of films produced with new technologies, but also as a reconstruction of the idea, perception and language of cinema.

Furthermore, the findings and conceptual framework of this study may serve as a basis for future explorations that engage with recent advancements in immersive storytelling platforms. In particular, emerging technologies such as artificial intelligence and virtual reality (VR) narratives could be employed to expand or simulate narrative flows in more interactive and personalized ways. These technologies offer new possibilities for rethinking digital narration beyond traditional screen-based formats.

With technologies such as artificial intelligence (AI), virtual reality (VR), and augmented reality (AR), the future of digital storytelling is being developed in a much more flexible, personalized, and user-friendly way. AI-powered narratives have the potential to reshape the scene, plot, and even character development in real time based on the viewer's previous preferences, emotional responses, or viewing performance. This dynamic structure largely invalidates the classical narrative approach, which relies on a fixed script; it transforms narrative from a static structure into a living, evolving organism. Meanwhile, virtual reality applications transform the viewer into a mere observer of the narrative, directly becoming a spatial component of the narrative. Thus, the viewer is no longer merely a mere spectator but an active subject within the narrative universe. These developments are transforming cinema beyond a mere audiovisual art form into a multidimensional and customizable experience platform. By rethinking the boundaries of classical cinema theory, they are laying the foundations for a new narrative paradigm.

REFERENCES

Aarseth, E. (1997). *Cybertext: Perspectives on Ergodic Literature.* The Johns Hopkins University Press. DOI: 10.56021/9780801855788

Aristoteles (1987). Poetika, (Çev: İsmail Tunalı), Remzi Kitabevi, İstanbul

Bal, M. (2009). *Narratology: Introduction to the theory of narrative* (3rd ed.). University of Toronto Press.

Baudrillard, J. (2003). *Simülakrlar ve Simülasyon.* Doğu Batı Yayınları.

Bordwell, D. (2008). *The Way Hollywood Tells It: Story and Style in Modern Movies.* University of California Press.

Bordwell, D. (2008). *Poetics of Cinema.* Routledge.

Bordwell, D., & Thompson, K. (2013). *Film Art: An Introduction* (10th ed.). McGraw-Hill.

Brecht, B. (2011). *Epik Tiyatro.* Agora Kitaplığı.

Brown, W. (2013). *Supercinema: Film-philosophy for the digital age.* Berghahn Books. DOI: 10.3167/9780857459497

Bruner, J. (1991). The Narrative Construction of Reality. *Critical Inquiry, 18*(1), 1–21. DOI: 10.1086/448619

Castells, M. (2010). *The rise of the network society* (2nd ed.). Wiley-Blackwell.

Chatman, S. (2009). *Öykü ve Söylem: Filmde ve Kurmacada Anlatı Yapısı, Çev. Özgür Yaren, 1.b.* De Ki Basım Yayın.

Elsaesser, T. (2015). *Digital Cinema: The Transformation of the Film Industry.* Routledge.

Ensslin, A., & Goggin, J. (Eds.). (2020). *Digital fiction and the unnatural: Transmedial narrative theory, method, and analysis.* The Ohio State University Press.

Erzurum, F. (2019). Transmedya Anlatım Sinema Filmleriyle Mi Başladı? *Dördüncü Kuvvet Uluslararası Hakemli Dergi, 2*(1), 32–42. DOI: 10.33464/dorduncukuvvet.542962

Field, S. (2005). *Screenplay: The foundations of screenwriting* (Rev. ed.). Delta.

Frank, A. W. (1995). *The Wounded Storyteller: Body, Illness, and Ethics.* University of Chicago Press. DOI: 10.7208/chicago/9780226260037.001.0001

Genette, G. (1980). *Narrative discourse: An essay in method*. Cornell University Press.

Gunning, T. (2004). The Cinema of Attractions: Early Film, Its Spectator, and the Avant-Garde. *Wide Angle*, *3*(4), 63–70.

Hutcheon, L. (2006). *A theory of adaptation*. Routledge. DOI: 10.4324/9780203957721

İspir, N., & Kaya, Z. (2011). Sinemada Postmodern Arayışlar. *Atatürk İletişim Dergisi*, (2), 81–99.

Jenkins, H. (2006). *Convergence Culture: Where Old and New Media Collide*. New York University Press.

Jenkins, H. (2016). *Cesur Yeni Medya: Teknolojiler ve Hayran Kültürü. (N. Yeğengil, Dü.)*. İletişim.

Jenkins, H., & Thorburn, D. (2003). *The New Media Reader*. MIT Press.

Kinder, M. (1999). Rewiring the spectator: The reception of narrative in digital cinema. In Sturken, M., Thomas, D., & Ball-Rokeach, S. (Eds.), *Technological visions: The hopes and fears that shape new technologies* (pp. 183–204). Temple University Press.

Kotan, S. (2022). *Korku filmlerinin ideolojisi ben'in içindeki öteki*. Eğitim Yayinevi.

Landow, G. P. (1997). *Hypertext 2.0: The Convergence of Contemporary Critical Theory and Technology, Johns Hopkins University Press, Manovich, L. (2001). The Language of New Media*. MIT Press.

McKee, R. (1997). *Story: Substance, structure, style and the principles of screen-writing*. HarperCollins.

McKee, R. (2011). *Öykü, (Çev. Nuray Yılmaz, Ertan Yılmaz ve Fatih Kınalı)*. Plato.

Metz, C. (1974). *Film Language: A Semiotics of the Cinema*. University of Chicago Press. DOI: 10.1515/9783110816044

Mulvey, L. (2006). *Death 24x a Second: Stillness and the Moving Image*. Reaktion Books.

Özdemir, E. (2005). *Anlatıbilim: Kuramsal bir giriş*. Phoenix Yayınevi.

Özturk, A. (2024). Sinemada Çok Katmanlı Anlatı: Inception (Başlangıç) Filmi Üzerinden Bir İnceleme. *Ahi Evran Üniversitesi Sosyal Bilimler Enstitüsü Dergisi*, *10*(3), 579–598. DOI: 10.31592/aeusbed.1505072

Pisters, P. (2012). *The neuro-image: A Deleuzian film-philosophy of digital screen culture*. Stanford University Press.

Randall, W. (1999). *Bizi Biz Yapan Hikâyeler. Çev. Şen Süer Kaya.* Ayrıntı Yayınları.

Ricoeur, P. (1984). *Time and Narrative* (Vol. 1). University of Chicago Press.

Ryan, M.-L. (2006). *Avatars of Story*. University of Minnesota Press.

Snyder, H. (2019). Literature review as a research methodology: An overview and guidelines. *Journal of Business Research, 104*, 333–339. DOI: 10.1016/j.jbusres.2019.07.039

Suner, A. (2006). *Hayalet ev: Yeni Türk sinemasında aidiyet, kimlik ve bellek.* Metis Yayınları.

Thakur, G. B. (2008). *Of Sture and Signifier in Michael Haneke's Cache. Psychoanalysys, Culture & Society. Sayı: 13.* Palgrave Macmillan.

Thompson, K. (2003). *Storytelling in the new Hollywood: Understanding classical narrative technique.* Harvard University Press.

ADDITIONAL READING

Bordwell, D., Staiger, J., & Thompson, K. (2003). Classical narration. In *The Classical Hollywood Cinema* (pp. 23–41). Routledge.

Culkin, N., & Randle, K. (2003). Digital cinema: Opportunities and challanges. *Convergence (London), 9*(4), 79–98. DOI: 10.1177/135485650300900407

Gaut, B. (2008). Digital cinema. In *The Routledge companion to philosophy and film* (pp. 75–85). Routledge.

Isaacs, B. (2005). Non-linear narrative. New Punk Cinema, 126-138.

Kim, N. W., Bach, B., Im, H., Schriber, S., Gross, M., & Pfister, H. (2017). Visualizing nonlinear narratives with story curves. *IEEE Transactions on Visualization and Computer Graphics, 24*(1), 595–604. DOI: 10.1109/TVCG.2017.2744118 PMID: 28866524

Lowe, N. J. (2000). *The classical plot and the invention of Western narrative.* Cambridge University Press. DOI: 10.1017/CBO9780511482281

Madej, K. S. (2008, August). "Traditional Narrative Structure"–not traditional so why the norm? In 5th International Conference on Narrative and Interactive Learning Environments (pp. 40-47).

Manovich, L. (1995). What is digital cinema. The visual culture reader, 2, 405-416.

Ryan, M. L. (2015). Transmedia storytelling: Industry buzzword or new narrative experience?. Storyworlds: A Journal of Narrative Studies, 7(2), 1-19.

Schasché, V. (2010). Non-Linear narrative structure in contemporary cinema. na.

Schiller, M. (2018). Transmedia Storytelling. Stories. Screen Narrative in the Digital Age, 97-108.

KEY TERMS AND DEFINITIONS

Breaking chronology: Breaking chronology is the deliberate disruption of the chronological order of events within a narrative. This technique is frequently used in non-linear narrative forms and directs the viewer's or reader's attention to the emotional or thematic effects of events rather than to resolve their cause-effect relationships. Methods such as flashbacks, flashforwards or parallel stories break the linear flow of time and offer different points of view and perspectives. This type of narration both adds depth to the story and immerses the viewer in the events, creating a more intense experience.

Digital cinema: Digital cinema is a form of cinema in which images and sounds are recorded, processed and distributed using digital technologies instead of analog film. With this method, films are shot using digital cameras, edited on computers and presented using digital projection systems. In addition to reducing production costs, digital cinema provides great flexibility in visual effects, color correction and post-production processes. It has also revolutionized the production and consumption of cinema by making it easier for films to reach wider audiences through the internet and digital platforms.

Narrative: The concept of narrative refers to the verbal or written transmission of events, situations or experiences with a specific plot and perspective. Narrative is not just a chain of events, but also how, in what way and by whom these events are told. Narrative, which we encounter in genres such as stories, novels, tales and epics, is shaped around basic elements such as time, place, character and plot. In this respect, narrative is a way of making sense of and sharing the experiences of an individual or society.

Non-Linear Narrative: Nonlinear narrative is a narrative form in which events are not tied to chronological order, but rather in fragmented, skipped, or flashbacks. This narrative style requires the viewer or reader to think more actively and piece together the flow of the narrative. Especially in cinema, scenes that go back and forth between the past and the future, different character perspectives, or parallel storylines are used to add depth and multi-layeredness to the narrative. Nonlinear narrative is often preferred as an alternative to the classical narrative structure, to reflect the subjective perception of time and the inner world of the characters.

Postmodern narrative: Postmodern narrative is a narrative form that questions and consciously breaks traditional narrative structures. In this approach, concepts such as reality, truth and authority are questioned, and stories are often fragmented, ambiguous and multi-layered. Instead of a linear timeline, techniques such as time-breaking, flashbacks, parallel stories or the merging of multiple perspectives are often used in postmodern narratives. Intertextual relationships, that is, references from one work to another, are also common. Postmodern narrative blurs the boundaries between reality and fiction, encouraging the viewer or reader to think more about the content of the story and actively construct meaning.

The alienation effect: The alienation effect is a technique that, especially in branches of art such as theater and cinema, distances the audience from being fully involved in events and from establishing a conventional emotional bond. This effect aims to prevent the audience from emotionally identifying with the work by placing them in the position of an "observer" at all times. In this way, the audience looks at the story from the outside and focuses more carefully on the structural features, social criticisms, or artistic aspects of the work. This technique, which holds an important place in Bertolt Brecht's understanding of theater, aims for the audience to actively interact with the work rather than remain passive spectators.

Traditional Narrative: In terms of cinema, traditional narrative is a film narrative that is based on classical storytelling rules and has a structure that the audience can easily follow. In this form of narrative, events usually proceed in chronological order, are based on cause-effect relationships, and characters have clear purposes. This structure, with clear beginning, middle, and end sections, is frequently seen in Hollywood cinema and aims to emotionally draw the audience in.

ENDNOTES

1 In postmodern narratives, traditional narrative structures are deliberately disrupted, and thus the viewer or reader is brought into a more active position. The prominence of purposeless or indecisive characters and the failure to establish a clear cause-and-effect relationship between events weakens the classical

integrity of the narrative, creating a more fragmented and scattered structure. While stylistic elements (visual style, camera use or editing techniques) become more decisive than the content, this situation makes the director's artistic identity visible. The fact that narratives are often concluded in an open-ended or ambiguous manner prevents the viewer from reaching a singular conclusion, creating a structure open to multiple meanings. Thus, meaning is constructed not in the text itself, but in the viewer's interpretive process, and this constitutes one of the most distinctive aspects of postmodern narrative. Gürata expressed this situation as "the prevalence of the aimless hero figure, the weakening of the causal link between narrative events, the foregrounding of stylistic elements and thus the demonstration of the director's artistic presence and intentions, the ambiguous ending of the narrative and thus the provision of speculation by the audience about the meaning of the film" (2004, p. 98).

2 The alienation effects used in epic theatre with the Brechtian approach are an important aesthetic and narrative strategy that feeds the non-linear narrative form. "To alienate means to strip it of its naturalness, familiarity and plausibility, and to give it a surprising and unexpected appearance in the audience" (Brecht, 2011, p. 89). This narrative technique prevents the audience from being completely emotionally involved in the narrative, and enables them to look at events with a more distant and critical eye. Thus, the audience enters a process of intellectual questioning instead of identification and has the opportunity to evaluate the social, political or philosophical layers contained in the narrative more consciously.

Chapter 7

Changing Perception if Space in Art With Digital Visual Culture and Television

Nursel Bolat
https://orcid.org/0000-0002-3986-7408
Ondokuz Mayıs Universty, Turkey

ABSTRACT

With the digital age, the ways of production, sharing and consumption of visual culture have radically changed. The perception of space in traditional arts has been reshaped with digital technologies, and representations of art and space form a more dynamic, interactive and multi-layered structure, especially with the influence of audio-visual media such as television. While space is generally represented with a static composition in traditional works of art, in the digital age the perception of space has become dynamic, variable and interactive with the viewer. Works of art move space from being just a physical phenomenon to digital environments with virtual reality, augmented reality or computer-based designs on digital platforms. Television has transformed works of art into visual experiences that merge with time and movement, rather than a static object. This transformation has allowed the work of art to be reproduced on digital screens, no longer dependent on physical space. Television is one of the most important tools that enable art to reach large audiences.

DOI: 10.4018/979-8-3373-1967-4.ch007

INTRODUCTION

Visual culture, which is the field that includes visuals produced by human beings, which carry aesthetic concerns and reflect the values of a society, has found the opportunity to reach wide areas with new technologies. Today, when visuality is intensely prominent and completely surrounds people, culture, unlike all other ages, acts through visuality in a way quite different from oral and written culture. Every visual image emerges as a product of visual culture. The fact that we are in the age of technology is one of the biggest factors that reveal this situation. Posters, computer-generated images, brochures, brochures, videos, moving pictures and many more are images that have a cultural background and carry more meaning than what we see. Visual culture comes into play in the interpretation of these meanings.

The concept of visual culture emerged for the first time in the 1980s. In a panel organized, the relationship between society and art was mentioned and visual culture was referred to. Especially in the context of developing technologies, advertising, toys, posters, CD covers, movies, video games, TV programs, computer graphics, internet networks, children's books and many other forms of visual culture that surround every moment of human life have an important place in the emergence of this concept. However, television emerges as a medium at the heart of the rise of visual culture and plays an important role in its spread.

It is known that digital technologies, which entered human life rapidly after television, are not only communication but also cultural tools that connect individuals and organizations around the world. The most important function of digitality is that it virtualizes objects, allowing them to be reproduced and transmitted over electronic networks. Digitalization causes changes in the formation process and quality of visual productions/communications. As individuals integrate with new technologies, they are surrounded by a new formation called "digital visual culture".

In this context, in the context of the transformation experienced with digital visual culture, it is seen that there have been changes in the forms of production, sharing and consumption of visual culture. In addition, digital technologies also have an impact on traditional arts. From painting to sculpture, from television programs to cinema, it is seen that the perception of space in many visual cultures has changed and transformed towards virtual spaces. The experience of space with virtual reality and augmented reality has started to bring virtual spaces to the forefront by neutralizing the concept of physical space with digitalization. It has moved from art galleries to television screens or computer screens.

The study focuses primarily on the process of change in visual culture in the digital age through a literature review. In this context, the process that visual culture has undergone from photography to the internet with the digital revolution is examined and the production and exhibition spaces of art are evaluated within the framework of the changing concepts of visual culture and art with digitalization. The transformation process of art from physical space to virtual space with technological developments is discussed. With digitalization, space breaks away from traditional space and shifts to very different dimensions. The concept of space has also become dynamic and interactive spaces with digitalization. Under the heading of digital platforms redefining space, the way art takes place in more fluid time and space channels instead of the time and space compression of digitalization is emphasized. In the process of individualization of technology and space, the transformations brought about by technology and digitalization in terms of the viewer are examined. In addition, the new unique presentation of the work of art and its transmission to the audience are discussed. Under the title of television's representation and mediation role in art, television art representation and digitalization process are evaluated.

The Transformation of Visual Culture in the Digital Age

The digital revolution process has features that affect the cultural structure in the social sense and enable the emergence of new imaginary forms, diversify people's worlds of meaning towards images and step into a new era (Burnett, 2012, p. 20). In contemporary communication practices, it is seen that visual images, which are far beyond words and borders, and visual discourse, which has taken over the power of communication and expression through digitalization, have been transformed through technology. As a result of this transformation, the distinction between what is real and what is visible becomes difficult (Lovejoy, 2004, p. 272). Cultural design, which is considered as digital visual culture, stands out with media messages and it is seen that this process has developed through new production possibilities, especially through visuals created through photography technologies.

Digital culture is understood as a set of products created through both creative and digital media. It is necessary to examine the concept in more detail in order to avoid confusion. In particular, technology and culture emerge as concepts that can be confused with each other. The social context of digital culture in the network society and the changes in media use enabled by the characteristics of digital media are emphasized. Charlie Gere states that digital culture is not as new as it seems and its development is not determined by technological developments. Digital culture refers to the options of thinking and doing business shaped by technology (Dijk, 2016, p. 293). Digital culture affects individuals' living spaces both through the networks they voluntarily participate in and in areas where institutional needs are met as citizens. People have to organize their lifestyles with digital culture. Digital culture also brings about change in visual culture in connection with technology.

Based on Barnard's definition of visual culture as "making the values and beliefs of a culture visible in various ways" (Barnard, 2002, p. 22), it is seen that today these features are realized through digitalization. The concept of digital visual culture should also be evaluated in terms of the aspect that goes beyond making a culture visible through digitized visuals with the use of digital technologies, causing cultural changes and creating new cultures by creating visuals. Digitalization brings new approaches through media with the development of internet technologies.

The Internet is basically a complex communication platform where new forms of media are emerging. The Internet combines multiple stages of communication processes such as storage, modification, interaction, reproduction, reproduction and modification of communication channels. New media have changed the ways in which information is collected, processed, produced, transmitted, stored, received and displayed (Pavlik, 1998, p. 80).

Looking at the changing communication, production, consumption, access to information, sharing habits and social life practices after digitalization, it is seen that all these technological, social and economic changes have had an impact on the field of art and television as much as in other fields.

The foundation of the new communication system was laid with television, but the emergence of decentralized and versatile new media with the creation of the multimedia system, which emerged as a new development in the 1990s, and the widespread use of the Internet have led cultural progress. New communication technologies have also had a major impact on social interests, government policies and the business world, leading to the emergence of a new culture (Castells, 2013, pp. 441-442). Although it is seen that the change in television after digitalization is closely related to technology, it is not enough to explain the existing change only in a technological context. In order to understand the change, it is necessary to look at the changing socio-cultural structures and the characteristics of the new audience, which has a more passive role in the new communication environment. The change experienced after digitalization has created a new communication model based on symbols and visuality that changes the roles of the source and receiver in previous communication models, where interaction is possible. The process of change that emerged with digitalization has brought about changes in the production, sharing and consumption forms in visual culture, as well as bringing a different dimension to art in the artistic field.

Digital art is generally created, stored and distributed through digital technologies; It is defined as a calculable art. This art combines the features of technology uses it as a tool. Digital artworks are digital. Process-oriented, time-based, dynamic and can be real time; participatory, collaborative and performative; modular, variable, productive and It has customizable features (Erdem, 2022, p. 58). Digital art uses the internet, network connections, original software, virtual reality, virtual environments, artificial life and organisms, databases, robots, body-worn headgear, prostheses, machine extensions, artificial intelligence, data imaging and mapping, hypertextual narratives and games are considered as digital artworks (Çokokumuş, 2012, p. 56). With this statement, it is possible to say that the field of digital art is very wide and is used in applications in both production and distribution. Therefore, it is seen that digital technologies have significantly changed the effects and spaces on traditional art and moved art from physical spaces to virtual spaces.

It can be said that people are undergoing an adaptation in order to adapt to the changes in the network of relationships where processes and visuals emerging with digitalization are effective, and in this process, they have started to move away from the real. Today, when people come together, instead of chatting and communicating face to face, they are distracted by the mobile technologies that are always with them, preferring the virtual to the real. Yaykın states that "as the landscape offered by virtual environments enters the new image space, we begin to think that we no longer need the real world" (Yaykın, 2010, p. 65). This situation is caused by digital technologies taking control of human life, detaching them from real life and imprisoning them in virtual spaces. Digitalization also brings about changes in the exhibition space as well as the production space of art.

Changes in the Perception of Space in Art

Traditional and digital perceptions of space have tried to survive in a continuous change in connection with technology from past to present. The relationship between art and technology has developed in direct proportion to each other throughout history. At the same time, these two concepts have existed as two basic elements that distinguish humans from other beings. Therefore, the presence of technology in the production process of a work of art has moved away from reflecting nature in art, giving it the appearance of an experiment. The artist has started to produce his artistic works, which cannot be produced with other tools or techniques, by using computer technology. With the introduction of the computer, the meaning, content and position of reality has changed almost completely. Virtuality has now replaced the established reality in every field and level, and has come to the forefront in artistic production (Sağlamtimur, 2010, p. 216). Art has evolved away from physical space and towards virtual spaces with new technologies. This evolution process has determined the new space of art as virtual spaces. This process has brought about a change in the perception of art and artistic space in relation to technology.

Baudelaire uses the expression "every age has its own attitude, its own look and stance" (Antmen, 2012, p. 15) when explaining art. Especially before industrial societies, it is possible to talk about an integrated common taste area that exists between the artist and the consumer and is guided by the social existence field. In the globalizing social structure and postmodern period with the new developments in communication technology, it is not possible to search for a single artistic movement in a certain style that reflects new perception norms, new feelings and thoughts in the artist's art (Antmen, 2012, p. 275). In the transition from traditional art to digital art, technologies also carry differences in the understanding of art and artists.

Sütçü states, "The most important function of digitality is that it virtualizes objects, allowing them to be reproduced and transmitted over electronic networks" (2012, p. 84). According to Christina Paul, digital environments are interactive, participatory, dynamic and change on demand. In addition to physical possibilities, the Internet enables the emergence of many new types of art forms. The computer is seen as an important medium in terms of both shaping and publishing many new art forms such as net art, network art, software art, cyber art, pixel art, new media art, multimedia art, interactive film, computer-based music. (Ürper, 2012, pp. 21-24): Artificial intelligence, artificial life, virtual reality, telerobotics, database aesthetics, data visualization, net activism, games, narrative hypermedia environments, body and identity are studied as themes in digital art.

Benjamin defines the use of technology for art as "the fact that social reality has matured enough to make technology an organ for its own purposes" and, in turn, "the fact that technology has become strong enough to correspond to the fundamental forces of society". He emphasizes that technology creates a new aestheticism that emerges with new developments. He states that a brand new aestheticism has emerged based on technological reproduction and mass media that can create its own public. The new means of communication offer the opportunity to film and record events and then reproduce them without the need for the function of reproducing and distributing them. Thus, it is possible to edit, reformat and correct the content. The transition to radio and sound film, in particular, both perfected and strongly disseminated the modern technologies of production (Benjamin, 1995, pp. 61-92). With these technological tools, art has found a place in places that reach and deliver to the masses from closed spaces that offer the opportunity for individual viewing. After radio and cinema, art has increased its dissemination through television, an electronic and audiovisual medium.

The space change of art, which started with traditional media, has reached different dimensions by finding new spaces with the digital process. The rapid introduction of the Internet into human life has enabled art, which has always progressed in connection with technology, to find a place for itself in this field. However, today it is not limited to the digitalization of artworks, many museums and art institutions have also shifted to the digital environment with their changing artistic practices. In addition to focusing on the acquisition and exhibition of art, the digital environment has also led to changes in documentation and archive studies, exhibition and presentation models.

With digitalization, art spaces have also become dynamic and interactive spaces. In addition to this, it is seen that the artists who are active in this field of net art exist with their artworks rather than their own identities. While these works meet with the audience in virtual space, they also enable artist, art and audience interactions. Yücel defines net art as "the existence of a globally accessible audience, its constant changeability, its ability to be copied and reproduced and easily eliminated if desired, its cheap production and distribution conditions, and thus its highly dynamic structure, have given it an almost unique character in the field of art until today" (Yücel, 2012, pp. 36-37). In parallel with the developments in internet technologies, the developments in camera technology and the reduction in the size of cameras, their integration into mobile phones and even becoming much smaller and turning into pens, buttons, keychains, glasses, etc. It is seen that their use hidden in different tools and objects enables more video activities (Akmeşe & Deniz, 2015, p. 502). In this rapid change process, spaces have begun to shift towards television screens and digital platforms in connection with technology.

Digital Platforms Redefine Space

With the digitalization of message sharing between individuals that started with digitalization, the time and space compression specific to traditional media is replaced by fluid time and space channels (Çakın & Bolat, 2018, p. 176). The Internet's initiation of the era of global communication has contributed to the development of new methods of intercultural dialog. For artists, the internet has emerged as a new medium of creation with the creation of new communication networks, a global sensitivity and the existence of an international audience" (Yücel, 2012, p. 31). In addition to the Internet creating new communication networks between artists, advanced approaches have emerged in the way the audience sees artworks and visits art environments with developing technologies. This process has been carried to higher levels with virtual reality and augmented reality applications.

Space experience with virtual reality and augmented reality Virtual reality applications, which first emerged with military technology, paved the way for the development of augmented reality applications and led to the emergence of these types as two separate fields today (Yılmaz, 2017, p. 265). Augmented reality is a system that allows users to interact with real and computer-generated objects by displaying 3D virtual objects stored in the user's natural environment (Bajura & Neumann, 1995, p. 189). In augmented reality, information that enhances or augments the real world is provided by the computer and incorporated into the user's reality, giving them a perception of the real world. This provides a different dimension from virtual reality where the user is completely immersed in the computer world. With augmented reality, the user interacts with the real world in a natural way, using the computer to simultaneously discover relevant information and interact with virtual objects (Ahlers et al., 1995, p. 1). This application therefore provides real emotions in the virtual environment and creates the impression of a physical environment.

Exhibition spaces designed in virtual spaces provide convenience in terms of accessibility regardless of social, cultural and economic differences. Making the artifacts in virtual exhibition spaces more readable contributes to increasing the curiosity and desire of visitors to physically visit the exhibition space. In contemporary museum designs, information transfer has become more effective with virtual reality (VR) and augmented reality (AR) technologies. This situation offers the opportunity to visit virtual museums, learn about the work and get to know it. With the development of technology, exhibition spaces have been positively affected by these developments (Yakışkan, 202224, p. 328).

As digital environments transcend the perception of physical space, the doors of the world of perception with the internet open to virtual individuals with "sights, smells and sounds" and thus to "virtual selves" (Ekim, 2022, p. 278). The virtual reality environment, designed with three-dimensional models and animations, provides interaction between the virtual space and the user. One of the main purposes of virtual space is to provide an experiential interaction, to care for the senses, to concretize abstract concepts. Virtual reality is defined as the technology that allows three-dimensional animations and shapes created using computers to interact with these objects in addition to giving the feeling of being in a real environment in the mind of the individual using technological tools (Franck, 2002, p. 35). In addition, these technologies affect the quality of communication between individuals and serve as virtual applications of communication. In addition to the diversity of presentation and participation, virtual media also include richness and abundance of space.

Individualization of Space with Technology: Art and Audience Interaction

Technology and art are not only a world constructed by electronic tools in the sense used today; in ancient Greece, art itself was perceived as an act of "techne", a process of making. In other words, art is a direct act of technology (Shanken, 2012, p. 11). From the first cave wall paintings to stone sculptures, all art products produced are reproductions and each of them is a technique. Art progresses with technology by using each new emerging technology. Although every art product is a technique, photography, which is one of the important inventions of visual culture and enables the recording of the image, can be defined as the art produced with the first technical visual tool of visual art.

"With photography, the human hand was for the first time relieved of its most important artistic obligations in the reproduction of painting; these obligations were now assumed solely by the eye looking into the lens. Since it takes much less time for the eye to perceive than for the hand to draw, the process of reproduction through painting changed at a speed that could go hand in hand with speech." (Benjamin, 2001, pp. 53-61). While the picture drawn on canvas is unique, photography has made this unique uniqueness of art reproducible. After the invention of photography and the camera, it is possible to evaluate video art in the perspective of digital art. Video art, which is among the branches of contemporary art, can also be considered as a part of the formation of digital art. Although video is used as a goal for artists who perform video art, the use of video as a tool has brought it to an effective position. After photography, the process of culture and art has also taken its place in television, an important visual communication tool.

Television is a means of standardization and adaptation to the world. While television has adapted to the world as a tool, it has also helped people adapt to the world order (Sirer, 2020, p. 52). As a modern storytelling and mythtelling medium, television offers a unique combination of the communication and art forms that preceded it (Tutar, 2020, p. 110). The role of television in transforming artworks starts with the transportation of artworks to homes through electronic media and is realized by delivering them to the audience in unique cultural forms.

When looking at the interaction between time and space in the context of the effect of time and motion on the perception of space in television and other electronic media, two assumptions underlie all approaches. The first assumption argues that time and space are not abstract concepts that can be handled separately. They cannot be treated as singular in pure form, but coexist together as a compound. According to the second assumption, even if they are considered singularly, they show two different characters in compound form. Space tends to be quantitative and time qualitative (Deleuze, 1986). Television, which tries to create a new reality by using time and space, provides existential environments of experience through the conceptualization of space and its representations.

Television carries art from galleries to its screens and realizes this process within the context of its own rules. "Television is both an industry and a technological tool; it is both a cultural and artistic production and a source of entertainment; it is also a political and social institution" (Mutlu, 1991, p. 24). As television entered the broadcasting life, it also offered an environment for art. To the extent that television is an effective means of mass communication, it is also an effective means of art. Parsa bases the fact that television is an art on the definition of art as "a series of movements made by human beings with unique tools and different materials in order to announce and show their inner appearances to others and to arouse special feelings in others". In this framework, he argues that television should be considered an art in terms of the creativity of the artist, as well as the camera and other technical tools, sound and light waves (Parsa, 1994, p. 8).

In this artistic process of television, the subjectivity of the viewer's spatial experience has a place in the understanding of art. Television and cinema, which are located at the intersection of the relationship between mental and physical space and produce various representations of existence, aim to reach the audience by manipulating the images and spatial fiction. Television narrative creates an experiential interaction with images and spatial constructions. While experiencing a television product, all that comes to the viewer's mind consists of images. In television narrative, spatial constructions are organized with elements such as light, shading, contrast, angles, frames, intonations, music, etc., and a film space perceived by the director completely under his control is created and the viewer's experience is thus directed (Sarıberberoğlu, 2019, p. 28). For this purpose, the locations can be unusual, secluded, dark, foggy or alien environments to support the narrative and direct the audience, or they can be constructed as mysterious or open in relation to the features of the narrative. These space creation features create familiar environments in the viewer's perception after a while.

This process, which turns into a phenomenon of representation by being expressed in the context of different stories/fictions according to the linguistic/visual possibilities of art, is usually handled with shocking, provocative images in order to evoke the dialectic between life and death; art creates spaces for the viewer to approach the phenomenon they avoid from different perspectives. In this sense, if the essence of the work of art involves a questioning of human existence, it can offer starting points for the viewer to think about the nature and existential meaning of human beings with an individual approach in nothingness, which is a controversial phenomenon with its unknownness like death or dying, through the act of art. One of the images that arouse curiosity for artists in creating these starting points is home images, which are effective in line with the meanings attributed to human life through space and are often analyzed as an unconscious image of past experiences (Barlas, 2024, pp. 191-192). In art narratives, houses and similar spaces are the places where art is realized, while the spaces used in the transfer of these arts to the audience are the places where they meet art lovers.

Digital culture refers to a new cultural structure articulated in today's technology-intensive climate. The cultural structure created through digital tools or as a result of the digitization of existing elements is reflected in the daily lives of individuals, especially in their consumption habits, as it affects almost every field (Erdem, 2017, p. 249). When it came to the 1990s, it offered important opportunities to artists who were interested in non-physical spaces as a place for the production and exhibition of art. Artistic and cultural products have come to the fore in digital spaces in terms of both production and exhibition.

Perception of Space in TV Series and Video Arts

Although areas such as video and cinema were designed as an industrial structure, they have developed in such a way that the individual can produce even alone as a result of technological developments. Considering this aspect, video is much more democratic than cinema (Kılıç, 2000, p. 2). The period when video created its own language independent of cinema and video art developed is seen as the 1960s. Unlike photography, video images are recorded on an electronic medium.

The perception of video images and the discovery by artists from different fields of art of the properties of video that can be used in art, its suitability for artistic works, and especially its features such as stopping, superimposing, image distortion, playing with color, sound recording, etc. have developed video art. At the same time, artists working with the video technique saw television as the representative of a new visual aesthetic and believed that they could show their work to millions of viewers through television channels and radically democratize art in this way. It is also seen that the documentary feature of video is used with the narrative types of conceptual art such as environmental art, pop art and minimal art.

The video image has started to lose its impact with digitalization since the 1990s (Arapoğlu, 2017). According to video art analysis, video artists aim to stimulate the senses, the mind and the creative spirit through the use of technological skills and imagination. The impact on viewers and those who can relate to video art is shaped by the methods used to challenge the discipline or language of media arts. Video art analysis focuses on artistic creation through technology or software to create different sensory and emotional experiences in the viewer or in the public sphere, to explore new forms of artistic expression-narrative, and to reflect on the relationship between technology and human beings (Bernaschina, 2023, p. 172). Developments in this field have led to the emergence of internet art or digital art with different works in this process. Therefore, video art has evolved from analog to digital and created a new video design.

The digital image is functionalized as a "reproduction" in the same way that the analog image was created. The image is re-visible and simulated on the screen according to coded commands (Vaughan, 2005, p. 6). In digital art, the re-visibility of the image according to coded commands, that is, its simulation through digitization and its transfer to the virtual environment, enables the desired changes to be made on it. In connection with this, it is concluded that new media is effective in the artist's forms of art creation, and that the object, subject, content and communication stages are also changed by artists in the "digital age" with new media (Güney, 2014, p. 134). With digitalization, the production and creation process of art has moved to virtual dimensions. In addition, significant changes have been experienced not only in the production but also in the exhibition of art with digitalization.

While the boundaries of digital art are constantly expanding with the internet, the museum concept has also been transformed in line with these developments. In recent years, the organization of noteworthy digital art exhibitions, which show that digital art is becoming more widespread, is a reflection of this. In addition, museums and galleries are also exhibiting their traditional art collections online, making it even easier to access them. The experience of art now extends to any private or public space with internet access (Wands, 2006, p. 8). While this new space brings about a new culture, it also brings about significant changes in visual culture. Digital visual culture is shaping the understanding of art and art spaces of today's generations with new spaces and new perceptions.

Digital art is intertwined with science and technology. This raises the question of whether this is a result of technological determinism or a reflection of modern culture (Wands, 2006, p. 11). The evolution of digital creativity, the development of technology and visuality, and the process of change and development in creativity with the development of technological advances and digital creativity, which includes new forms of artistic and visual expression, including virtuality and media art, have revealed the process of change and development in creativity. The understanding of digital art has also had an impact on the art and art format of television as an electronic and technical device.

When analyzed in terms of how television formats based on continuity change the perception of space, television shapes the perception of space according to its formats with features such as the immediacy of the event, spatial proximity, the elitism of those involved in the event, being out of the ordinary, being conflictual, containing tension, excitement and consistency.

Modern people find themselves in them and identify themselves with them while watching a worker who rebels against his boss, an actor who beats the person who insults him, lovers who live their love against many obstacles. Popular culture characters, which have structures based on fiction through television program types, create a dramatic structure by using narrative method with tools such as time and space. In addition, the audience's need for dreams and fantasies is satisfied at some point (Geraghty et al. 1998, p. 65). In all these formats and storytelling, television defines space in the viewer's perception by creating a built-in space specific to certain formats. Television itself represents the space in which the narrative is developed, fictionalized and presented in a unique framework. With the digital transformation and digital television platforms, formats have become more flexible and interactive.

The study of television in the digital age focuses on whether the advent of digital technology has changed the social practice of watching television. With digital television broadcasting, television content is no longer only transmitted linearly, but can also be used on different platforms and technical devices. In the digital age, visual content can distribute its messages not only through television, but also through more and different tools than ever before, often reaching the audience in better ways. Today, these tools are different technical devices such as laptops, tablets and phones (Mikos, 2016, p.159; Karaduman & Karataş, 2022, p. 991). The development of digital broadcasting and the ability of viewers to access content from many tools has led to the need to create messages for the tool, and has even changed the time and duration of the time that viewers spend in front of the screen. Digital platforms, which are accessed independently of time, space and device used, can be seen as the new form of television broadcasting (Karaduman & Karataş, 2022, p. 991). While different platforms have created new spaces for television broadcasting, they have changed the perception of space of television formats based on continuity.

One of the artistic approaches related to television broadcasting is video arts and the conceptual transformation of space. Especially experimental video arts use a digital space approach in terms of space. It is not possible to evaluate the relationship between video art and space only as the space surrounding the moving image. The relationship between video art and space reveals a much broader concept of space, including the space where the images are shown and the organization created within this space.

Representation and Mediation Role of Television in Art

Culture constitutes the most important differentiation item of cities undergoing change in the globalizing world, and accordingly, globalization also leads to serious changes in the fields of culture and art. Many factors such as education, social and economic environment, information technologies, etc., which change with the changes in cities, directly and indirectly affect individuals engaged in art. This situation has led to the birth and change of contemporary art by affecting the art world and its products. In the process of globalization, television reshapes the function of art in reaching large masses.

In the process of globalization, Featherstone also attributes importance to the disappearance of the boundaries between art and everyday life and the collapse of the hierarchical distinction between popular culture and high culture due to post-modernism (2013, p. 30). In connection with the spread of mass communication culture, art has become uniform and ordinary by intertwining as a model and series of each other. Artists who produce similar mass productions and the works produced have become almost monophonic. Free art, which criticizes and opposes, has been replaced by an art product and mass that acts together with society and walks side by side with it. Artists are now included in the system instead of taking an attitude against the system. This situation has enabled art to be treated like a commercial enterprise. The motive to make a profit from the artworks, the effort to become famous and rich has now become a priority for the artist.

In the process of art's profit-making, the support and development of contemporary art, photography and performance arts by corporations has brought the arts to an important focal point in both economic and political terms (Stallabrass, 2016, p. 23).

In another aspect, art is considered as an important tool for establishing social reputation for individuals, various commercial and non-commercial organizations. An individual's artistic endeavor and sympathy in the society creates a sense of having an intellectual capital and a way of life. Some of the individuals desire to gain a social position in the society by using art as a tool (Dimaggio & Useem, 1978). Individuals use art and artworks as a kind of identity, believing that art will add features such as being more enlightened, culturally knowledgeable and emotionally oriented to their personality. The mass media also enables reaching wider areas through dissemination in this process of acquiring a social identity.

Among the media arts, which have the widest range in terms of artistic production reaching the masses, TV is the most important tool that comes to mind. This is because TV is involved in the marketing of artistic products both as a tool and as a medium of artistic production. Initially used more as an advertising and promotion tool, TV has now become the world of all kinds of drama plays (series), programs and media users (Başaran Alagöz & Ekici, 2016, p. 196). With the effect of globalization, there is a fusion between different cultures today. The development of technology also reveals the reality of the commercial process by sending the art produced all over the world. The same development has led to the fact that people can reach anywhere in the world and this accessibility has led to the rapid spread of different cultures (Jenkins, 2016, p. 264). This commercialization process has also enabled art to take its place in the marketing of art through mass media.

Artworks, like products in a typical market, are offered to the market through various campaigns, take part in competition and go through the stages in the product life cycle (Başaran Alagöz & Ekici, 2016, p. 191). One of the examples of the use of the product life cycle in the artistic field is the career development of artists. Each period in the product life cycle corresponds to some areas in the artist's career development. These are; the "pre-development" stage is paired with the "unknown visual artist", the "entry" stage with the "emerging visual artist", the "growth" stage with the "established visual artist" and the "maturity" stage with the "famous visual artist" (Lehman & Wickham, 2014, p. 680). Especially in what is considered the modern period, mass media and especially television have enabled this process to spread faster and wider.

Today, with the opportunities offered by communication technologies, the modern individual has entered a period in which the influence of words and writing has weakened and he/she is confronted with more images. In the process of technological development, people are bombarded with images at any time and anywhere, first through cinema, television, billboards and smartphones. Thus, while texts lose their impact in the encompassing power of images, they are also devalued. While the image was once seen as a method to explain the word, today the word (text) has turned into a tool to explain the image (Ellul, 2004, p. 158). Therefore, with technological developments, visual images have spread to every field with the tools where the image produces content intensively. Especially with digitalization and the inclusion of the internet in the age of visual culture, visuality has carried art and information to the highest point in terms of production, distribution and space.

With the dizzying development of technological tools and the superiority of the image, reading culture and the future of the printed book are being questioned. As the most valuable cultural medium and the main source of knowledge, the book is losing its value day by day, replaced by the rapid flow of images and image exchange on social media (Lauristin & Vihalemm, 2014, p. 7). As he emphasized in the context of the art of cinema, art history has witnessed similar debates for photography and film before digital games, especially with the proliferation of mechanically reproducible mass media. In fact, every 'new media' was initially met with skepticism by the dominant media of the era into which it was born. For example, just as there was a period of primary oral culture (Ong, 2007) in which memory-based text production was at the center, today we are living in a period in which visual culture is at the forefront.

With the rise of visual culture, in addition to the ever-changing and controversial nature of the definition of art over time, the interaction relationship between art and the audience, digital art forms, genres and methods, and their position in the context of contemporary art have also added new dimensions to the debate. In the aftermath of the digital revolution, new technology has begun to open up new channels for art production ; but at the same time, it has brought similar debates on the forms, genres and methods of art back to the agenda. The 21st century has welcomed a succession of technological developments, accompanied by new materials, techniques and methods for art production. Digital technology has opened an integrated field of artistic expression with materials such as film, video and photography.

According to Wands, the author of The Art of the Digital Age, digital art includes digital animation and video, digital installation and virtual reality, software, database, net art and game art among many other art forms. During this period, museums became more interested in organizing digital art exhibitions and started to publish their traditional art collections on the Internet (Wands, 2006, p.8). In addition to the realization of image creation using digital technology and the re-presentation of previous forms with digital technology, the element of interactivity, which can be used as a possibility with digital technology, has also opened up a new field of expression.

As a representation medium of digital space, television has kept pace with the developing technologies by keeping pace with the developing technologies in television. The phenomenon of digitalization, which affects artistic and cultural production in many areas, has brought innovations in media arts and led to the emergence of new TV-derived mixed technology broadcast media (cable TV, digital TV, satellite TV, interactive TV, web TV, direct broadcast satellites, etc.). Thanks to these broadcast mediums, both documentaries and dramas produced by channel producers and artistic productions such as movies, dramas, documentaries, concert recordings whose broadcasting rights have been purchased have had the opportunity to meet their target audience through subscription systems. Dramas that are appreciated by the audience can also bring drama music together with enthusiasts if they are appreciated, thus creating new demands for musical art presentations. Likewise, for archive-savvy art lovers, the sale of acclaimed dramas in formats such as vcd, dvd, dvix, etc. can also be shown as one of the marketing efforts (Başaran Alagöz & Ekici, 2016: 196). Therefore, television, as a medium that closely follows technology, both digitalizes itself and sets an example for other digital media in terms of program type and content.

Digital art mimics the appearance of traditional tools and genres. That is, it is often difficult to tell to what degree digital tools play a role in the production of an image and when working digitally is organized. Furthermore, it is widely believed that it is impossible to know where traditional and digital art begin and end (Nalven and Jarvis, 2005, p. 8). Arguments in favor of technological determinism in art have proclaimed that this interconnectedness is due to the development of technology that allows artists to create the works they do. Nevertheless, if we look at art as a creative reflection of modern culture, digital art can be considered as a subset of contemporary art (Wands, 2006, p.11). It is observed that numbers, symbols, virtual forms, expressions and fictions are proliferating in recent artworks that have been shaped by digital technology, and works such as software and animation are characterized as digital art.

The Internet can be considered an advanced example of mass media, the best examples of which are radio and television (Wands, 2006, pp.11-12). One of the most important features of contemporary art is that the distinction between artistic disciplines has disappeared. While movies can contain photographs, videos can make use of film content and photographs, the internet can link text, image, sound and time-based communication tools such as video and animation.

Another field of study for artists and scientists working digitally is virtual life, virtual touch, tactile sensation, virtual smell and taste. They carry out their work and research in order to give the feeling that artistic works that are forbidden to touch can be touched, and objects that cannot be smelled can be smelled. In this process, television is one of the indispensable display areas since it is available in every home. Television, which has made a rapid transition to the digital process, maintains its place as an important venue for digital arts while progressing in harmony with technology with digital platforms as well as its traditional features.

CONCLUSION

Visual culture exhibits a dynamic structure that is constantly transforming with the development of technology. Especially the process of digitalization has radically changed the ways in which art is produced, consumed and exhibited. While the physical boundaries of traditional art spaces have expanded through digital platforms, television has become one of the most important tools of this transformation. Although television was initially seen only as a source of entertainment and information, over time it has functioned as a space where art is both produced and exhibited.

New forms of expression offered by the digital age have changed the perception of art and moved the concept of space from physical reality to the virtual world. Especially television, digital effects, augmented reality (AR) and virtual reality (VR) technologies transform traditional narrative patterns and reshape the viewer's perception of space. The evolution of art from being merely an exhibited object to an interactive structure that the audience can directly experience has paved the way for television to transform into an art platform.

According to the findings of the study, the digitalized understanding of art and the content produced through television reduce the need for traditional art spaces and increase the accessibility of art. Artworks produced in the virtual environment contribute to the democratization process by reaching large masses through television and digital media platforms. While this process offers new creative spaces for artists, it also makes the relationship with artworks more personal and interactive for the audience.

However, these new spatial experiences offered by digital art and television bring with them some controversies. The weakening of art's ties with traditional spaces raises debates on authenticity and aesthetic value between physical artworks and digital art. The digitalization of art and the transformation of television into an art platform necessitate new analyses in terms of the content of art, its relationship with the audience and its economic dimensions.

In conclusion, digital visual culture is radically changing the perception of art and space. Television has become not only a narrative medium but also an important center of artistic production and experience. In the future, television and digital art are expected to become even more intertwined, making it possible for new forms of artistic expression to emerge. In this context, interdisciplinary research on the relationship between digital art and television is of great importance for understanding the transformation of art.

REFERENCES

Ahlers, K. H., Kramer, A., Breen, D. E., Chevalier, P. Y., Crampton, C., Rose, E., Tuceryan, M., Whitaker, R. T. ve Greer, D. (1995). Distributed Augmented Reality for Collaborative Design Applications, *European Computer-Industry Research Centre Gmbh, Proceedings of Computer Graphics International*, Maastricht, Munich, 1-21.

Akmeşe, Z., & Deniz, K. (2015). *Dijital Aktivizm Olarak Video Aktivizm: Redhack Belgeselleri. Yeni Medya Çalışmaları II. Ulusal Kongresi*. Kadir Has Universty.

Antmen, A. (2012). *20. Yüzyılın batı Sanatında Akımlar: Sanatçılardan Yazılar ve Açıklamalarla*. Sel Publication.

Arapoğlu, F. (2017). *Video Sanatının Öyküsü*. [Belgesel / Kısa Film / Video Sanatı]. (N. Ç. Bikiç ve F. Özgür, Ed.). İstanbul: Doruk Publication.

Bajura, M. ve Neumann, U. (1995). Dynamic Registration Correction in Augmented Reality Systems, *Virtual Reality Annual International Symposium*, March 11-15, 1995 Research Triangle Park, North Carolina, 189-196. DOI: 10.1109/VRAIS.1995.512495

Barlas, M. (2024). Hiçlik Ve Ölümün Gölgesinde Bir Hayaleti Hatırlatmak: Kayıp Ev/ La Maison Manquante. *SDÜ ART-E Güzel Sanatlar Fakültesi Sanat Dergisi*. 17/33, 190-207. ISSN 1308-2698

Barnard, M. (2002). *Sanat Tasarım Ve Görsel Kültür* (Korkmaz, G., Trans.). Ütopya Publication.

Başaran Alagöz, S., & Ekici, N. (2016). Tartışmalı bir konu olarak sanat pazarlama: Kavramsal bir değerlendirme. *Kahramanmaraş Sütçü İmam Üniversitesi İktisadi ve İdari Bilimler Fakültesi Dergisi*, *6*(2), 189 202.

Benjamin, W. (1995). *Estetize Edilmiş Yaşam*. Der Publication.

Benjamin, W. (2001). *Tekniğin Olanaklarıyla Yeniden Üretildiği Çağda Sanat Yapıtı, Pasajlar* (Cemal, A., Trans.). Yapı Kredi Publication.

Bernaschina, D. (2023). Screen Video Art: A New Experimental Proposal for Creative Study. *Art Vision*, *29*(51), 170–177. DOI: 10.5152/ArtVis.2023.23202

Burnett, R. (2012). *İmgeler Nasıl Düşünür* (Pusar, G., Trans.). Metis Publication.

Çakın, Ö., & Bolat, N. (2018). *Sosyal Medya Reklamcılığı ve Dijital Medya Etkileşimleri. Dijital Kültür ve İletişim* (Çakmak, V., & Çavuş, S., Eds.). Literatürk Publication.

Castells, M. (2013). *Ağ Toplumunun Yükselişi*. İstanbul Bilgi University Publication.

Çokokumuş, B. (2012). Dijital Ortamda Kültür ve Sanat. *İnternational Journal of New Trends in Arts, Sports &. Science Education, 1*(3).

Deleuze, G. (1986). *Cinema I: The Movement Image, Sinema I: Hareket İmge* (Özdemir, S., Trans.). Norgunk Publication.

Dijk, J. V. (2016). *Ağ Toplumu.* Kafka Publication.

Dimaggio, P., & Useem, M. (1978). Social Class and Arts Consumption. *Theory and Society, 5*(2), 141–161. DOI: 10.1007/BF01702159

Ekim, B. (2022). The Museum Of Me: A Visual Presentation Of Your Online Social Life On Facebook. Visualist 2012, Uluslar Arası Görsel Kültür Kongresi, İletişimde, Sanatta ve Tasarımda Yeni Yaklaşımlar. In Zeybek, I., & Yengin, D. (Eds.), *Diijitalleşme* (pp. 273–280).

Ellul, J. (2004). *Sözün Düşüşü* (Arslan, H., Trans.). Paradigma Publication.

Erdem, M. N. (2017). Dijital Çağda Kültür, Yeni Tüketici Kapitalizmi ve Reklam Anlatısı. *1st International Conference on New Trends in Communication.* İstanbul: Ticaret Universty.

Erdem, S. (2022). Görsel İletişim Tasarımında Dijital Dönüşümü Nft Üzerinden Tartışmak.

Franck Atalay, O. (2002). *Düşünce İçin Mimarlık: Sanallığın Gerçekliği, Mimarlık ve Sanallık. Çağdaş Mimarlık Sorunları Dizisi.* Boyut Yayın Grubu Publication.

Geraghty, C., & Lusted, D. (1998). *The Television Studies Book.* Arnold Publication.

Güney, E. (2014). Dijital Görsel Kültür ve Yeni Medya Ekseninde Sanatın Değişen Rolü. *PhD Thesis.* İzmir: Ege Universty.

Jenkins, H. (2016). *Cesur Yeni Medya" Teknolojiler ve Hayran Kültürü* (Yeğengil, N., Trans.). İletişim Publication.

Karaduman S., & Ünlü Karataş N. (2022). Dijital Yayıncılık ve İzleme Pratikleri: "Gain" Üzerine Bir Analiz. *İstanbul Ticaret Üniversitesi Sosyal Bilimler Dergisi*, 21(45), 982-1007. DOI: 10.46928/iticusbe.1001254

Kılıç, L. (2000). *Görüntü Estetiği.* İnkılap Publication.

Lauristin, M., & Vihalemm, P. (2014). *Reading in Changing Society. Studies in Reading and Book Culture.* University of Tartu press. DOI: 10.26530/OAPEN_496790

Lehman, K., & Wickham, M. (2014). Marketing Orientation and Activities in The Arts Marketing Context: Introducing A Visual Artists' Marketing Trajectory Model. *Journal of Marketing Management, 30*(7–8), 664–696. DOI: 10.1080/0267257X.2013.838987

Mikos, L. (2016). Digital Media Platforms and the Use of TV Content: Binge Watching and Video-Ondemand in Germany. *Media and Communication, 4*(3), 154–161. DOI: 10.17645/mac.v4i3.542

Mutlu, E. (1991). *Televizyonu Anlamak.* Gündoğan Publication.

Nalven, J. ve Jarvis J.D. (2005). Going Digital: The Practice and Vision of Digital Artists. *Thomson Course Technology.* USA.

Ong, W. (2007). *Sözlü Kültür, Yazılı Kültür, Sözün Teknolojileşmesi* (Postacıoğlu Banon, S., Trans.). Metis Publication.

Parsa, S. (1994). *Televizyon Estetiği.* Ege Üniversitesi İletişim Fakültesi Publication.

Pavlik, J. (1998). *New Media Technology: Cultural and Commercial Perspectives* (2nd ed.). Ally and Bacon.

Sağlamtimur, Z. Ö. (2010). Dijital Kültür. *Anadolu Üniversitesi Sosyal Bilimler Dergisi, 10*(3), 213–238.

Sarıberberoğlu, M. T. (2019). Sinematik Kurgunun Bilinçaltı Mekânları-Tekinsiz Mekânlar. *Journal of Architecture and Life* 5(1), 27-39. ISSN: 2564-6109 DOI: DOI: 10.26835/my.645548

Sirer, E. (2020). *Zamanın Ruhu: Eşzamansızlık. Yeni İzlence Yeni İzlerkitle* (Sirer, E., Ed.). Literatürk Publication.

Stallabras, J. (2016). *Sanat A.Ş. Çağdaş Sanat ve Bienaller* (Soğancılar, E., Trans.). İletişim Publication.

Sütçü, C. S. (2012). *Sosyal Medyaya Girmeden Önce Bilinmesi Gerekenler. Yeni Medya ve...* (Yengi, D., Ed.). E Publication.

Tutar, C. (2020). Postmodern Kentsel Mekanda Kültürel Belleğin İnşa Süreci. *Akdeniz Üniversitesi İletişim Fakültesi Dergisi, 34*(34), 102–121. DOI: 10.31123/akil.781252

Ürper, O. (2012). *Dijital Teknoloji Çağında Reklam Fotoğrafçılığı.* Say Publication.

Vaughan, W. (2005). *Digital Art History. (Eds: Anna-Bentkowska-Kafel, Trish Canhen, Hazel Gardiner) History of Art in the Digital Age: Problems and Possibilities.* USA: Intellent Ltd.

Wands, B. (2006). *Dijital Çağın Sanatı* (Akınhay, O., Trans.). Akbank Sanat Publication.

Yakışkan, N. N. (2024). Sergileme Mekânlarının Sanal ve Fiziksel Ortam Arasındaki Farkının İncelenmesi. *İstanbul Modern örneği. yedi: Sanat, Tasarım ve Bilim Dergisi*, Sanatta Dijitalizm Özel Sayısı, 327-337. DOI: 10.17484/yedi.1458408

Yaykın, M. (2010). *Sanat, Teknoloji, Bilim ve Fotoğraf.* Kalkedon Publication.

Yılmaz, R. (2017). Türkiye'de Reklam Anlatısının Dijital Çağdaki Yapısal Dönüşümü. *1st International Conference on New Trends in Communication*. İstanbul: Ticaret Universty.

Yücel, D. (2012). *Yeni Medya Sanatı ve Yeni Müze*. İstanbul Kültür Üniversitesi Publication.

ADDITIONAL READING

Ede, S. (2005). *Art and Science*, New York: I.B.Tauris & Co Ltd.

Fischer, E. (2012). *Sanatın Gerekliliği* (Çapan, C., Trans.). Sözcükler Publication.

Gombrich, E. H. (1999). *Sanatın Öyküsü, (Çev. Erol Erduran - Ömer Erduran).* Remzi Kitapevi Publication.

Hopkins, D. (2000). *After Modern Art 1945-2000.* Oxford University Press.

Kagan, S. M. (2008). *Estetik ve Sanat Notları* (Çalışlar, A., Trans.). Karakalem Kitabevi Publication.

Mattick, P. (2003). *Art in Its Time, Theories and Practices of Modern Aesthetics.* Routledge.

Schiller, D. (1999). *Digital Capitalism: Networking The Global Market System.* The MIT Press. DOI: 10.7551/mitpress/2415.001.0001

Storey, J. (2009). *Cultural Theory and Popular Culture.* Pearson Longman.

Touraine, A. (1992). *Modernliğin Eleştirisi.* Yapı Kredi Publication.

Türkoğlu, T. (2010). *Dijital Kültür.* Beyaz Publication.

Ulusoy, M. D. (2005). *Sanatın Sosyal Sınırları.* Ütopya Publication.

Yamaner, G. (2007). *Postmodernizm ve Sanat.* Algı Yayın.

Yılmaz, M. (2012). *Sanatın Günceli, Güncelin Sanatı*. Ütopya Publication.

KEY TERMS AND DEFINITIONS

Augmented Reality (AR): Augmented reality creates a new perception environment by placing digital elements such as computer-generated sound, image, animation, and hologram onto our environment in real time through smartphones, tablets, and virtual reality glasses.

Digital Art: Digital art, or computational art, is generally defined as a form of art that involves the production of non-physical objects, in which a computer plays a role. In this process, the computer can be found anywhere on the spectrum, from a traditional auxiliary tool to an indispensable co-creator.

Digital Culture: It is a concept that defines the idea that the internet significantly shapes the way individuals communicate and interact, think and behave. It points to the development in information and communication technologies and the changing cultural and social structure with the introduction of the internet into individuals' lives.

Digital Visual Culture: As individuals integrate with new technologies, they are surrounded by a new formation called "digital visual culture". It can be said that visual discourses realized in the virtual environment, in which messages are encoded through technology, are effective in the formation of individuals.

Virtual Reality (VA): Technology is one of the information technologies that has been frequently mentioned recently, allowing users to experience the virtual world in the real world, and in this context, allowing users to participate in a fictional or simulated environment, allowing them to experience images in three-dimensional reality instead of perceiving them on two-dimensional screens.

Virtual Selves: The self is the characteristics and experiences that enable a person to have internal awareness and define themselves as a separate person. With the emergence of new communication technologies, a digitalized and new type of self, the virtual self, has emerged. In this context, the virtual self is a type of self that people need in virtual worlds.

Virtual Space: With the developing computer technologies, the world of the products of the human mind, referred to as the third world, and the cyber world formed by mental schemas, are giving rise to a new type of space, namely the virtual space. Throughout human history, this virtual space has already existed with different variations.

Chapter 8
Television, Streaming, and Audience

Recep Bayraktar
https://orcid.org/0000-0002-6285-8059
Giresun University, Turkey

ABSTRACT

Each innovation that emerged in the field of television broadcasting has transformed the audience's experience of consuming media content, and today streaming platforms have become functional as the main pillar of this transformation. Thanks to streaming, content, audience, viewing experience and television broadcasting have incorporated new definitions. These definitions, which are intertwined with the characteristic features of computer and internet technologies, make today's television understandable.

INTRODUCTION

Since the early days of television broadcasting, the audience's experience of watching programs has also undergone various transformations in parallel with the technological features of television, broadcasting applications and signal transmission tools. First of all, the linear interaction form between television and the audience has gained flexibility with video recording devices, and over time, the trends of the on-demand video phenomenon have developed. While each technology that has emerged in the field of television broadcasting has made the audience autonomous from the program broadcast flow, increasing mobility and content diversity have allowed the audience to spend more time in front of the television. Computer, internet technology, miniature screens and new media platforms, which are considered

DOI: 10.4018/979-8-3373-1967-4.ch008

alternatives to television, have integrated with television, traditional television has repositioned itself within the internet ecosystem.

Today, streaming platforms are among the main environments where this repositioning and television content is consumed intensively. In addition to multiplying the content diversity exponentially, streaming platforms redefine the assumptions and viewing practices of the audience through algorithmic systems. The algorithm, which develops content suggestions by making predictions based on the expectations, interests and desires of the audience, strengthens the bond between television and the audience through the characteristic features of individuals. With the phenomenon of binge- watching, television screens have evolved into the viewing practice desired by broadcasters since the early days of traditional television. This study examines the stages that television broadcasting and the audience have gone through over time through the relevant literature, and explains the effects of streaming platforms on television.

FROM TERRESTRIAL TO INTERNET TELEVISION

Various experts claim that television was killed many times over in the 1980s by cable TV and VCRs; in the 1990s by the internet and video games; in the 2000s by TiVo and iPod; and in the 2010s by smartphones and iPads. While television continues to be the medium through which most people obtain visual entertainment and information, and through which advertisers reach their target audiences, the number of traditional TV audience is declining as other screens integrated into the internet ecosystem become more attractive, and new technologies that allow greater control over programming can also be a concern for advertisers (Butler, 2018). Over time, the television industry has evolved from a practice where broadcasters send content to audience to a process where audience receive the content they desire (Taşdelen ve Kesim, 2020: 707). In TV broadcasting, programmers have sought to reach and retain audience in different ways. Over time, programs have been transmitted via terrestrial cable and satellite transmitters ite (Lotz, 2018). With the increasing availability of cable services (Lotz, 2014), the television viewing experience was transformed by video cassette recorders (VCRs), which were developed after Sony and JVC introduced Betamax and VHS tapes in the mid-1970s. This allowed television audience to record their favorite programs and do other things while they were broadcasting; the TV system evolved from a one-way broadcast stream to a video-on-demand (VOD) system (Marshall, 2009, p. 83).

One of the most important changes in television broadcasting standards after the addition of color images to television in the 1960s was the transition from analog to digital television (Seel, 2018, p. 69). In the digital transformation process, media consumption habits are constantly changing (Erdem, 2024). While digital video recorders (DVRs) allow audience to record programs in a qualitative way, cable providers began introducing VOD services in the early 2000s. DVRs and VOD services are the precursors to Internet-distributed television. (Lotz, 2017, p. 2). VOD also marks the beginning of streaming television, with audience receiving content from a central server rather than storing it locally (Marshall, 2009).

The changes in television are part of broader changes in the communications sector. It has become increasingly difficult to distinguish between television sets and computers, and the two devices have become increasingly similar, increasing their common functions. As television has become integrated with computers, its characteristics have also changed (Hughes, 1990, p. 170). Thanks to digitalization, broadcasters have the opportunity to transmit their programs beyond traditional television broadcasting, through Internet Protocol (IP)-based distribution mechanisms (Demetriades, 2003, p. 177). The online distribution of digital video has led the way in the distribution of television programs over the internet, and many broadcasters have not remained indifferent to this tendency of audience who want to access the content they want for free or through subscription. Thus, television broadcasts have moved to an online environment where audience can watch on demands (Seel, 2018, p. 69).

Internet-distributed TV (ITV) enables personalized content delivery independent of program schedules (Lotz, 2017, p. 3). In this ecosystem, television content is transferred to the internet via traditional media websites or social networking sites (Turner & Tay, 2009, p. 2). Thus, in the interconnected ecosystem of the Internet with social networks, television content is restructured to adapt to the streaming logic of mobile media devices such as smartphones (Marshall, 2009, p. 9). In the 1980s, with the widespread acceptance of cable and satellite distribution, it was important to send audience a program at a time and keep them in front of the screen during the broadcast. In the distribution of TV content over the internet, the feature that allows audience to watch programs whenever they want comes to the fore. These two types of viewing experiences are described as linear and non-linear television broadcasting, and while the former presents audience with consecutive broadcasts that are part of a sequential program at a specific time, the latter is understood as programming presented in no particular order on VOD services such as YouTube and Hulu (Butler, 2018).

The audience's experiences with computers and the portable screen technology of the mobile phone have also necessitated the redefinition of television. In this context, Lotz discusses the stages that television has gone through as the network era, the transition to multi-channel, and the post-network era. While the network era includes the implementation of traditional broadcasting practices, the post-network era refers to television that includes digitalization, mobile devices, smartphones, the internet, and streaming services (Lotz, 2014). ITV refers to content packaged for direct transmission over the internet, where audiovisual quality may vary depending on internet speed (Marshall, 2009, p. 46). On the other hand, Internet Protocol Television (IPTV) technology, which is distributed using internet protocol, guarantees the secure delivery of high-quality television broadcasts, on-demand video and audio content over a broadband network. IPTV generally operates like a standard pay TV service and data is streamed over the network infrastructure belonging to the service provider. While a PC is usually used to access Internet TV services, a digital set-top box is used to access and decode video content transmitted over the IPTV system. The International Telecommunication Union defines IPTV as follows:

> *IPTV is defined as multimedia services such as television/video/audio/text/graphics/ data delivered over IP based networks managed to provide the required level of quality of service and experience, security, interactivity and reliability (O'driscoll, 2008).*

STREAMING

Streaming media began with the proliferation of personal computers in the 1990s (Shattuc, 2020). Among the elements that have fundamentally changed the internet paradigm in these years is P2P-based streaming technology (Li & Yin, 2007, p. 94). P2P first gained popularity through file sharing applications such as BitTorrent. Since then, many commercial P2P-based live streaming systems have been designed and implemented to provide live TV programs over the Internet, including PPLive (Tang at al., 2007).

Streaming is the transfer of data such as text, graphics, audio and video at appropriate quality and speed levels across increasingly crowded communication channels. Streaming has a variety of uses, from entertainment to business applications (Demetriades, 2003, p. 3-4). Historically, streaming audio and video content over the Web began in the mid-1990s when Progressive Networks introduced RealAudio. Microsoft introduced Windows Media Player shortly thereafter, followed by Apple introducing Quicktime, an MPEG-based codec. In 1999, Apple introduced QuickTime TV using Akamai Technologies' streaming media distribution service. A few years later, Flash, DIVX, and MPEG players were developed. Streaming video systems offer live, time-shifted, and on-demand services to audience (Wilkinson, 2010, p. 118). Video streaming systems have evolved into a variety of business models, including SVOD, which relies entirely on subscription revenue, AVOD, which relies entirely on advertising revenue, and hybrid-financed, which combines the two. These models emphasize different audience strategies, ranging from entirely audience-financed to advertiser-financed (Tang & Wei, 2023, p. 4).

In the streaming system, TV broadcasts are transferred directly to the user's computer over the internet without the need for any recorder. By 2018, Hulu, SlingTV, FuboTV, DirecTV, Now, Roku, Apple TV and YouTube platforms were offering live multi-channel TV broadcasts. This service, which is distributed by the content provider over the internet, also represents a major technological change in the content and format of television since the cable era (Shattuc, 2020). In this context, the popular video streaming platform Netflix is among the leading platforms that have transformed the previous form of television and the connection audience have with the content they consume, with terms such as original content, binge-watching and cord-cutting. Instead of storytelling in episodic formats week by week, Netflix's content distribution model allows for long-term, uninterrupted programming, with narratives presented in eight or thirteen hour-long episodes that are supposed to be watched quickly. This also eliminates the need for audience to rely on recaps and recap tools to recall previous episodes and events (Buck & Plothe, 2019, pp. 1-2).

With the availability of broadband and the emergence of YouTube as the leading hub for user-generated video content, the online video phenomenon has multiplied exponentially (Barnhart, 2022). Netflix's move into streaming was also inspired by YouTube's success as a video streaming service (Shattuc, 2020, p. 153). Netflix has introduced an entirely new way to experience television content to a market that has been accustomed to consuming television via cable and antennas. Streaming providers such as Netflix, Hulu, Amazon Prime, and YouTube offer content for web-enabled smart TVs, tablets, computers, and smartphones using real-time interfaces that are more like a website than traditional broadcasting. This streaming practice delivers content in real time over the internet without downloading. These platforms are subscription-based and typically release new programming series all at once. This is a major change from the ad-supported weekly distribution model associated with traditional television. As a result, the experience of consuming television content has changed dramatically over the years thanks to these platforms (Tefertiller & Sheehan, 2019, p. 595).

Netflix categorizes certain elements of series with more than 2000 audience groups with similar tastes worldwide and designs its programs, production, dubbing, marketing and advertising according to the groups it categorizes. Netflix labels all its content with hundreds of general terms. This allows it to develop marketing recommendations for individual users. Netflix offers content to niche-audience groups with niche programs and movies, thus appealing to the specific interests of these communities. In addition, among Netflix's most important strategies is the concept of original content. Netflix Original is content specially produced by Netflix. (Shattuc, 2020). In 2013, Netflix released three original series that changed the paradigm of television and streaming. Although Hulu began airing original television series as early as 2011, before Netflix, these series did not receive the audience number and Emmy Awards of Netflix series (Arditi, 2021, p. 81).

For an increasing segment of the global audience, television is understood as an online service environment dispersed across an ecology of websites, portals and applications. The key elements of this distribution ecology include:

- Online TV portals such as BBC iPlayer, ABC iView, NRK TV and Toggle, which offer a combination of live TV broadcasts, new releases and library content via the websites and applications of the major broadcast networks;
- SVOD services such as Netflix, Amazon Prime Video, Hulu, HBO Now, Hayu and CBS All Access;
- Transactional video-on-demand (TVOD) services that offer content for purchase or rental, such as iTunes, Google Play Store, Microsoft Films & TV and Chili;
- Portals that include user-uploaded content, which is generally free, and professional content with premium content tiers available through subscription or direct purchase, such as YouTube, Youku and Tencent Video;
- Video-sharing platforms such as Daily Motion, which offer free, ad-supported amateur and professional content, which is generally uploaded informally;
- On-demand download services such as BitTorrent, Popcorn Time file hosting sites and illegal streaming sites;
- Unlicensed TV streams offered through pirate websites, illegal TV streaming boxes and streaming services such as Periscope;
- Recommender apps such as JustWatch that tell you what content is available on various streaming platforms (Lobato, 2019, p. 8).

ALGORITHM AND AUDIENCE

Lotz describes the broadcasting practice in which television content is streamed as post-network, and describes the type of television content that this period makes possible as prized content. Prized content refers to programming that audience seek out and specifically desire. Prized content is not a matter of watching whatever is on TV. Prized content is determined by the audience and is a post-network phenomenon that constitutes the norm of content streaming, which acts as the dominant organization for television. Prized content encourages audience to follow news about content, read endless conversations on social networks and news sites, search for missed episodes, and pay for the most prized content (Lotz, 2014, p. 12).

While watching television used to be a social and cultural activity, today it is an individualized experience (Arditi, 2021, p. 99). In the multiplatform media environment, audience can simultaneously take on multiple roles such as content user, content contributor, content distributor, and content producer. For example, after a person watches a funny pet video shared by a friend on Instagram, they can like and comment on it. They can share this video with their followers. They can respond to this video by making another funny video of their own pet (Ha, 2021). Thus, the active audience interacts with the content (Yu, 2023).

Platforms like Netflix have changed traditional linear television in two ways. The first of these is the original content phenomenon related to the content, and the second is the algorithm system that includes personalized recommendations and promotions related to the audience. Thanks to the algorithm, data including features such as how long audience watch a program, when they pause the program and when they stop watching is obtained. With this data, interest categories are created for audience and general audience profiles are determined. The algorithm system can make recommendations by predicting how much the audience will enjoy certain programs (Van Esler, 2020). For example, YouTube's algorithm automatically creates a suggested list based on the audience's viewing history and selects the most-watched videos that are similar to the videos watched in the past. YouTube also has an autoplay feature that plays a similar video after finishing a video. YouTube can also use an ad algorithm to ensure that the same ad is not shown to the same person twice. Each person's YouTube viewing experience is different from other users because each person receives different ads and different recommendations (Ha, 2021).

Our online and offline lives are increasingly shaped by algorithms as we are exposed to algorithm-based web search results, targeted ads, personalized social media content, and consumption recommendations. The relationship between algorithms and audiences reveals a new dimension of the information communication process that fundamentally transforms the relationship between media and the "audience. Netflix recommends movies, Pandora music, Amazon recommends consumption objects. The AI-powered algorithmic communication model behind these applications revolutionizes the news and information consumption behavior of its users and restructures the relationship between media and audience. In this context, the audience, which is intertwined with the algorithmic system, symbolizes a completely new concept of the information distribution process, where the wishes and desires of individual audience are not embedded in the wholesale operations of professional media organizations and are not subject to them. Individual preferences become a priority in the communication process (Yu, 2023).

Navigating through hundreds of channels on satellite and cable TV can take as much time as a program on air. Similarly, the same applies to online environments where thousands VOD content are available. Using information about audience, context, demographic information, and on-platform browsing behaviors are obtained for targeted content, and filters are created for audience. In this sense, the paradigm in the digital ecosystem has shifted from user-controlled browsing to algorithm-controlled ranking (Andrejevic, 2009, pp. 35-36). Streaming services need algorithms that will generate effective analytics to help stream brands retain subscribers and attract more, while also identifying audience's interests in data creation, content creation and distribution, and creating user-centric recommendation systems (Tang & Wei, 2023). Data obtained from audience is processed through active data production processes that are generated by their actions, such as searching, subscribing, and commenting, and through passive data production processes that audience generate within the platform, such as pausing, leaving the page, and accepting recommendations. Both active and passive data production are important for the commercial success of streaming media (Marr, 2015). Audience's digital footprints and the amount of time they spend on a digital media outlet are automatically processed. Streaming services like Netflix and Hulu have much more detailed information about their subscribers. E-commerce company Amazon even integrates online video audience data with shopping behavior (Ha, 2021).

According to Strangelove, television becomes more seductive and addictive as it becomes more interoperable with platforms (Strangelove, 2015, p. 173). Jenner argues that there are two important elements to the way streaming technologies are marketed and viewing practices have developed. These are linked to the concepts of choice and control. On the one hand, the audience's design of their own programmes rather than being dictated by broadcasters, and on the other hand, the audience's expectation of watching individual episodes or entire programmes again, operationalises choice and control over television content. The different practices in how audience consume television, and the idea of taking control of television, meant that programmed platform television existed in parallel with linear television, starting in the late 1940s with the introduction of technologies such as the remote control, and streaming offered audience more choice and control. Choice and control, the freedom to choose and the power to control what programmes are played on the TV set, intrinsically connect the audience to the content. The more choice audience are given, the more control they are assumed to have over their own experience. In this context, choice and control have been ever-expanding throughout the history of television, and binge-watching is the current culmination of both (Jenner, 2021).

Binge-watching a TV program is watching it faster than the program schedule allows. A TV series is a dramatized action divided into episodes. In this sense, most episodes of a series end at a climax, leaving the viewer eagerly waiting for the next episode. Binge-watching has been fueled by Netflix's simultaneous release of all episodes of a single season. Binge-watching TV series changes the cultural position of television in our daily lives, and binge-watching emerges as a cultural dimension of television (Marshall, 2009). Binge-watching comes from the selective, controlled and indiscriminate viewing that gradually became part of the television discourse through different technologies adopted as part of the television culture (remote control, VCR, DVD, etc.), from the processes of previous media practices and recording technologies that allowed for more viewing. In this context, Netflix's contribution to binge-watching lies in being a platform that successfully targets the assumed desires and expectations of binge-watching audience in its marketing, interface and algorithm (Jenner, 2021).

CONCLUSION

Since the invention of television, which has made moving images widespread worldwide, it has also affected the viewing practices of audience with its technological transformations. While the first applications in the traditional sense allowed for one-way, linear transmission of television programs, it was a necessity for audience to stick to the program flow. The video recorder devices that were developed have eased this limitation for audience, and thanks to the recorded television programs, audience have had a flexible viewing experience in terms of time. While the first examples of video on demand applications emerged with the cable broadcasting phenomenon developed to overcome the geographical difficulties of television broadcast signals, content that appealed to the specific interests of audience has developed as a result of the exponential increase in the variety of television channels with satellite broadcasting.

The integration of computer and internet technology into the broadcasting field and the convergence of these devices with television have enabled the evolution of analog content into digital, while the interaction of audience familiar with computer and internet technology with media content has increased, and the orientation of media content has taken on a structure where ordinary users as well as professional producers can intervene and produce. The integration of television into the internet ecosystem makes it possible to access TV broadcasts anytime, anywhere with miniature screens. Traditional television, which is under the threat of new media platforms, incorporates the characteristic features of new media and reaches audience without time and space restrictions with the streaming system.

Streaming media increases the targeting ability for the content it delivers to the audience, and with algorithmic systems, the content expectations of the audience can be analyzed with their behavior patterns in the digital ecosystem, and content suggestions are developed accordingly. Streaming systems allow audience to consume media content uninterruptedly and for a long time, and unlike traditional television broadcasting, all episodes of a series can be consumed one after another with the streaming system. After discovering what the television audience desires in terms of content with algorithms, it is time for the binge-watching act to take place within streaming media, which allows them to consume this content breathlessly. While streaming platforms that define their audience through algorithms manage to keep the audience in front of the screen with the suggestions they offer far beyond traditional television broadcasting, the possibility that algorithmic systems that surround the audience in every aspect with their suggestions within streaming media may imprison the audience in filter bubbles may be among the topics of discussion for future studies.

REFERENCES

Andrejevic, M. (2009). The twenty-first-century telescreen. In Turner, G., & Tay, J. (Eds.), *Television Studies After TV: Understanding Television in the Post-Broadcast Era* (pp. 31–40). Routledge.

Arditi, D. (2021). *Streaming culture: Subscription platforms and the unending consumption of culture*. Emerald Publishing. DOI: 10.1108/9781839827686

Barnhart, A. (2022). *The primetimer guide to streaming TV*. Primetimer.

Buck, A. M., & Plothe, T. (2019). Introduction. In Plothe, T., & Buck, A. M. (Eds.), *Netflix at the Nexus Content, Practice, and Production in the Age of Streaming Television* (pp. 1–9). Peter Lang Publishing.

Butler, J. G. (2018). *Television visual storytelling and screen culture*. Routledge. DOI: 10.4324/9781315181295

Demetriades, G. C. (2003). *Streaming Media building and ımplementing a complete streaming system*. Wiley.

Erdem, S. (2024). Çapraz medya stratejileri: Markaların çoklu platformlarda etkileşim sanatı. In M. Nur ERDEM & Cansu MAYADAĞLI (Eds.), *Marka İletişimi Tüm Boyutlarıyla Marka ve Tüketici Etkileşimi* (pp. 131-142). Palet Yayınları.

Ha, L. (2021). The changing audience. In Mahoney, M., & Tang, T. (Eds.), *Handbook of media management and business* (pp. 21–38). Rowman & Littlefield.

Hughes, P. (1990). Today's television, tomorrow's world understanding television. In Goodwin, A., & Whannel, G. (Eds.), *Understanding Television* (pp. 170–187). Routledge.

Jenner, M. (2021). Introduction. In Jenner, M. (Ed.), *Binge-Watching and Contemporary Television Research* (pp. 1–20). Edinburgh University Press.

Li, B., & Yin, H. (2007). Peer-to-peer live video streaming on the internet: Issues, existing approaches, and challenges [Peer-to-Peer Multimedia Streaming]. *IEEE Communications Magazine*, *45*(6), 94–99. DOI: 10.1109/MCOM.2007.374425

Lobato, R. (2019). *Netflix Nations: The geography of digital distribution*. New York University Press. DOI: 10.18574/nyu/9781479895120.001.0001

Lotz, A. D. (2014). *The television will be revolutionized*. New York University Press.

Lotz, A. D. (2017). *Portals: A treatise on ınternet-distributed television*. Michigan Publishing. DOI: 10.3998/mpub.9699689

Marr, B. (2015). *Big Data: Using SMART big data; analytics and metrics to make better decisions and improve performance*. Wiley.

Marshall, P. D. (2009). Screens: television's dispersed 'broadcast. In Turner, G., & Tay, J. (Eds.), *Television Studies After TV: Understanding Television in the Post-Broadcast Era* (pp. 41–50). Routledge.

O'Driscoll, G. (2008). *Next generation IPTV services and technologies*. Wiley.

Seel, P. B. (2018). Digital television & video. In Grant, A. E., & Meadows, J. H. (Eds.), *Communication Technology Update and Fundamentals* (pp. 67–76). Routledge. DOI: 10.4324/9780203702871-6

Shattuc, J. (2020). Netflix, Inc. and online television. In Wasko, J., & Meehan, E. R. (Eds.), *A Companion to Television* (pp. 145–164). John Wiley & Sons. DOI: 10.1002/9781119269465.ch7

Strangelove, M. (2015). *Post-TV: Piracy, cord-cutting, and the future of television*. University of Toronto Press. DOI: 10.3138/9781442666184

Tang, W., & Wei, M. (2023). Streaming media business strategies and audience-centered practices: A comparative study of Netflix and Tencent Video. *Online Media and Global Communication*, *2*(1), 3–24. DOI: 10.1515/omgc-2022-0061

Tang, Y., Luo, J. G., Zhang, Q., Zhang, M., & Yang, S. Q. (2007). Deploying P2P networks for large- scale live video-streaming service [Peer-to-Peer Multimedia Streaming]. *IEEE Communications Magazine*, *45*(6), 100–106. DOI: 10.1109/MCOM.2007.374426

Taşdelen, B., & Kesim, M. (2020). User central content design in the age of interactive television. *Erciyes İletişim Dergisi*, *7*(1), 703–722. DOI: 10.17680/erciyesiletisim.612545

Turner, G., & Tay, J. (2009). Introduction. In Turner, G., & Tay, J. (Eds.), *Television Studies After TV Understanding Television in the Post-Broadcast Era* (pp. 1–6). Routledge. DOI: 10.4324/9780203878316

Van Esler, M. (2020). Reproducing television canons: Streaming services and the legacy of linear TV. *Journal of Popular Culture*, *53*(4), 946–966. DOI: 10.1111/jpcu.12940

Wilkinson, J. S. (2010). IPTV: Streaming media. In Grant, A. E., & Jennifer, H. M. (Eds.), *Communication technology update and fundamentals* (pp. 117–131). Focal Press. DOI: 10.1016/B978-0-240-81475-9.50008-6

Yu, R. (2023). *Algorithmic audience in the age of artificial intelligence: Tailored communication, information cocoons, algorithmic literacy, and news literacy*. Peter Lang Publishing.

ADDITIONAL READING

Arrojo, M. (2015). Social television as a new relationship between conventional TV and the audience an analysis of its aims, processes, and results. *International Journal of Social Science Studies*, *3*(4), 37–49. DOI: 10.11114/ijsss.v3i4.854

Ferchaud, A. (2021). *Binge and bingeability: The antecedents and consequences of binge- watching behavior*. Routledge.

Hallinan, B., & Striphas, T. (2014). Recommended for you: The Netflix Prize and the production of algorithmic culture. *New Media & Society*, *18*(1), 117–137. DOI: 10.1177/1461444814538646

Keltie, E. (2017). *The Culture ındustry and participatory audiences*. Palgrave Macmillan. DOI: 10.1007/978-3-319-49028-1

McDonald, K., & Smith-Rowsey, D. (Eds.). (2016). *The Netflix effect: Technology and entertainment in the 21st century*. Bloomsbury Publishing USA.

Nieborg, D. B., & Poell, T. (2018). The platformization of cultural production: Theorizing the contingent cultural commodity. *New Media & Society*, *20*(11), 4275–4292. DOI: 10.1177/1461444818769694

Radošinská, J. (2017). New trends in production and distribution of episodic television drama: Brand Marvel-Netflix in the post-television era. *Communication Today*, *8*(1), 4–29.

Scheibe, K., Fietkiewicz, K. J., & Stock, W. G. (2016). Information behavior on social live streaming services. *Journal of Information Science Theory and Practice*, *4*(2), 6–20. DOI: 10.1633/JISTaP.2016.4.2.1

Starosta, J. A., & Izydorczyk, B. (2020). Understanding the phenomenon of binge-watching-a systematic review. *International Journal of Environmental Research and Public Health*, *17*(12), 4469. DOI: 10.3390/ijerph17124469 PMID: 32580289

Wasko, J., & Meehan, E. R. (2020). *A companion to television*. John Wiley & Sons. DOI: 10.1002/9781119269465

KEY TERMS AND DEFINTIONS

Binge-watching: Binge-watching is the process of audience watching entire seasons of a series or movie in succession without interruption within streaming media systems. This viewing behavior is among the marketing strategies of digital streaming platforms. This strategy strengthens the audience's connection with the media content and the media content is consumed intensively.

Internet Protocol Television (IPTV): IPTV is a multimedia service that transmits text, audio, video, graphic data and television broadcasts to users via broadband internet technology, ensuring service quality.

Internet-distributed television (ITV): Internet-distributed television allows television content to be digitally encoded and transmitted via internet protocol-based distribution mechanisms. TV content distributed over the internet may vary depending on the receiver's internet speed and quality.

Multimedia: Multimedia refers to computer-based applications in which text, data, images, animation, audio and video content can be digitally stored, processed and transmitted.

Recomendation Systems: Recommendation systems are algorithmic structures that analyze the behavior of different users in the digital ecosystem and then recommend the right content and products in line with their interests.

Streaming: Streaming is a system that transfers data over the internet not as a single unit, but gradually through a continuous flow and without the need for any storage. Platforms such as Netflix and Hulu offer media content to their subscribers through streaming.

TV Broadcasting: Broadcasting is the transmission of audiovisual media content to audience via signals. In the early years of television broadcasting, these signals, which were transmitted terrestrially, have transformed into broadcasting via cable, satellite and internet broadcasting with the development of broadcasting technologies and the digitalization of analog systems. Today, television broadcasting has taken on the feature of being able to offer audience the content they demand in line with their interests, anytime, anywhere, through streaming platforms and increasing mobility opportunities.

Chapter 9
Transformation of Content Consumption in the Context of Digitalization:
A Comparative Analysis on Traditional Television Broadcasting and Online Platforms

Aslı Yurdigül
https://orcid.org/0000-0001-5333-9751
Atatürk University, Turkey

Yusuf Ayaz
Artvin Çoruh Unversity, Turkey

ABSTRACT

This study examines the transformative effects of digitalization on content consumption through traditional television broadcasting and online video platforms. To this end, interviews were conducted with 22 participants with different sociodemographic characteristics to understand the changes in media usage practices. The research findings reveal that traditional television viewing is based on habits and socialization. In contrast, it is understood that the diversity of content, ease of access, time-space flexibility, and personalization of digital platforms form the basis of new viewing habits. In conclusion, although television and digital platforms reveal a hybrid media consumption that complements each other, it is understood that digitization is the cause of change in viewing habits. This study reveals that the

DOI: 10.4018/979-8-3373-1967-4.ch009

transformation in media consumption is not only technological but also has cultural and social dimensions; it contributes to the development of user-oriented strategies in the digitization process of the broadcasting sector.

INTRODUCTION

As in many other areas, digitalization has brought significant changes in the field of content consumption. Television, one of the most traditional ways of content consumption, has gradually lost its former importance and online platforms, which have emerged as a product of new communication technologies, have become more and more widespread. Digitalization factors such as access, interactivity and personalization have played a major role in this. Based on this, the main purpose of this study is to reveal the content consumption experiences of users shifting from a traditional medium such as television to online platforms from a user perspective. For this purpose, 22 users with different demographic characteristics, who were determined by purposive sampling method, were interviewed and data were collected through semi-structured interview forms. The data obtained were evaluated within the framework of certain themes ussing thematic analysis method. As a result of the evaluation, it was observed that users' video consumption has undergone a transformation on the axis of content, technology and cost.

THE TRANSITION FROM TELEVISION TO ONLINE PLATFORMS IN THE DIGITALIZATION PROCESS

The superiority, quality, and usability of a new technology are generally determined by the standards set by existing technologies. This comparison also reveals how the features offered by the new technology improve the previous technology in terms of form and function. When the evolution of communication tools is examined in this context, a similar process is evident. Each new communication tool tends to offer a superior experience in terms of usability and functionality compared to its predecessor. This evolutionary process in communication tools spans from the camera to cinema and then to television, continuing to evolve with each new step.

In this sense, the invention of the camera made it possible to record details, while the discovery of cinema took the camera a step further by presenting moving images in an impressive way. Television continued this technological advancement by bringing the cinema experience into homes, making it more accessible and continuous. (Başer, 2010, p.46-47). Postman (Postman, 2020, p.108) also emphasizes this evolutionary process, stating that television is a continuation of the tradition initiated by the telegraph and photography. Indeed, with the emergence of internet technology, this evolutionary process has taken on a new dimension and continued its course. Television has now been replaced by online video platforms that enable content consumption independent of time and space constraints.

Therefore, the point reached at the end of this historical process should not be regarded as a mere technical transformation. This transformation has also profoundly affected and changed viewer habits, content consumption patterns, and the relationship established with media tools and media environments. This study evaluates the relationship established by viewers with the media environment and the effects on content consumption in detail in light of user experience as a result of this transformation.

The harmonious development of internet and video technology has significantly popularized online video broadcasting at both professional and amateur levels. Parallel to this, there has been a significant increase in the number of people using the internet to watch videos According to We Are Social's (2024) report, the percentage of individuals worldwide who use the internet to watch online videos, TV shows, and movies is 52.3%. The same report indicates that this percentage is 54.6% in Turkey. The report also shows that video streaming platforms rank among the most frequently visited websites in Turkey. This highlights a shift in content consumption preferences.

Especially among young people, the tendency to watch traditional television broadcasts has gradually decreased, and interest in online platforms has increased day by day. According to a 2022 study by RTÜK examining young people's television viewing habits, a significant decline in television viewing time has been observed both in Turkey and worldwide. In particular, 44.7% of individuals aged 15-21 stated that they watch television every day, but this viewing time is limited to an average of 1 hour and 4 minutes per day. When asked about the reasons for not watching television, 27% of these young people cited "finding the content inappropriate, unhelpful, and unrealistic" as the primary reason. Additionally, 13.1% mentioned "finding other platforms (the internet) more enjoyable and preferring them" as a reason. (RTÜK, 2022, p.103). Based on these data, it can be concluded that there is significant social momentum in the changing television viewing habits of young people and that there is an increasing tendency to prefer digital media.

Apart from the reasons mentioned above, various factors may influence individuals' preference for internet videos over traditional television broadcasts. However, at this point in the study, defining television as a means of communication and determining its place in society would be beneficial to the research as a whole. This would allow for a better understanding of the changes that have taken place in media usage among individuals.

The word "television," which means "seeing from a distance," is a combination of the Latin words 'tele' and "visio." (Kırık, 2010, p.23). This definition reduces television to a mere technological tool, reducing it to the level of a machine, and generally characterizes it in terms of typical capitalist relations such as production, labor, surplus value, and profit. However, what is essential is the social and intellectual capital that television offers through its program content. Therefore, it is necessary to focus not on what television shows or reflects as a mere tool, but on its underlying ideology and the narratives and symbolic representations that constitute this ideology. (Postman, 2020, p.112). Thus, rather than equating television with any particular mode of production, it is possible to evaluate it by focusing on its potential in terms of the images, worldviews, symbols, visions, and values it presents. (Meehan, 1986, p.448-449). Television technology has created a complex industry behind the scenes. The most important output of this industry is cultural production. Television has become an important part of political and social institutions by continuously engaging in cultural production. (Mutlu, 2005, p.22-28).

Television was designed to create a cultural impact by enriching its technological features with dramatic elements. (Esslin, 2001, p.17-21). The effective use of these dramatic elements has also created social and psychological effects on television viewers over time. (Williams, 2003, p.10-12). This situation is not a temporary phenomenon. Television has occupied a central place in people's lives since childhood. Despite all efforts to turn it off, children place this ever-present device at the center of their emotional and cognitive activities. Television has therefore taken on an important role in people's lives. (Silverstone, 1994, p.8-14). Television already plans the viewer's entire day. Television content is determined in advance, including what time it will be broadcast, what kind of storyline it will have, and what characters will appear, and viewers are expected to conform to this schedule. (Mutlu, 2005, p.91).

Television is still one of the most widely used mass communication tools by a large segment of the population. (We Are Social, 2024). However, the spread of the internet and new communication technologies that developed alongside it has happened much faster than in the early years of television. Devices such as laptops, cell phones, and tablets, along with increasing internet speeds, have made it possible to watch television broadcasts anywhere. This rapid development has led to the formation of approaches in broadcasting that have not yet matured. (Demirkıran, 2010, p.74). One of the fundamental misconceptions here is the assumption that live broadcasts accessible from anywhere are of great importance. However, despite the idea of live broadcasts accessible from anywhere, users find the ability to consume content independently of place and time much more appealing. (Lotz, 2014, p.54).

In order to respond to changes in viewer habits, video streaming service providers have adapted their content delivery methods to this new system. As a result, content is now classified according to viewer habits, recommendation systems have been developed, and technological features such as subtitles and dubbing have been introduced to facilitate viewing. (Sarı & Türker, 2021, p.61). This adaptation process has transformed video streaming platforms into a media ecosystem integrated with all the features of the internet. With its ability to offer a wide variety of content and closely track viewer habits, the internet has a significant advantage over television broadcasting. Unlike traditional television broadcasting, internet-based broadcasting allows for viewer interaction. This allows viewers to be identified more clearly and transparently, and even enables content producers to increase viewer satisfaction by producing content tailored to their preferences. In order to respond to changes in viewer habits, video viewing service providers have adapted their content delivery methods to this new system. As a result, content is now categorized based on users' viewing habits, recommendation systems have been developed, and technological features such as subtitles or dubbing are provided to enhance the viewing experience (Sirer, 2020, p.1164-1665). On the other hand, the internet is extremely advantageous compared to the high costs of television because it is based on a low-cost and fast distribution structure (Özel, 2015, p.1-3).

This framework is also linked to the development of online video broadcasting. Today, the internet has expanded the boundaries of the media industry with its global accessibility, interactivity, and user-focused content selection, emerging as an advanced form of television with new qualities.

In line with these advantages, traditional television broadcasting has undergone a multi-layered and user-centered transformation alongside new broadcasting technologies such as Internet TV, IPTV, OTT, and mobile TV. When it comes to Internet-based broadcasting, the first thing that comes to mind is Internet TV, or more commonly known as Web TV. Web TV refers to an environment where individuals, groups, or organizations broadcast video content they produce through their websites. (Sarı, 2021, p.177). The fundamental difference between Internet TV and traditional television broadcasting is the medium of transmission and, consequently, the flexibility offered to viewers. In this regard, the fact that Internet TV is generally free, accessible from anywhere with an Internet connection, and allows viewers to consume content independently of location makes it a distinct model of broadcasting (Demirkıran, 2010, p.78).

Another form of television broadcasting on the Internet, IPTV (Internet Protocol Television) emerged with the integration of the Internet and television. The basic technology of IPTV broadcasting is the transmission of content to subscribers via a digital data network using the Internet protocol (Kandemir, 2013, p.18). The storage capacity and high transmission speeds offered by digital technology have increased the variety of content available on IPTV and enabled the transmission of the same content with much less loss compared to traditional broadcasting. (Levy, 1999, p.4). In this publication format, content is delivered to end users in a highly controlled and secure manner within a closed network via Internet IP (Internet Protocol) packets (Kırık, 2015, p.147). In addition, the IPTV service offers temporal flexibility in content consumption, such as "watch again," "pause," "select," and "rewind," and also allows users to recommend content to each other (Kırık, 2015, p.155).

As an alternative to traditional broadcasting, another media platform that offers users greater freedom is OTT, which stands for "Over The Top." Although IPTV and OTT TV broadcasting are often considered to be the same thing, they differ in terms of how they are used. While IPTV uses a closed network structure, OTT TV enables content to be viewed on any screen with internet access, at any time and place. Among the most prominent examples of OTT TV, which offers a media experience tailored to user preferences, are YouTube, Netflix, Exxen, Hulu TV, Gain, Amazon Prime Video, and Disney+ (Aslan & Öztekin, 2023, p.856). OTT TV offers a more interactive and content-based service compared to traditional television broadcasting. In this way, it has transformed the viewing experience by going beyond the linear content model offered to viewers (Li & Sun, 2019, p.165). In traditional television broadcasting, viewers' freedom of movement is limited to changing channels. In contrast, OTT services offer a much wider range of freedom by allowing viewers to categorize content into genres such as crime, comedy, and drama (Sarı & Türker, 2021, p.73). In addition, OTT services have enabled viewers to influence the flow of content by adding interactivity to storytelling, thereby removing viewers from a passive position and actively involving them in the content consumption process (Ünlü & Akyol, 2025, p.51-52).

The economic advantages that users gain by adopting innovations also apply to the media sector. In parallel with developments in the information technology sector, the importance of the Internet network for the media continues to grow day by day. Undoubtedly, economic factors that will reduce costs lie at the root of this importance. As media conglomerates discover the opportunities offered by the internet, they increasingly believe that this field is more profitable for them. For this reason, major newspapers, television channels, and other media organizations have quickly established their own online publishing platforms (Coşkun, 2015, p.68).

Economic concerns have always been a fundamental motivator for changes in user behavior. Social media offers users the opportunity to communicate at a much lower cost and with higher efficiency compared to traditional communication tools (Kaplan & Haenlein, 2016, p.370). In the media sector, the relative price-performance ratio of alternative services also influences user preferences. Almost none of the major online media outlets charge for access. This price difference between online and offline media directs readers toward affordable options. (Ahlers, 2006, p.33).

In addition to these economic advantages, the main reason why Internet-based communication environments have become so tightly integrated into our daily lives is that they have different characteristics from other communication tools. One of the most prominent of these characteristics is interaction. Social media platforms enable mutual interaction. In traditional television broadcasts, it is difficult to receive feedback on the message, however, in social media, there is the possibility of responding quickly to messages. This demonstrates that social media is a communication tool that enables two-way interaction rather than being one-way (Güngör, 2018, p.398).

Therefore, in the digital age, as communication tools have developed, interactive features have become one of the most important elements. Here, interaction refers to the user's ability to directly intervene in the content. While content is published in traditional media according to the decisions of editors and editors-in-chief, this process is controlled by users in Internet-based environments (Aydın, 2021, p.38). Therefore, the most important aspect of communication is not the type of medium used, but rather the interaction and response between the source and the recipient. In this context, digitization has significantly improved television broadcasting (Kırık & Karakuş, 2013, p.65). In addition, social networking sites have transformed the way users interact with media content and paved the way for individuals to take on more active roles in media consumption (Çil & Yurdigül, 2018, p.455).

This new trend in communication tools has also changed the framework within which interactions take place at the individual and social levels. A completely different form of interaction has emerged compared to the face-to-face interaction that has long dominated human relations. Although this new form of interaction varies depending on the structure of the communication tool, it has added a new dimension to interaction by transcending time and space (Thompson, 2019, p.129-131). Indeed, this transformation has led to the adoption of a more interactive and engaging communication model, replacing the one-way structure of traditional media by prioritizing user participation in new media environments (Yurdigül & Yüksel, 2012).

One of the important factors influencing users' media preferences is content diversity. (LaRose & Atkin, 1991, p.9-10). For example, online video streaming platforms offer much higher quality productions than traditional television broadcasts thanks to their technical capabilities and the variety of content they offer. This is seen as one of the main reasons why these platforms are preferred by viewers (Aslan & Öztekin, 2023, p.1889). For this reason, online video platforms with extremely diverse content that transcend time and space constraints and offer a personalized experience are important in terms of ease of use.

This new media environment is expanding the range of content available on traditional television platforms and eliminating constraints such as time and place. Unlike television, consumers can access customized content tailored to their personal preferences via the internet, smartphones, and other mobile devices anytime, anywhere. Naturally, this has led to changes in people's media usage habits (Kuyucu & Karahisar, 2013, p.173). Research conducted on this topic also emphasizes that the ability of viewers to store content on their mobile devices and watch it at any time, as well as applications such as in-site search and personalization, are important factors influencing media choice (Akıncı & Başer, 2020, p.481). With this digital transformation, it is clear that television viewing habits have evolved and the role of traditional broadcasting has diminished.

Internet broadcasting services, social media platforms, and similar online tools are increasingly pushing traditional broadcasting into the background. Online publishing offers a significant advantage due to its ability to provide personalized services. As a result, anyone with an internet connection can access content from anywhere and at any time, even through their home television. Additionally, internet-based platforms offer a personalized experience tailored to individual preferences through mobile devices, overcoming time and location constraints. (Aydın, 2021, p.53-54).

This transformation clearly demonstrates the impact of digital technology on communication forms. With the integration of digital technology into communication, different forms of communication such as mass communication and face-to-face communication have merged onto the same infrastructure. While this signifies a qualitative change in communication brought about by digital technology, new communication tools and infrastructures have led to quantitative changes such as the production, processing, storage, and distribution of messages, reduced costs, and increased capacity (Aydoğan & Başaran, 2012, p.219).

This change has affected not only communication methods but also archiving and information storage methods. Before the development of Internet technology, archiving was carried out in a bureaucratic manner. For example, in the Roman Empire, all weapons, materials, and official documents were kept in a specific place. Such a bureaucratic archiving system was abandoned with the advent of the Internet. Now, anyone can organize and protect their own archive according to their personal importance. Especially on social media, something that is insignificant to one person may be valuable to another. Thus, every user can create a personal archive based on their own preferences (Parikka, 2017, p.146-147). This transformation has increased the accessibility of television content, blurring the rigid boundaries of media and offering a more dynamic consumption environment.

From this perspective, television is now everywhere. The convergence of amateurs and professionals, i.e., producers and consumers, has eliminated the rigid boundaries of television. Different tools and content are meeting people's needs in a practical way. In addition, this new environment has given users significant mobility. Content consumption on computer, phone, and tablet screens is becoming the new form of television (Özel, 2015, p. 22).

Television can now be consumed individually through devices with different screen sizes. It can be watched anytime, anywhere, shared, discussed, and rewatched in new media environments (Özsoy, 2011, p.124-125).

The idea that viewers are changing and actively participating in production is increasingly emphasized. Television has been replaced by the growing and rapidly spreading Internet network. However, the fact that the habits and behaviors we have acquired through television continue in new media environments should not be overlooked (Özel, 2015, p.13). From this perspective, it is important to evaluate traditional television broadcasts and online video platforms from the consumer's point of view.

When reviewing previous studies in the literature, it is observed that they generally focus on changes in television viewing habits and the potential of digital platforms to replace television. (Genel, 2018; Kuyucu, 2021; Söğüt, 2020). Some studies, however, point to differences in the motivations of Generations Y and Z for watching television and using internet media (Guo, 2022; Kuyucu, 2019).

Therefore, existing studies mostly focus on the rise of online video platforms, the decline in television viewership, and the characteristics of this new audience. This study, however, differs from previous research in that it takes a micro-level approach to media environment preferences, offering a qualitative perspective that focuses on the viewer under the influence of factors such as individual experience, technological factors, emotional tendencies, viewing comfort, and economic access. In this context, semi-structured interviews were conducted with viewers to obtain first-hand, detailed, and qualitative information on the subject. The findings of this study differentiate it from previous studies by providing television broadcasters with the opportunity to understand the changing user and by providing data for the development of innovative strategies. Additionally, considering the user experience is important in terms of identifying user expectations in digital media production and consumption and guiding the adaptation of effective traditional broadcasting to the digital environment.

In this context, the study aims to guide the digital transformation of broadcasting and assist in the development of user-focused strategies by revealing the media consumption habits of viewers/users on online video sharing platforms and television broadcasting with qualified data. Based on this, the objective of the research is to compare television broadcasts and online video platforms from the user's perspective in terms of content, technology, and cost, and to understand changing viewer habits and expectations.

RESEARCH METHODOLOGY

Since this study examines the reflections of changing content consumption patterns due to digitalization on user experience, a qualitative case study was deemed appropriate as the research design. The primary objective of a case study is to conduct an in-depth examination of the characteristics of a specific situation within its own context, focusing on factors such as the environment, individuals, events, and processes. This allows the changes and processes occurring within the specific situation to be analyzed in a multidimensional manner (Şimşek & Yıldırım, 2016, p.73) In addition, case studies enable the acquisition of qualified information about events and facts, thereby laying the groundwork for future studies (Punch, 2011, p.149). In this sense, case studies are defined as an approach that provides a deep understanding of complex situations that cannot be understood with superficial techniques. Therefore, the phenomenon under consideration can be structured in an extremely broad manner in the form of a person, place, group, institution, or event. In addition to all this, case studies can be supported by single and multiple data collected from various sources such as interview data, observations, personal documents, letters, photographs, and notes. Furthermore, case studies can lay the groundwork for future studies by obtaining qualitative information about events and phenomena (Kümbetoğlu, 2020, p.44).

Within this scope, the interview technique was chosen as the qualitative data collection method in line with the research design. The interview technique was chosen because it allows participants to express their thoughts openly and naturally, and provides detailed and meaningful data through qualitative feedback related to the research topic. The interview technique is an effective qualitative data collection tool that goes beyond superficial methods and enables a deep understanding of the reasons behind emotions, thoughts, and behaviors (Türnüklü, 2000, p.544).

Semi-structured interviews are the most commonly used interview technique in qualitative research (Polat, 2022, p.168). In semi-structured interviews, even though questions are prepared in advance, the interviewer can ask new questions to increase the scope and depth of the interview depending on how the interview is going (Bal, 2016, p.163). In this technique, interviews are conducted face-to-face, by telephone, or electronically with the individuals to be interviewed.

The semi-structured interview form developed within the scope of this study aims to understand the transformations brought about by digitalization in content consumption practices in a multidimensional way through the experiences of participants. The interview form prepared in this direction consists of a three-stage structure. The first section of the form includes information about the participants' demographic characteristics. The second section focuses on participants' television viewing practices, while the third and final section examines the user and media consumption relationships transformed by digitalization. Thus, the interview form is thematically organized to explore many dimensions of this transformation.

In this regard, the research team consists of individuals selected from different age and occupational groups in Turkey. Therefore, the study was conducted using a non-random sampling method in line with qualitative research, taking into account certain criteria and ensuring that participants shared similar characteristics (Şimşek & Yıldırım, 2016, p.115). In this regard, a total of 22 participants with different socio-demographic characteristics were contacted, and interviews were conducted with those who were deemed suitable. The interviews were conducted face-to-face and individually.

Research Model: This study aims to reveal users' content consumption practices and experiences as they shift from traditional media such as television to online platforms from the users' perspective. In this regard, a descriptive research model was used to examine the effects of digitalization on content consumption habits. For this purpose, a semi-structured interview technique was used.

Working Group: The study was conducted with a total of 22 participants selected through random sampling from different age, education, and occupational groups in order to gain a deep understanding of the impact of digitalization on media consumption practices. Face-to-face and individual interviews were conducted with the participants in appropriate settings. Through these interviews, participants' media usage habits, content preferences, and viewing practices were examined. As a result, the diversity of the research group enabled the collection of multidimensional data.

Development and Implementation of the Interview Form: In this study, a semi-structured interview form was used to examine changes in media usage habits in line with the digitalization process. The interview form consists of three main sections. The first section includes information about the participants' demographic characteristics. The second section collects data on participants' viewing preferences and media usage habits. The third and final section includes in-depth questions about changes in individuals' media usage habits and their expectations for the future of broadcasting.

Data Collection and Analysis: In the study, one-on-one interviews were conducted with each participant in an environment where they felt comfortable. Prior to the interviews, the purpose of the study was explained to the participants, and it was emphasized that the study was voluntary and that confidentiality would be maintained. With the participants' consent, the interviews were recorded. During the data analysis phase, descriptive analysis was used, and the data obtained were coded and interpreted within the framework of pre-determined themes. Each participant was coded with letters and numbers (K1-K2). Percentages and frequencies were used to analyze demographic data.

RESEARCH FINDINGS

This study, which was prepared based on the transformation in content consumption as a result of digitalization, was conducted with twenty-two participants. Care was taken to ensure a wide age range among the participants. The findings section presents the analysis of the participants' personal information, media usage habits, media consumption preferences, differences between content, and usage experience.

Demographic Information of The Participants

Data on the gender, age, education level, and financial status of the participants are presented in Table 1.

Table 1. Personal Information Regarding Research Participants

		Frequency	Percentage
Gender	Male	13	%59,1
	Female	9	%40.9
Age	20-30	7	%31,8
	31-40	9	%40.9
	41-50	2	%9.1
	51-60	4	%18,2
Education Level	Secondary School	1	%4,5
	High School	6	%27,3
	Associate Degree	1	%4,5
	Bachelor's Degree	10	%45,5
	Postgraduate	4	%18,2
Financial Status	Insufficient	6	%27,3
	Equal to Expenses	9	%40.9
	Able to Save	7	%31,8

When looking at the gender of the participants in the study, 59.1% were male and 40.9% were female. When examining the age range, it was observed that the majority of participants were between the ages of 20 and 40. In terms of education levels, 45.5% were university graduates and 27.3% were high school graduates. According to an assessment of participants' financial situations, 40.9% reported that their expenses were equal to their income, 31.8% stated that they were able to save, and 27.3% indicated that their income was insufficient.

Information On Participants' Media Use

This section of the study includes information about the media environment preferred by participants, their usage times, and frequency of use.

Figure 1. Participants' Television Usage Times

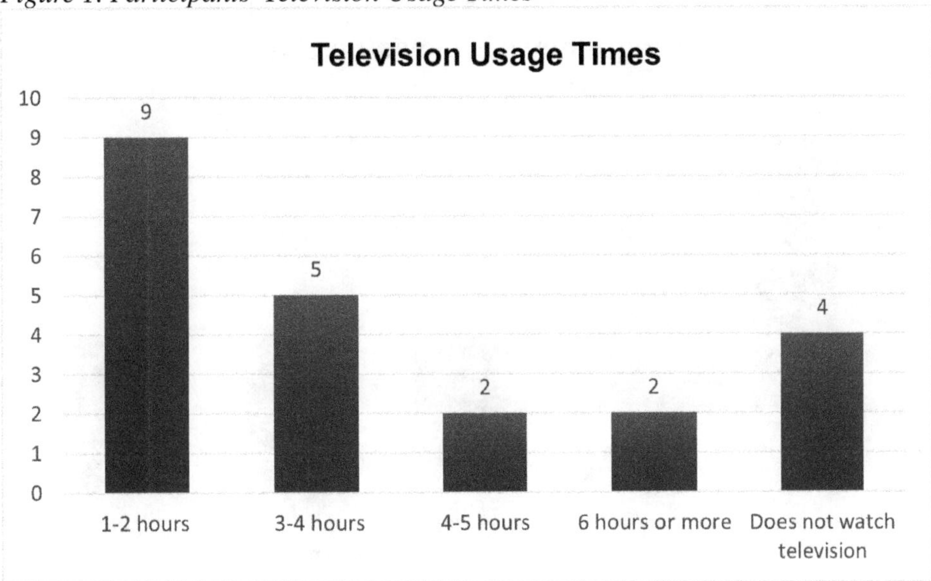

The graph shows data on participants' television viewing times. As shown in the graph, most participants (9 people) watch television for 1-2 hours per day. In addition, while 5 people watch for 3-4 hours, the viewing times of participants who watch for 4-5 and 6 or more hours are limited. The number of participants who do not watch television at all is 4.

Figure 2. Participants' Television Usage Intensity

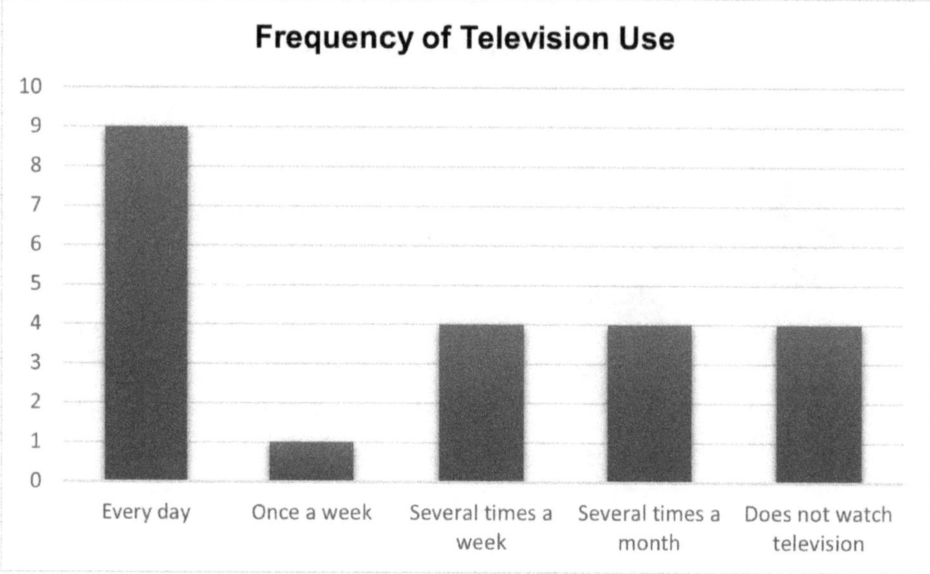

As can be seen in the graph, when looking at the television viewing habits of the participants, it is seen that most of the participants (9 people) watch television every day. The number of participants who watch television several times a week, several times a month, and never watch television are evenly distributed (4 people each). Only 1 person watches television once a week.

Figure 3. Participants' Digital Environment Usage Times

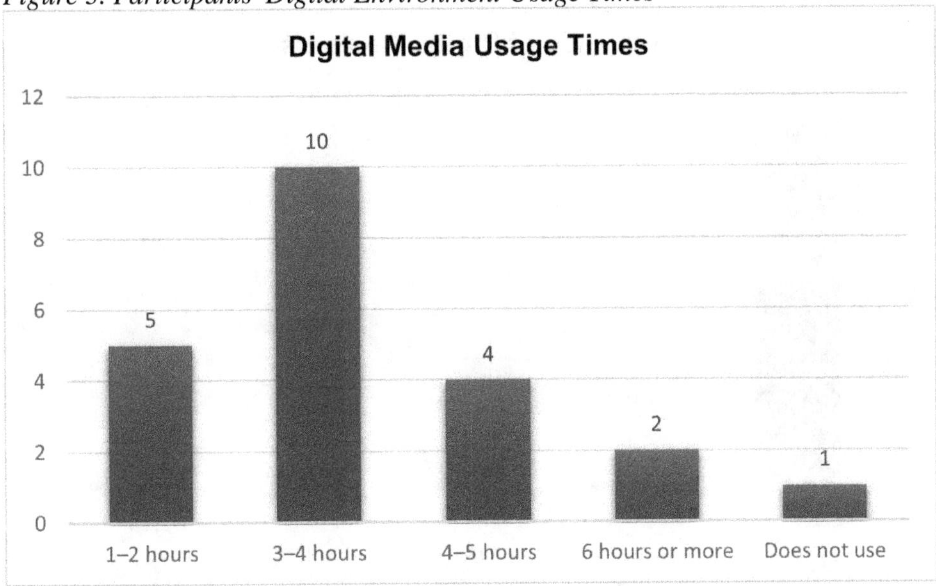

As shown in Figure 3, the majority of participants (10 people) use digital platforms for 3-4 hours to view content. Five participants use them for 1-2 hours, and four participants use them for 4-5 hours. Two participants use them for 6 hours or more, and one participant does not use them at all.

Figure 4. Participants' Intensity of Digital Environment Use

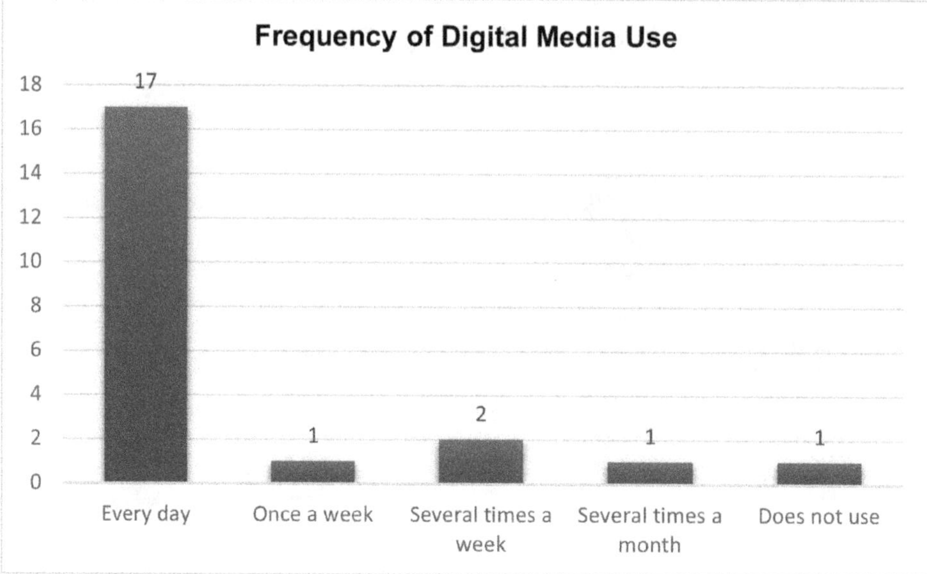

Graph 4 shows data on the duration of use of digital media by participants while viewing content. According to this data, it can be seen that the vast majority of participants use digital media every day. In addition, the number of participants who use it several times a week is 2, once a week, several times a month, and the number of participants who say they never watch is 1.

Findings On Users' Media Consumption and Media Preferences

In this section of the study, a series of questions were asked to identify changes in media consumption habits in line with the digitization process from the participants' perspective. These questions were related to topics such as viewing habits, content preferences, personalization and interaction experiences, content quality (script, filming, acting, etc.), content diversity, and viewing comfort.

Media Preferences and Viewing Habits

In this section of the study, questions were asked about which environments participants preferred for viewing content, the reasons for their preferences, and the factors influencing their preferences, in order to reveal the factors affecting participants' environment preferences while viewing content and their changing viewing habits.

When examining participants' content consumption habits and environment preferences, it can be understood from their statements that for some participants, television is not just a viewing device, but has become a habit that is part of their daily lives.

"I watch it on TV. But I also go on YouTube sometimes. (...) When Müge's program starts, I don't even go outside. I watch TV at that time. And when MasterChef is on, I'm glued to it." (K2)

"I turn on the TV every day. I especially watch the news after seven o'clock, comparing different channels (...) Watching TV is a habit. I want to have some background noise." (K4)

"I love TV. I feel relaxed when I watch it. The sound makes me happy. If I'm home around 5.30 or 6 in the evening, the TV stays on until I go to bed. Do I watch it? That's debatable, but it stays on." (K6)

"I watch television for an hour a day. I usually watch news and talk shows (...) The television is always on while I cook." (K10)

"I watch television... My family watches mostly news, so I join them. The television is always on at home, so when I sit down in front of it, I start watching out of habit." (K17)

It has been observed that participants' interests and moods are among the factors that determine their media preferences. Some participants expressed their reasons for their preferences as follows.

"Online video platforms are better because I can choose the type of content I want. If I want to be happy, I can find content that makes me happy." (K7)

"I watch YouTube. I really enjoy watching vlogs. I like watching cooking videos. Recipe videos (...) I watch Turks traveling abroad. I feel like I'm learning in my own language while also seeing other countries. (K11)

"YouTube, cakes, recipes... It educates me. I watch it to learn." (K13)

"I don't watch TV. I watch vlog videos. I like videos of people traveling around cities." (K20)

For some participants, watching is more than just entertainment, it is seen as a means of socializing, facilitating family communication and spending time together. It's nicer when someone is with me because I can comment (...) I enjoy watching with my friend more." (K12)

"When we stay with my family, we like to cook a meal we enjoy and watch it together." (K22)

"I don't really like watching alone. I usually like watching with my spouse or children." (K6)

Time Management And Ease Of Access

With digitalization, one of the factors affecting people's media usage habits is the change in the way they access content and manage their time. In this context, participants were asked questions about the flexibility of time and space, content control, and ease of access to content offered by online media environments to users.

Online media environments offer the freedom to select content as well as the ability to customize the viewing experience. This increases the viewer's control over content, transforming the traditional viewer profile into a more active one. Participants' statements clearly show that digitization has led to a trend toward independence from time and place in media use.

"I used to watch whatever was on TV, I had no choice. Now I can watch whatever I want whenever I want on online platforms. (...) I want to watch a camping video at night, but I can't watch it on TV, I can find it online right away." (K1)

"On TV, I'm glued to the screen for two hours. With streaming, I can watch the show on those platforms whenever I want. If I'm not available, I can pause it and watch it later when I am. If I'm tired today, I can just watch it tomorrow or the next day." (K8)

"At least we can pause it (...) and watch from where we left off. I can't do that on TV when I'm with my family or alone. I pause it, then watch from where I left off or rewatch it." (K19)

"There are two reasons why I prefer online platforms. The first is that they allow me to watch the content later. The second is that I can access it whenever I want when I'm out or watch content I've downloaded beforehand." (K14)

There are serious, fundamental functional differences between online media and television in terms of flexibility, speed, and control. While television broadcasts are independent of user preferences, in the digital environment, users can directly access the content they want at any time. Additionally, online media platforms have evolved beyond being mere entertainment spaces to become important sources of information. As a result, many users, particularly on platforms like YouTube, have indicated that they use these platforms to gain knowledge, satisfy their curiosity, and find solutions to problems they encounter in their daily lives.

"I can access anything I want instantly. It takes seconds to get there. You have to turn on the TV and fiddle around with it. Here (online media), it's more immediate." (K4)

"(...) I can access the video I want or the information I need much faster online. After all, you can't get that instantly on TV. For example, you turn on a news channel, you want to hear the news, and you have to wait for that news item to come up or wonder if it will be mentioned at all. But online, I can access it easily." (K5)

"To learn, to get information, for example, if I want to make a cake or cook a meal, I immediately look at YouTube. Because they provide recipes." (K9)

"(...) because it shows things practically, like YouTube. Recently, I had a problem with an outlet, and I fixed it by watching how to do it there." (K15)

"I watch it to learn the information I want, not just to watch movies or something, but to watch the things I need." (K16)

Interaction and Personalization Experience

One of the important aspects of digital content consumption that has transformed media consumption is its interactive and personalized nature. In order to identify the differences this creates in terms of usage, participants were asked to compare the interactive capabilities of television broadcasts and internet-based video platforms based on their own experiences, whether this has a role in media consumption preferences, how personalization in online media affects viewing decisions, and whether they have had an experience that made them feel interactive online.

In content consumption, situations such as sharing, liking or disliking, and commenting offered by the internet can be considered direct feedback. This is among the factors that affect user satisfaction. Participants stated that, unlike television, which leaves viewers passive, digital platforms are multifaceted tools that increase participation.

"I definitely look at the comments on my favorite TV shows. I look at at least ten or fifteen comments. I also comment myself. I especially comment on sports-related things." (K4)

"I don't think there's much interaction on TV. In modern environments, especially online, there's more interaction, and I use it too. It allows me to directly access content creators or communicate with them." (K12)

"There are more opportunities online, and the networks are broader. It's hard to share something you see on TV with someone else, but here I can send it and share it." (K19)

"When we watch a series on digital platforms, there are like buttons. When we click on the like buttons, it shows us content based on that. On TV, whatever they show us is what we get. We can't comment or like things on TV. But on digital platforms, I can both like the show and instantly share it with the people around me." (K8)

Recommendation systems shaped by algorithms based on user habits directly influence preferences such as content type and viewing time by keeping users engaged. According to participants, recommending content tailored to their interests is beneficial, but it can also lead to prolonged viewing sessions. Participants expressed their thoughts on this matter as follows.

"The recommendations catch my attention, so I continue watching. The type of content I watch, my preferences, and my timing can be influenced from start to finish." (K17)

"For example, when I watch entertainment content, it shows me similar content. I start watching those too. Even if not all of them catch my attention, I like that there is a certain limitation." (K8)

"Personalized content makes things easier for me. I get to watch content based on my preferences. Algorithms are important in this regard on internet platforms." (K6)

"The recommendations catch my attention. I think YouTube is quite advanced in this regard. It draws us in, sometimes we watch for hours without even realizing it." (K3)

Online media platforms create a space for direct interaction between users and content and content producers. In this regard, participants described their interactions with content and producers and their opportunities for feedback as follows.

"When I watched Black Mirror, there were choices. You could direct the characters and the story based on those choices. You even determined the ending of the series. Of course, I felt a direct connection with the content." (K8)

"It allows me to access content creators online or communicate with them. In fact, once a comment I made on a video led me to directly communicate with an artist." (K12)

"I sometimes go live. I criticize things, for example. Some people say 'good job,' others kick me out of the stream. But it's cool there's immediate interaction." (K10)

The Determining Factors of Content Quality and Diversity in Media Preferences

With the development of broadcasting technologies, the diversity of content and the quality of content production can also be influenced by users' preferences in content consumption. For this reason, participants were asked questions about the differences between content on television and video platforms, content diversity, content production quality, and image and sound quality.

In their assessments of the differences between content published in both media environments, participants highlighted topic diversity, originality, and freedom of expression. Participants used the following statements in relation to this situation:

"Television content is either news, Turkish series, or talk shows. On digital platforms, however, I can find all kinds of content. I can choose content based on my mood, such as horror or entertainment." (K8)

"I can't find everything on TV, and I can't access the content I want instantly. On the internet, however, I can easily find the movie I want or the type of content I want. I only use TV to watch the news." (K10)

"There is usually censorship on TV, and I don't like that. Internet platforms, on the other hand, offer more advanced and uncensored content." (K11)

"The variety of content on the internet is countless. TV, on the other hand, is subject to certain rules such as RTÜK. Compared to this, the internet is a more free and open environment." (K12)

"Internet content is up-to-date and follows trends that's why it's more appealing. TV caters to the popular audience, so I can't find content that suits me." (K16)

"Censorship is contrary to human nature. Reactions that are part of life's natural flow are censored this is not seen on online platforms. On television, programs are made with advertising concerns in mind, targeting the general audience." (K17)

"TV content repeats itself and is generally lacking in originality. However, the internet offers a more free and open environment in this regard." (K22)

Evaluations of content quality are not limited to technical aspects. Participants' comments highlight various elements that shape the perception of quality, such as the way content is presented, the cinematographic structure of scenes, and image and sound quality.

"There are some series on digital platforms that are really well made. The actors, scenes, and camera angles are great." (K19)

"Some content on YouTube is really high quality. It looks professional. For example, some cooking videos have such good visuals and lighting that it feels like you're watching a series." (K20)

"I don't watch TV because most of the content is misleading. I prefer digital platforms because I can find series that reflect real life." (K5)

"While I can watch 4K videos on YouTube or other platforms, I can't watch content of that quality on TV." (K1)

"I find TV to be of higher quality in terms of content. Because there's a control mechanism in place, and content is presented in a more regulated manner. However, online platforms are more advantageous in terms of content variety and ease of access." (K21)

The Transformation of User Experience and Monitoring Comfort

The digitization process has replaced the structural limitations of traditional television broadcasting with the flexible access opportunities offered by online platforms, which are independent of time and place. Consequently, with this transformation, factors such as platform interface, viewing comfort, and ease of access, which play a major role in the viewer experience, have come to the fore. In this section of the study, questions were posed to determine the impact of these factors on participants.

When participants were asked questions about the convenience provided by new technologies, they stated that factors such as portability in content consumption, ad-free viewing, no time and place restrictions, and the physical comfort provided by portability had a positive impact on their user experience.

Digital platforms are not restrictive. We can access them wherever we want. That's why they're much more convenient, but television isn't like that. (K22)

"When I watch online, i.e. on my phone, I can watch in whatever position I feel comfortable in. I don't get tired, but when I sit in front of the TV, I get tired more quickly because I'm always looking at the same place." (K2)

"(...) so now cell phones are like mini TVs. That's why I can access anything similar to TV wherever I want, whether I'm in the car, at home, or even lying on the grass in the street (...) online platforms are becoming more appealing." (K16)

"(...) Internet platforms are more advantageous because you can watch whatever you want. With television, there are commercials, something happens, and you can't get up. But that doesn't happen with the internet. You can pause it, do what you need to do, and continue watching. Plus, we can watch anywhere." (K10)

"Online platforms are more accessible. I can access them from anywhere with my phone, I can even watch them in bed. Sitting in front of the TV feels much more comfortable to me. Plus, I don't have to deal with ads, but that's not the case with TV. We have to be more patient. We have to wait for the ads." (K20)

Fee and Value Perception in Access to Content

Participants were asked questions aimed at understanding the psychological and emotional effects of paying to view content on online media platforms. As a result of the interviews, it was observed that participants adopted different attitudes toward paid content viewing from economic, emotional, and practical perspectives.

"I think people get used to everything. For example, it feels like something has always been there, like it's something that needs to be done... Maybe at first there is some hesitation in the decision-making process; we think, 'Should we

do this?' But once we do, things change. There are automatic payments now, so it feels like it's always been there." (K5)

"If I really want to use a video or platform, I'll get a premium subscription even if there are a lot of ads. I still don't give it up. For example, if there's a show I love and watch all the time, after a while I can't resist and switch to a premium membership." (K20)

"My choices usually depend on my financial situation. If the service is expensive, I won't pay for it that month and won't use it. But if my financial situation allows it later, I'll buy it again." (K11)

"Since I don't pay, this situation doesn't affect me. If I did pay, it would probably affect me. I'd probably prefer to watch TV." (K9)

"I prefer that platform because of the advantages it offers me. So even if it's paid, I'm aware of that and make a conscious choice. Because it offers me more options and the content I'm looking for is there." (K15)

General Assessment of Digitalization and Perceptions for The Future

In order to understand participants' awareness of the transformation caused by online media platforms in the media ecosystem and their predictions about the future of this transformation, opinions were gathered on the effects of digital platforms on television broadcasting, the future of television, and how content consumption will shape up in the future.

"People in our age group still watch television. But the younger generation isn't very interested in television. I would say they don't watch it at all. People our age have been watching television for years, so they continue this habit. But the younger generation spends more time on the internet and on their phones, they are hardly interested in television." (K2)

"I don't think these media will completely disappear, but I think they will evolve. Actually, they are already becoming very similar. In fact, some content is broadcast on both internet platforms and television. Perhaps in the future, these two media will become even closer, or even equal. I think that's what will happen." (K3)

"I think the rise of the internet has greatly reduced television broadcasting. I don't think the younger generation is very interested in television anymore, in fact, television can even be cringeworthy to them. I also think this has become a trend. It's almost like saying, "Are you watching television?" has become a status symbol among them. Watching series on platforms like Netflix is seen as more prestigious for the new generation." (K5)

"Now everyone has the internet at their fingertips and can find whatever they want in seconds. That's why no one wants to stick to a plan like, 'My show starts

at 8:30 on TV, I'll sit down and watch it.' Some old habits seem like they'll disappear completely over time." (K6)

"Television will never completely disappear. In fact, television is like a member of the family. Even if no one watches it, having it on in the background is like it's part of the house, an accessory. I don't think the television age is coming to an end it will simply continue to exist in a different form. It will evolve over time, but it won't disappear completely." (K16)

"I can't say anything definite about the future of television right now. Because the programs and movies are always the same, they don't offer anything new to people. That's why no one is interested in television anymore. On the other hand, the internet offers users many more options and alternative worlds." (K13)

CONCLUSION

This study examines the transformative effects of digitalization on viewer habits. To this end, semi-structured interviews were conducted with 22 participants with different demographic characteristics. Through these interviews, the transformative impact of digitalization on viewers was evaluated in terms of their attitudes toward television and online media platforms, content preferences, access to content, interaction elements, broadcast quality, and user comfort.

The findings indicate that users tend to prefer online media platforms in their media consumption preferences (Allam & Dinana, 2021; Aslan & Öztekin, 2023; Papathanasopoulos & Varoutas, 2022; Yurdigül et al., 2024). However, the findings reveal that users are not entirely shifting toward online platforms rather, participants utilize traditional television broadcasting and digital platforms not as mutually exclusive but as complementary functions. While participants benefit from the content diversity, ease of access, and personalization offered by online media platforms, they also prefer television's habit-based viewing patterns and family time spent together. Therefore, despite their tendency to use online media environments, participants exhibit a hybrid media usage pattern in their content consumption preferences. This indicates that, in addition to technical capabilities, social opportunities also influence users' viewing and preference choices (Nick, 2012; Sarı & Türker, 2021).

The flexibility of time and space offered by online media platforms to users has brought about a profound change in media consumption practices. As can be understood from the participants' statements, this transformation is not so much a technological advancement as it is a shift away from the passive nature of the user. Now, every user can shape their media consumption according to the rhythm of their own life. Features such as pause, rewind, and resume are redefining the viewing experience and giving the viewer an active role. In addition, participants emphasize that media is not just an environment for passing the time but also serves a function in accessing everyday information. This indicates that the relationship with media has become multidimensional. Media is no longer just something that is watched but has become an experience that is shaped by its audience through the many features it offers. The findings of this study support theoretical approaches in the literature that suggest the time and space flexibility offered by online media platforms transforms the audience from passive recipients into active agents who actively direct their media experience (Jenner, 2017; Lobato, 2019; Napoli, 2011).

It is observed that the opportunities offered by online video platforms for liking, commenting, and sharing content among users play a decisive role in participants' transition from passive viewers to active participants. This finding demonstrates that media interaction enriches the user experience (Çolak & Zinderen, 2023; Wulf et al., 2020). Additionally, the recognition of user preferences through algorithms and the provision of personalized content recommendations emerge as elements that strengthen both social interaction and individual satisfaction. In this context, online media platforms are no longer merely tools that connect viewers with content but are increasingly recognized as dynamic, participatory platforms that customize the media experience (Taylor & Choi, 2022).

Interviews with participants have shown that two main factors determine viewers' media preferences: content diversity and production quality. Traditional television broadcasts create a perception of high technical quality thanks to control mechanisms, but censorship, repetition, and limited subject selection cause them to fall short of viewer expectations. In contrast, online media platforms can offer content that appeals to viewers' moods and interests through topic diversity, narrative originality, and recommendation systems based on user preferences (Sarı & Türker, 2021).

In this regard, the portability, location independence, and user-friendly interfaces offered by online media platforms enable viewers to break free from all physical constraints. According to participants' statements, this also leads to a significant increase in viewing comfort. Considering that traditional television broadcasting increases physical fatigue by requiring viewers to watch from a fixed position, the emphasis on "we can watch anywhere" highlights the importance of the uninterrupted and liberated streaming experience offered by online platforms. As Simons (2009) noted, viewers today want the freedom to create their own television and decide what, where, and when to watch (Simons, 2009).

Participants' statements indicate that perceptions of content viewing fees tend to normalize over time. Although initial hesitation and financial concerns were expressed, automatic payment systems and subscription models softened these perceptions by creating a sense that they had always been there. On the other hand, while economic conditions remain one of the primary factors influencing preferences, users are increasingly willing to pay for content they enjoy and consume regularly. In this context, the extensive content pool and ease of use offered by digital platforms enable participants to justify the cost based on the “price-performance” criterion. Thus, a multi-layered value perception encompassing economic, emotional, and practical dimensions emerges for online platforms.

Underlying all these views of the participants, assessments of what users can expect from themselves and television broadcasting in the future are important for determining the future state of television broadcasting. Participants have pointed out that younger generations in particular do not have television viewing habits and that their interest in television has decreased significantly. Studies on intergenerational media consumption habits also support the participants' views (Hasançebi Ayaz & Çelik Yılmaz, 2025). Among the most important predictions is the view that television will not continue to exist in its current form. Participants have stated that the integration of television broadcasts into the digital environment is necessary. It is believed that television can only survive by adapting to changing conditions in this way.

In conclusion, it is evident that digitalization has a transformative effect on individuals' media consumption habits. It is understood that the features offered by online media platforms, such as ease of access, personalization, and content diversity, have led to significant changes in media consumption. Flexibility, interaction, and individual experience are now at the forefront of media usage. Thus, it has been concluded that digitalization has transformed media consumption not only technologically but also socially and culturally.

REFERENCES

Ahlers, D. (2006). News Consumption and the New Electronic Media. *The Harvard International Journal of Press/Politics, 11*(1), 29–52. DOI: 10.1177/1081180X05284317

Akıncı, S., & Başer, E. (2020). Reklamdan Kaçınma Bağlamında Geleneksel ve Modern Film İzleme Ortamlarının Genç İzleyiciler Üzerinden Karşılaştırılması: Netflix ve Sinema Salonları Örneği. *Erciyes İletişim Dergisi, 7*(1), 1. Advance online publication. DOI: 10.17680/erciyesiletisim.622176

Allam, R., & Dinana, H. (2021). The Future of TV and Online Video Platforms: A Study on Predictors of Use and Interaction with Content in the Egyptian Evolving Telecomm, Media & Entertainment Industries. *SAGE Open, 11*(3), 215824402110408. DOI: 10.1177/21582440211040804

Aslan, Ö., & Öztekin, H. (2023). İnternet Tabanlı Televizyona Geçiş Sürecinde İzleme Pratiklerinde Yaşanan Değişim veslie Dönüşüm: Erciyes Üniversitesi Öğrencileri Örneği. *Erciyes Akademi, 37*(4), 4. Advance online publication. DOI: 10.48070/erciyesakademi.1394517

Aydın, A. A. (2021). *Yeni Televizyon Teknolojileri ve Değişen Habercilik.* Kriter Yayınevi.

Aydoğan, A., & Başaran, F. (2012). Yeni Medyayı Alternatif Medya Bağlamında Anlamak. İn Ö. Özer (Ed.), *Alternatif Medya Alternatif Gazetecilik.* Literatürk.

Bal, H. (2016). *Nitel Araştırma Yöntem ve Teknikleri.* Sentez Yayınları.

Başer, A. (2010). Yeni İletişim Teknolojilerinin Evriminde Televizyonun Yer. İn F. Aydoğan & A. Akyüz (Eds.), *İkinci Medya Çağında İnternet.* Alfa Yayınları.

Çil, S., & Yurdigül, Y. (2018). Sosyal Paylaşım Siteleri ve Demokratikleşme İlişkisi: Facebook Örneği. *Uluslararası Kültürel ve Sosyal Araştırmalar Dergisi, 4*(2), Article 2.

Çolak, Y., & Zinderen, A. (2023). YouTube Haberciliği Üzerine Dijital Etnografik Bir Araştırma ve Gazeteci Görüşleri: Cüneyt Özdemir ve Nevşin Mengü YouTube Haber Kanalları Örneği. *Gümüşhane Üniversitesi İletişim Fakültesi Elektronik Dergisi, 11*(1), 1. Advance online publication. DOI: 10.19145/e-gifder.1194597

Coşkun, M. (2015). Sayılsal İletişim Teknolojileri Bağlamında Tv Haberciliği. İn S. Özel (Ed.), *Yeni Medya Çağında Televizyon.* Derin Yayınları.

Demirkıran, C. (2010). Geleneksel Televizyonun Sanal Modeli Olarak Web Tv Ve Livestream Portalında Tv Yönetimi. *Akdeniz Üniversitesi İletişim Fakültesi Dergisi, 14*, Article 14.

Esslin, M. (2001). *Televizyon Çağı*. Pınar Yayınları.

Genel, M. (2018). *Online Medya Mecrasında Video İçerik Üretimi ve Reklam Geliri Elde Etme: "Youtube Örneği*. 491–500.

Güngör, N. (2018). *İletişim Kuramlar-Yaklaşımlar*. Siyasal Kitabevi.

Guo, W. (2022)... *The Comparative Analysis of the Utilization of Traditional Broadcast TV in the UK and China, 65–69*. Advance online publication. DOI: 10.2991/assehr.k.220502.015

Hasançebi Ayaz, E., & Çelik Yılmaz, N. (2025). Dijital Yerliler ve Dijital Göçmenlerin Sosyal Medyadaki Haberlere Karşı Güven Algıları. İn E. Küçük Durur, A. Yılmaz, & Z. Cinisli (Eds.), *İletişim Bilimlerinde Yenilikçi Araştırmalar I* (pp. 59–74). Gazi Kitabevi.

Jenner, M. (2017). Binge-watching: Video-on-demand, quality TV and mainstreaming fandom. *International Journal of Cultural Studies, 20*(3), 304–320. DOI: 10.1177/1367877915606485

Kandemir, C. (2013). *IPTV Yayıncılığının Sorunları ve Geleceği: Sektörün Aktörleriyle Derinlemesine Mülakatlar*. Derin Yayınları.

Kaplan, A. M., & Haenlein, M. (2016). Dünyanın Bütün Kullanıcıları Birleşin- Sosyal Medya Zorlukları ve Sağladığı FırsatlarDünyanın Bütün Kullanıcıları Birleşin-Sosyal Medya Zorlukları ve Sağladığı Fırsatlar (D. Işık, Trans.). İn H. Hülür & C. Yaşın (Eds.), *Yeni Medya Kullanıcının Yükselişi* (2016th ed.). Ütopya Yayınevi.

Kırık, A. M. (2010). *Etkileşimli Televizyon*. Anahtar Kitaplar Yayınevi.

Kırık, A. M. (2015). İnternet Teknolojisi ve Sayısal yayıncılık Bileşkesinde Gelişen IPTV'nin Günümüzdeki Durumu. İn S. Özel (Ed.), *Yeni Medya Çağında Televizyon* (pp. 129–169). Derin Yayınları.

Kırık, A. M., & Karakuş, M. K. (2013). Sosyal Medya ve İnternet Teknolojisi İle Yöndeşen Televizyon Yayıncılığı: Sosyal TV. *AJIT-E: Academic Journal of Information Technology, 4*(11), 61–73. DOI: 10.5824/1309-1581.2013.2.003.x

Kümbetoğlu, B. (2020). *Niteliksel Araştırmalarda Analiz*. Bağlam Yayıncılık. https://www.baglam.com/home/book/niteliksel-arastirmalarda-analiz

Kuyucu, M. (2019, August 11). *The Differences between Television Watching and Digital Media Using Habits of Y and Z Generations*.

Kuyucu, M. (2021). Geleneksel Televizyon Mecrasının Yaşadığı Dijital Dönüşüm ve Bunun Mecraya Olan Etkileri. *Avrasya Sosyal ve Ekonomi Araştırmaları Dergisi, 8*(3), Article 3.

Kuyucu, M., & Karahisar, T. (2013). *Yeni İletişim Teknolojileri ve Yeni Medya*. Zinde Reklam Yayıncılık.

LaRose, R., & Atkin, D. (1991). Attributes of movie distribution channels and consumer choice. *Journal of Media Economics*, *4*(1), 3–17. Advance online publication. DOI: 10.1080/08997769109358200

Levy, D. A. L. (1999). *Europe's Digital Revolution: Broadcasting Regulation, the EU, and the Nation State*. Psychology Press. DOI: 10.4324/9780203278284

Li, Z., & Sun, Y. (2019). OTT Industry Development in China. *Journalism and Mass Communication*, *9*(4), 164–175. DOI: 10.17265/2160-6579/2019.04.002

Lobato, R. (2019). *Netflix Nations, The Geography of Digital Distribution*. New York University Press. DOI: 10.18574/nyu/9781479895120.001.0001

Lotz, A. D. (2014). *The Television Will be Revolutionized*. NYU Press.

Meehan, E. R. (1986). Conceptualizing culture as commodity: The problem of television. *Critical Studies in Mass Communication*, *3*(4), 448–457. DOI: 10.1080/15295038609366675

Mutlu, E. (2005). *Globalleşme Popüler Kültür ve Medya*. Ütopya Yayınevi.

Napoli, P. M. (2011). *Audience Evolution: New Technologies and the Transformation of Media Audiences*. Columbia University Press.

Nick, C. (2012). *Media, Society, World: Social Theory and Digital Media Practice* (Vol. 30). Polity.

Özel, S. (2015). Televizyon Ekseninde İzleyici ve İnternet Üzerindeki Video Hizmetlerinin İzleyiciye Etkisi. İn S. Özel (Ed.), *Yeni Medya Çağında Televizyon* (pp. 1–31). Derin Yayınları.

Özsoy, A. (2011). *Televizyon ve İzleyici: Türkiye'de Dönüşen Televizyon Kültürü ve İzleyici*. Ütopya Yayınevi.

Papathanasopoulos, A., & Varoutas, D. (2022). On the competition between Video OTTs Platforms vs Traditional TV: A Niche Case Study in Greece. *31st European Regional ITS Conference, Gothenburg 2022: Reining in Digital Platforms? Challenging Monopolies, Promoting Competition and Developing Regulatory Regimes*, Article 265664. https://ideas.repec.org//p/zbw/itse22/265664.html

Parikka, J. (2017). *Medya Arkeolojisi Nedir?* (Kılıç, E., Trans.). Koç Üniversitesi Yayınları.

Polat, A. (2022). Nitel Araştırmalarda Yarı-Yapılandırılmış Görüşme Soruları: Soru Form ve Türleri, Nitelikler ve Sıralama. *Anadolu Üniversitesi Sosyal Bilimler Dergisi, 22*(Özel Sayı 2). *Article Özel Sayı, 2*(Özel Sayı 2), 161–182. Advance online publication. DOI: 10.18037/ausbd.1227335

Postman, N. (2020). *Televizyon Öldüren Eğlence* (Akınhay, O., Trans.). Ayrıntı Yayınları.

Punch, K. F. (2011). *Sosyal Araştırmalara Giriş Nicel ve Nitel Yaklaşımlar (Z. Etöz* (Bayrak, D., Arslan, H. B., & Akyüz, Z. Eds. & Trans.). Siyasal Kitabevi.

RTÜK. (2022). *Gençlerin Medya Kullanımı ve Dijital Okuryazarlık Araştırması 2022* (p. 216). Radyo ve Televizyon Üst Kurulu. https://www.rtuk.gov.tr/genclerin-medya-kullanimi-ve-dijital-okuryazarlik-arastirmasi-2022/4402

Sarı, Ü. (2021). Televizyon Yayıncılığının Dijital Dönüşümü. İn H. S. Türkoğlu (Ed.), *Dijitalleşme Çağında Değişen İletişim Dinamikleri* (pp. 171–197). Beta Basım Yayım.

Sarı, Ü., & Türker, H. (2021). Dijital Platform Kullanıcılarının İzleme Alışkanlıklarına Yönelik Bir Araştırma: Netflix Örneği. *Anadolu Üniversitesi Sosyal Bilimler Dergisi, 21*(1), 1. Advance online publication. DOI: 10.18037/ausbd.902558

Silverstone, R. (1994). *Television and Everyday Life* (Routledge). Routledge.

Simons, N. (2009). "Me TV": Towards changing TV viewing practices? *Proceedings of the 7th European Conference on Interactive TV and Video*, 219–222. DOI: 10.1145/1542084.1542128

Şimşek, H., & Yıldırım, A. (2016). *Sosyal Bilimlerde Nitel Araştırma Yöntemleri (Hasan Şimşek-Ali Yıldırım)*. Seçkin Yayıncılık.

Sirer, E. (2020). Televizyon Yayıncılığında Büyük Veri. *Atatürk Üniversitesi Sosyal Bilimler Enstitüsü Dergisi, 24*(4), Article 4.

Söğüt, F. (2020, December 11). *Dizilerin Yeni Mecrası Olarak İnternet: Televizyon Dizileri İle İnternet Dizileri Arasında Karşılaştırmalı Bir Analiz.*

Taylor, S., & Choi, M. (2022). An Initial Conceptualization of Algorithm Responsiveness: Comparing Perceptions of Algorithms Across Social Media Platforms. *Social Media + Society, 8*(4), 205630512211443. DOI: 10.1177/20563051221144322

Thompson, J. B. (2019). *Medya ve Modernite* (Öztürk, S., Trans.). Kırmızı Yayınları.

Türnüklü, A. (2000). Eğitimbilim Araştırmalarında Etkin Olarak Kullanılabilecek Nitel BirAraştırma Tekniği: Görüşme. *Kuram ve Uygulamada Eğitim Yönetimi, 24*(24), Article 24.

Ünlü, T. T., & Akyol, O. (2025). Türkiye'de Dijital Platformlarda Yayınlanan İnteraktif Yapımların İncelenmesi. *Communicata, 29*(29), 29. Advance online publication. DOI: 10.32952/communicata.1619972

We are Social. (2024). *Digital 2024 Global.* https://wearesocial.com/uk/blog/2024/10/digital-2024-october-global-statshot-report/

We are Social. (2024). *Digital 2024 Turkey.* https://datareportal.com/reports/digital-2024-turkey

Williams, R. (2003). *Televizyon, Teknoloji ve Kültürel Biçim* (Türkbağ, A. U., Trans.). Dost Kitabevi.

Wulf, T., Schneider, F., & Beckert, S. (2020). Watching Players: An Exploration of Media Enjoyment on Twitch. *Games and Culture, 15*(3), 328–346. DOI: 10.1177/1555412018788161

Yurdigül, Y., Bayraktar, R., & Çil, S. (2024). A study on Kahramanmaraş Earthquake survivors: "Social media platforms were more effectively used than the traditional media.". *Connectist: Istanbul University Journal of Communication Sciences, 66*(66), 66. Advance online publication. DOI: 10.26650/CONNECTIST2024-1460282

Yurdigül, Y., & Yüksel, H. (2012). Gazeteciliğin Dönüşümü: Yeni Medyaya Entegrasyon Sürecinde Değişen Habercilik Pratikleri. *Akdeniz Üniversitesi İletişim Fakültesi Dergisi, 18*, Article 18.

ADDITIONAL READING

Couldry, N. (2012). *Media, Society, World: Social Theory and Digital Media Practice.* Polity Press.

Deuze, M. (2007). *Media Work.* Polity Press.

Ellis, J. (2000). *Seeing things: Television in the age of uncertainty*. I. B. Tauris. DOI: 10.5040/9780755604784

Evens, T., Henderickx, A., & Conradie, P. (2023). Technological affordances of video streaming platforms: Why people prefer video streaming platforms over television. *European Journal of Communication*.

Grainge, P., & Johnson, C. (2015). *Ephemeral media: Transitory screen culture from television to YouTube*. Palgrave Macmillan.

Holt, J., & Sanson, K. (Eds.). (2014). *Connected viewing: Selling, streaming, & sharing media in the digital era*. Routledge.

Jenkins, H. (2006). *Convergence culture: Where old and new media collide*. New York University Press.

Meikle, G., & Young, S. (2012). *Media convergence: Networked digital media in everyday life*. Palgrave Macmillan. DOI: 10.1007/978-0-230-35670-2

Şeker, N. T. (2009). *Teknoloji ve televizyon*. Literatürk Yayınları.

Van Dijck, J., Poell, T., & de Waal, M. (2018). *The platform society: Public values in a connective world*. Oxford University Press. DOI: 10.1093/oso/9780190889760.001.0001

KEY TERMS AND DEFINITIONS

Audience Culture: This refers to how individuals relate to media content, how they consume and interpret it, and how they integrate it into their daily lives. With the digitalization process, the audience has transformed from merely watching to becoming an active subject that selects, interprets, and shares content, and the media experience has become intertwined with social, cultural, and individual habits.

Content Consumption: Content consumption is the process of interaction between users and the information, audiovisual content, and messages presented in media environments. With digitalization, this process has gone beyond mere viewing or listening to become a personalized, flexible experience that takes place across multiple platforms. Users have evolved into active actors with the power to select, schedule, and format content.

Digital Transformation: Digital transformation is not merely the use of digital technologies as tools, but rather a structural and cultural change that fundamentally reshapes the way individuals, institutions, and social practices operate. In the context of media, this transformation has led to the emergence of a new communication structure focused on interaction, speed, and personalization at every stage, from content production to the viewer experience.

Digitalization: Digital transformation is a process that goes beyond the use of technological tools and leads to fundamental changes in many areas of social life. It has a wide-ranging impact, from the production and transmission of information to the ways in which individuals communicate and their habits of consuming media content. In this process, analog systems are being replaced by digital infrastructure, with elements such as access, speed, interaction, and personalization coming to the forefront. Digitalization transforms individuals from mere content consumers into content producers. In this sense, digitalization is regarded not merely as a technical advancement but as a socio-technical phenomenon that reshapes the communication language and cultural structure of the era.

Interaction: This refers to the user's ability to communicate with content and content creators in digital media environments. Unlike the one-way communication model in traditional media, digital platforms offer users opportunities for active participation, such as commenting, sharing, liking, and directing content flow. This removes the viewer from a passive position and makes them part of the experience.

Media Transformation: Media transformation refers to fundamental changes in content production, consumption, and audience roles that accompany digitization. Users move away from being passive viewers to becoming selective and interactive in their engagement with content. This transformation creates a new media experience that, unlike traditional media, focuses on flexibility, personalization, and participation.

Online Platform: Online platforms are digital media environments that deliver content via the internet infrastructure, allowing users to view, share, and interact with content. Unlike traditional broadcasting, these platforms offer time- and location-independent access to content, personalization, and multi-faceted interaction. They remove the user from the passive viewer role and involve them as active participants in the content process.

User Experience: User experience encompasses the emotional, functional, and cognitive dimensions of individuals' interactions with content in digital media environments. This experience is shaped by factors such as content access, ease of use, level of interaction, and personalization. On digital platforms, users are not just spectators, they are active participants who guide and transform the experience.

Chapter 10
Colors, Places, and Games:
Reading the Metaphorical World of the Netflix Series "Squid Game"

Gözde Sunal
https://orcid.org/0000-0002-9535-5714
Istanbul Commerce University, Turkey

Gamze Nil Arkan
https://orcid.org/0000-0002-6637-5329
Beykent University, Turkey

Gül Gürsoy
https://orcid.org/0009-0004-3136-1578
Istanbul Commerce University, Turkey

ABSTRACT

This chapter explores how Squid Game uses color, spatial design, and childhood games as visual metaphors to critique capitalist structures and power dynamics. The stark contrast between the green tracksuits of players and the pink-red uniforms of guards reflects rigid social hierarchies. Labyrinthine staircases and playful color palettes evoke disorientation and nostalgia, while intensifying the brutality of the games. Innocent childhood contests are transformed into violent spectacles, revealing the psychological toll of systemic inequality. Through visual analysis, the chapter highlights how the series blends aesthetic contradiction and narrative tension to expose moral compromises in contemporary society. It situates Squid Game within broader debates on surveillance, social stratification, and the ethics of competition.

DOI: 10.4018/979-8-3373-1967-4.ch010

INTRODUCTION

In twenty-first-century late capitalist societies, individuals are subjected not only to the pressures of economic precarity but also to the pervasive influence of invisible surveillance mechanisms and the hegemonic operations of mediatized representation. Neoliberalism has instituted a sociopolitical order in which perpetual competition is normalized, solidarity is delegitimized, and personal failure is recoded as individual deficiency (Brown, 2015, pp. 9–10). Among the most insidious instruments of this regime is the culture of spectacle, which aestheticizes and legitimizes violence while rendering it consumable. Contemporary media functions as a producer of commodified visual regimes that capitalize on suffering and survival, thereby transforming spectacle into a site wherein the neoliberal system is both reproduced and normalized (Debord, 1994; Foucault, 1977).

In this context, the 2021 Netflix series *Squid Game* can be interpreted not merely as a survival narrative, but as a striking allegory of neoliberal order, class-based inequalities, and the structure of the surveillance society. The fictional universe in which indebted individuals are compelled to engage in deadly competition exposes both the desperation of those marginalized by the system and the mechanisms through which human life is commodified. The games depicted in the series evolve into experiments wherein individuals are tested not only physically, but also morally and psychologically—transforming violence into both a method and a spectacle (Dolgun, 2008; Debord, 1994).

This study has approached the series *Squid Game* as a multilayered representation of neoliberal political economy, class-based violence, and the culture of surveillance. The series interrogates not only the competitive order in which individuals are positioned against one another but also the ways in which this violence is both spectatorship and internalized. In this regard, *Squid Game* serves as a compelling narrative that enables an analysis of contemporary capitalism's domination over the individual through the lenses of the aesthetics of spectatorship and body politics (Foucault, 1977).

This chapter aims to explore *Squid Game* as a critical cultural text that unveils the visual and narrative strategies through which late capitalist ideologies are disseminated and normalized. By examining the series through the lenses of surveillance society, visual culture, and neoliberal subject formation, the study seeks to demonstrate how symbolic imagery, spatial design, and aestheticized violence contribute to the internalization of systemic oppression. In doing so, the chapter not only investigates how media representations function as ideological apparatuses but also how they shape spectatorship, agency, and ethical boundaries in contemporary societies. The purpose is to critically engage with the intersection of visual codes, class metaphors, and mechanisms of control in order to reveal the latent discursive operations that underpin the spectacle of survival and commodified suffering in *Squid Game*.

THE SEMIOTICS OF COLOR: VISUAL CODES AND CLASS-BASED METAPHORS

Cinema, as a predominantly visual art form, employs color as a potent semiotic resource in the construction of meaning and affect. Colors operate not solely as aesthetic components but as complex visual codes that communicate nuanced information concerning characters' psychological dispositions, spatial atmospheres, and broader sociocultural structures (Bordwell & Thompson, 2010, p. 184). Each color embodies distinct semiotic weight and layered significations (Kırık, 2013, p. 72). This semiotic potency is particularly salient in the cinematic articulation of class distinctions and social hierarchies, wherein color symbolism assumes a critical role (Arnheim, 1977, p. 59). Accordingly, the deployment of color in cinema transcends individual character representation, functioning as a discursive mechanism that renders visible the ideological and hierarchical underpinnings of society (Metz, 1974, p. 67; Bazin, 1967, p. 83). Moreover, the efficacy of such visual signification is further accentuated through analyses attentive to the intersectionality of gender and class (Mulvey, 1989, p. 45).

A critical analysis of color utilization in the early stages of cinema indicates that a principal impetus was the imperative to faithfully replicate natural reality. Within this framework, early cinematic productions privileged the direct, objective reproduction of the external world over the deployment of colors as carriers of subjective or symbolic meaning. Accordingly, instances wherein color assumes a deliberately codified symbolic function or where chromatic harmonies are orchestrated with conscious aesthetic intent remain relatively infrequent during this period (Koca, 2019, p. 231).

Color usage in cinema is recognized as a pivotal visual apparatus that significantly augments the emotional and psychological complexity of narrative structures (Bordwell & Thompson, 2010, pp. 170–185). The culturally and symbolically charged meanings of color function critically in articulating characters' affective states and developmental trajectories (Gage, 1999, pp. 50–75). For example, the color red is conventionally coded to signify passion, danger, or violence, whereas blue chromatic registers are associated with notions of tranquility, detachment, or melancholia (Arnheim, 1977, pp. 120–135). Such chromatic semiotics assume a determinative role in the construction of cinematic atmosphere and in eliciting affective responses from the spectator (Bazin, 1967, pp. 45–60).

The introduction of color usage in cinema initially emerged as a desire to faithfully reproduce the colors of nature on film. However, over time, filmmakers, much like painters, began to reorganize natural colors according to their personal aesthetic sensibilities to enrich their narrative worlds. Within this framework, a broad and multilayered domain of color usage developed in cinema, based on the qualities of certain colors evoking specific emotions or carrying symbolic meanings (as cited in Özön by Koca, 2019, p. 231). Theoretically, these aesthetic and meaning-oriented approaches to color usage are commonly classified under two primary categories: the first focuses on "realistic color usage," aiming to depict reality as it is; the second regards colors as active tools in meaning production, termed "expressive color usage." Both approaches address the function of color aesthetics in cinematic narrative on different planes. Accordingly, color aesthetics is evaluated along three fundamental dimensions: its visual impression (visually), its role as an emotional expression tool (emotionally), and its symbolic meanings (symbolically). These three intertwined elements sometimes operate separately, sometimes synergistically, revealing the depth and diversity of color's semantic production in cinema (Sözen, 2003, p. 10).

Moreover, colors serve as significant visual devices that reinforce spatial and temporal narration. Cool and dark tones typically create atmospheres of threat and uncertainty, while warm and bright colors evoke feelings of safety and comfort in the viewer (Mulvey, 1989, pp. 80–95). Directors and cinematographers strategically select color palettes by considering these psychological effects to visually enrich the narrative's layers of meaning (Monaco, 2009, pp. 210–230).

Additionally, the function of color within narrative structure extends beyond aesthetics; it is an effective tool for emphasizing transitions in time and space. Variations in color saturation and tone clarify chronological distinctions or reality-fiction divides within the narrative, facilitating the viewer's positioning in the story (Metz, 1974, pp. 100–115). Peter Greenaway's film *The Cook, The Thief, His Wife & Her Lover* (1989) exemplifies successful visualization of color usage and spatial transitions. The film employs distinct color palettes specific to each main setting: red dominates the restaurant scenes, symbolizing Albert's violent authority; white appears in the restroom scenes, creating a temporary sense of refuge; the kitchen, designed in green, subverts green's usual connotations of freshness, health, and tranquility, instead provoking discomfort in the viewer. Here, the director employs green ironically, juxtaposing the color's conventional meaning with an unsettling effect. This deliberate color orchestration is not merely a formal choice but also a cinematographic reflection of class and sexual tensions (Demirer, 2020, p. 577). Thus, color usage contributes meaningfully to the film's structural composition and deepens viewer engagement.

Visual codes employed in films are not merely aesthetic preferences but also crucial indicators of social class and identity representation. Visual composition directs viewer perception through the organization of form and color, serving as foundational elements in meaning production (Arnheim, 1977, pp. 120–135). The arrangement of space, lighting, color palettes, and framing techniques in cinema facilitate the visual transmission of social hierarchies and class distinctions (Bordwell & Thompson, 2010, p. 205). As André Bazin notes, characters from upper classes are typically situated in symmetrical, spacious, and well-lit environments, whereas lower-class individuals occupy confined, cluttered, or dimly lit spaces, producing powerful visual cues that evoke class divisions (Bazin, 1967, p. 80).

A contemporary example of such visual strategies is Bong Joon-ho's *Parasite* (2019). The film starkly contrasts the modern, bright, and geometrically ordered villa of the upper-class Park family with the cramped, damp, and chaotic basement dwelling of the lower-class Kim family. The color usage reinforces this opposition: pastel and cool tones prevail in the Park household, while dark and saturated colors characterize the Kim family's surroundings. Similarly, Hirokazu Kore-eda's *Shoplifters* (2018) symbolizes class fractures through dark, crowded, and narrow spaces.

Cinema, as a visual narrative medium, employs color not only as an aesthetic element but also as a multilayered language imbued with profound social and ideological meanings. Colors symbolically encode characters' psychological states, spatial atmospheres, and broader social structures, offering viewers subconscious yet significant cues. Particularly in depicting class differences and social statuses, the function of color becomes a critical tool in visualizing the film's ideological message.

As demonstrated through examples from diverse periods and genres, the selection and application of colors do not merely enhance the emotional and psychological depth of narratives; they also render social hierarchies and power relations visible. From Kieslowski's *Three Colors* trilogy to Bong Joon-ho's Parasite, and from Peter Greenaway's *The Cook, The Thief, His Wife & Her Lover* to Todd Haynes's *Far from Heaven*, color palettes articulate the social positions and conflicts of characters and spaces through class metaphors and visual codes. This evidences that color's ideological function in cinematic language plays a central role alongside cultural coding and meaning production.

In conclusion, the language of color in cinema, as a narrative tool where visual codes intersect with social class and ideology, enriches the film's formal complexity and semantic layers. The conscious and strategic use of color intensifies the ideological weight of cinematographic storytelling and renders the viewer's experience multidimensional. Therefore, the study of colors remains an indispensable domain within film studies, especially in the context of social analysis and visual culture.

Color as Power in *Squid Game*: Visual Semiotics, Class Stratification, and Disciplinary Mechanisms

Netflix's *Squid Game* (2021) constructs a multilayered critique of systemic structures not only through its narrative framework but also via the metaphoric dimensions embedded in its visual language. Among the series' most potent symbolic tools is color. Colors transcend mere aesthetic purposes, becoming deliberately instrumentalized codes that visualize power relations, class distinctions, and the individual's positionality within the system. This instrumentalization of color unveils the sociological potential of visual culture to generate meaning (Bellantoni, 2005).

One of the most salient colors in the series is red. The masked officials who administer the games wear uniforms in dark pinkish-red hues, a color historically associated with power, violence, and oppression. Within the context of *Squid Game*, red signifies the enforcers of the system—the coercive face of power that exercises control through violence. These characters are devoid of individual identity; their faces concealed behind masks, their bodies homogenized through uniformity, rendering them literal "extensions of the system." Here, color simultaneously symbolizes threat and the authority of the functionaries.

The shades of red deployed in the series are directly linked to power and surveillance. The game supervisors' dark pink uniforms mark them as embodiments of violence and order. Historically, red has been a signifier of danger, strength, and discipline (Pastoureau, 2017, p. 142). Within *Squid Game*, this aligns with Michel Foucault's analyses of disciplinary societies, wherein mechanisms continually monitor and regulate individual behavior (Foucault, 1977).

Conversely, the green tracksuits worn by the contestants symbolize both hope and repression. While green frequently signifies life and renewal in nature, within the series it denotes the coerced belonging of indebted individuals selected by the system. The uniformity of the contestants' attire creates an illusion of equality by erasing interpersonal differences. However, this "equality" masks structural inequalities explicable through Pierre Bourdieu's concepts of habitus and social capital (Bourdieu, 1984). The apparent homogeneity conceals class-based experiences and cultural codes, failing to transform individuals' positions within the system, merely rendering them invisible.

The metaphorical potency of color extends beyond characters' costumes to the chromatic design of the game arenas, which similarly carry ideological connotations. The pastel palette in the "Red Light, Green Light" game—blue skies, yellow floors, pink walls—evokes children's playgrounds. This scene illustrates how an ostensibly "innocent" environment can transform into a lethal space. The contrast between this aesthetic and the violence enacted therein creates an "aesthetic estrangement," unsettling viewers' emotional expectations. Interpreted through Guy Debord's (1994) concept of the Society of the Spectacle, these scenes metamorphose reality into simulation; the audience consumes stylized violence as spectacle. Particularly in the first game, the atmosphere of childhood play abruptly morphs into mass execution. The deliberate use of soft colors juxtaposed with ruthless scenes symbolizes the normalization and quotidianization of violence.

Another significant color usage in the series is green. According to Gamze N. Arkan (2024),

> *"For many, green has strong associations with nature, evoking lush grass, trees, and forests. In cinema, green is often described as a revitalizing and calming color. Other common associations include money, luck, health, and jealousy. Green is frequently employed in two distinct ways. On one hand, it represents living plants in nature, signaling growth and hope. Conversely, green-tinted scenes can appear diseased or lifeless; at times, green may even appear ominous or dangerous on screen."*

However, in *Squid Game*, these positive connotations are used ironically. Contestants enter the game wearing green uniforms, ostensibly signaling "a new beginning," yet their lives are imminently threatened. This irony exposes how "hope" is distorted within a merciless system. The uniformity of the green tracksuits obliterates individuality, rendering all contestants "equal." Class, profession, and background differences are masked; everyone becomes a mere cog in the game's machinery. This functions as a metaphor for capitalist systems' treatment of individuals as mere numbers. The green tracksuits, in both color and design, resemble prison garments, symbolizing the contestants' lack of freedom and their systemic control and confinement. Participants become "prisoners" who have surrendered their liberty and comply with the rules to survive. Thus, the green attire in *Squid Game* embodies layered meanings of hope and new beginnings' irony, de-individualization and imprisonment, and the tension between nature and systemic control, visually reinforcing the series' socio-psychological themes.

The "pink" color usage in the guards' overalls is another notable chromatic choice. The bright pink hue worn by these mid-tier operatives symbolizes figures who occupy lower rungs in the hierarchical order yet ruthlessly enforce the system's regulations. The stark contrast between the pink uniforms and the overall dark, somber atmosphere heightens this effect. While pink is generally associated with softness, innocence, and warmth, here it ironically signals a harsh authoritarian presence. This unexpected dissonance generates unease and tension for the viewer.

Culturally, pink is linked to childhood, play, and innocence. In *Squid Game*, however, children's games have deadly consequences, and the use of pink amplifies the tension between appearance and reality. Furthermore, the traditional feminine and passive connotations of pink are subverted, as it represents a strict, active authority—a commentary on the system's unpredictable and complex nature.

Additionally, the elite class represented by VIP characters adopts a deliberately opulent color scheme. Golden masks, dark luxury clothing, and lavish settings symbolize their class supremacy. These colors signify not only wealth but also inaccessibility and absolute power. In this chromatic coding, officials in red represent the system's executors, contestants in green the system's victims, and VIPs in gold the system's owners. This visual stratification facilitates the audience's cognitive construction of class hierarchies.

The VIPs are affluent individuals who have experienced every conceivable luxury and have become indifferent to all pleasures. They observe the competition in exclusive spaces unseen by contestants and gamble on the players' fates. Each VIP's golden mask is adorned with an animal figure emblematic of their traits or positions: the owl mask of wisdom worn by the game's creator and organizer; tiger symbolizing courage and nobility; bear for healing and wisdom; deer denoting kindness and divine power; bull representing strength, death, and fertility; lion symbolizing power and leadership; and eagle as the all-seeing messenger. These masks arguably reflect the wearers' professions and their wielded powers.

In conclusion, *Squid Game* employs color not merely as a visual element but as an ideological instrument that carries and shapes the narrative. Colors become part of a symbolic system through which social hierarchies are reproduced, individuals' positions within the system are defined, and viewers' emotional responses are guided. In this respect, the series harnesses the communicative power of visual representation to construct a critical sociopolitical discourse.

SPATIAL CONFIGURATIONS: PRISON, PLAYGROUNDS and DYSTOPIAN ARCHITECTURE

Squid Game offers the viewer a powerful ideological narrative not only through its characters and plot but also through the spatial universe it constructs. The enclosed, constrained, and labyrinthine spaces employed throughout the series serve as critical narrative devices that illustrate the individual's entrapment within the system and the extent of their alienation from freedom. Space here functions not merely as a backdrop but as an apparatus of power that directly shapes both the storyline and the social order (Lefebvre, 1991).

Although the spatial configuration of the series initially appears to consist simply of game arenas and dormitories, the architectural and aesthetic design of these areas is highly deliberate. The dormitory where contestants reside is conceived as a high-ceilinged, light-permeated yet confined space. This setting evokes not only a classical prison but is also constructed according to a panoptic surveillance logic. Michel Foucault's panopticon model, as elaborated in Discipline and Punish (1977), describes a disciplinary mechanism in which individuals are perpetually observed yet cannot see the observer. Within *Squid Game*, the contestants exist under constant surveillance by guards and cameras, producing a psychological regime of control that transcends mere physical boundaries.

The design of the game arenas strikingly recalls dystopian childhood spaces. The expansive pastel-colored area used in "Red Light, Green Light," the massive walled room for the "sugar honeycomb" game, and the towering structures for "tug-of-war" function both aesthetically and psychologically to heighten tension. This architectural staging may be read through Giorgio Agamben's (2005) concept of the "state of exception," wherein individuals are removed from ordinary life conditions and placed in spaces exempt from legal and ethical norms. Consequently, the architectural form engenders a "play universe" where law is suspended and the boundary between life and death is blurred.

Another notable spatial element is the presence of color-coded corridors and geometrically complex staircases reminiscent of the impossible architectures of Spanish artist M.C. Escher. The vibrant walls and intersecting pathways create a visual illusion while simultaneously disorienting contestants, eroding their sense of direction and belonging. This labyrinthine structure can be interpreted as a representation of Zygmunt Bauman's (2000) "liquid modernity," reflecting the dissolution of stability and identity characteristic of late capitalist society. Space here operates not for the individual but for the system: the contestant loses spatial orientation, while the system fully surveils and locates them.

The VIP sections mark a clear architectural hierarchy. While contestants struggle for survival in the lower levels, VIPs observe the game from elevated, soundproofed, luxuriously appointed spaces. This spatial segregation embodies not only a physical height differential but also the embodiment of class privilege within architecture. The "view from above" constitutes more than a mere vantage point; it symbolizes the all-seeing gaze of sovereign power controlling the world from the system's center. Considering Foucault's analyses of spatial forms of power, the VIPs' position is not simply a viewpoint but a structure where sovereignty is materially inscribed (Foucault, 1977).

In conclusion, the spatial design of *Squid Game* functions as a narrative plane that materializes the individual's vulnerability before the system, their existence under surveillance, and the reproduction of class divisions. The architectural constructs of the series stage the functioning of capitalist order not only representationally but as an experiential reality. Thus, space becomes not merely a locus of movement but an ideological structure wherein power manifests, class becomes visible, and the individual is effaced.

Dormitory: The Aesthetics of Prison and the Internalization of Surveillance

From the very early episodes, the dormitory space, which draws significant attention, emerges as the sole communal living area where the contestants spend their days. This large, high-ceilinged room—with its top-down lighting, stacked beds, and blank walls surveilled by cameras—embodies a carceral aesthetic. This spatial arrangement can be analyzed through Foucault's (1977) concept of the "panopticon." Similar to prisoners, contestants behave under the awareness of constant observation, though the observer remains unseen. This regime of surveillance instills a psychological self-discipline that transcends physical restrictions. In Episode 4 (Gganbu), the outbreak of chaos and violence in the dormitories following the lights being turned off exposes the inherently repressive nature of the system, precisely through the temporary withdrawal of surveillance. The rapid descent into disorder in the absence of oversight suggests that the system does not serve to sustain order, but rather operates as a mechanism of suppression and a regulator of violence.

Staircases and Colorful Corridors: Disorientation and Alienation

The spiral staircases and narrow pastel-colored corridors through which the contestants pass to reach the game arenas resemble the optical illusions in the works of Spanish artist M.C. Escher. These transitional spaces lack a clear sense of direction or exit; whether one is ascending or descending remains uncertain. While their playful, toy-like aesthetic may initially evoke a sense of childlike innocence for the viewer, for the contestants, they evolve into spaces of existential entrapment. This scene can be interpreted through Bauman's (2000) concept of "liquid modernity", in which the individual loses their sense of orientation within the system and becomes increasingly passive: in a world where everything is in motion, the individual belongs nowhere.

Playgrounds: The Fatal Transformation of Childhood Spaces

Each game in *Squid Game* unfolds within a uniquely designed architectural space. The open area used for the first game ("Red Light, Green Light") resembles a fabricated version of nature, with an artificial sky, cartoonish trees, and a dusty yellow ground. This architectural illusion mimics the seemingly safe spaces of childhood. However, the moment the game begins and players are shot and collapse, the innocence of the setting is shattered—the safe space becomes a site of execution. In this scene, a wide-angle overhead shot captures the fallen bodies scattered across the field, creating a spatial choreography of mass slaughter. The juxtaposition of play and death reveals that the space operates not merely as a physical construct, but as an ideologically charged structure.

Similarly, the setting for the "tug-of-war" game takes place on a narrow platform suspended dozens of meters above the ground. Contestants compete in a confined space where the losing team is plunged to their death. The terror of this space is not only due to its physical elevation but also the immediate and fatal consequence of falling. This scene metaphorically underscores the stakes of the capitalist system: falling does not simply mean losing a game—it equates to losing one's right to exist.

VIP Areas: Architectural Hierarchy and Class Surveillance

The areas where VIP characters are positioned in the series are organized with a distinct architectural hierarchy compared to the competition spaces. VIPs watch the games from isolated, luxurious, and aesthetically lavishly decorated rooms located on the upper floors of the competition venues. This physical elevation and visual grandeur symbolize class superiority and control. Foucault's (1977) analyses on the spatialization of power are directly observable here: VIPs are not merely spectators but figures of authority who regulate the world from the highest points of the space. The poor, who confront death on the lower floors, are watched by the powerful above, making the system's hierarchy spatially manifest.

CLASS CONFLICT, VIEWING CULTURE and THE SOCIETY of SURVEILLANCE: SQUID GAME'S ALLEGORY of THE SYSTEM

Surveillance and Postmodern Power: A Foucauldian Perspective

Surveillance, as one of the most significant mechanisms of social control, has retained its importance from the past to the present day. This enduring significance has necessitated its adaptation to new conditions and contexts. In traditional and modern societies, surveillance manifests in different forms. While in the former, it involves structures governed by binding internal rules in which God and group members watch one another, in the latter, power relations and the dynamics of authority come to the forefront (Gücüyener, 2001, p. 118).

According to Foucault, the dominant model of surveillance society is the "panopticon" designed by the English social reformer Jeremy Bentham in the last quarter of the eighteenth century (Laughey, 2010, p. 113). Bentham's 1791 design features a semicircular building plan centered around a "control space," surrounded by cells in which prisoners are unable to see their observer. With no area left outside the observer's gaze, prisoners cannot determine whether or not they are being watched at any given moment, which compels them to comply (Lyon, 1997, p. 93). The term "Panopticon," derived from Greek and meaning "seeing the whole," effectively captures this condition.

The dictionary meaning of surveillance includes "the act of watching, supervision, and protection." It is also used as a substitute for the legal term "custody." The verb "to observe," from which surveillance is derived, carries meanings such as "to protect, watch over, care for, safeguard; to give importance to, take into account, keep separate; to guard, wait for; to consider all details and factors in reaching a conclusion; to favor" (Turkish Language Association, 2005). Thus, the term embodies both positive and negative connotations.

Giddens approaches surveillance in the context of modernity and bureaucracy by distinguishing between surveillance that "collects data" about individuals and surveillance that "controls" individuals. This dual distinction encompasses both the accumulation of citizens' information by institutions or communities, which constitutes the basis for the formation of "power," and the monitoring of the activities of those lower in the hierarchy by superiors or, more generally, by the state (Dolgun, 2008, p. 115).

Within the Foucauldian theoretical framework, surveillance is not merely a technique or tool; it must be understood as a practice of power that institutionalizes the domination of bodies and minds by modern and postmodern forms of power. Surveillance permeates all social spheres not only through micro-power relations experienced by individuals in their interactions with one another but also through macro-control mechanisms operating continuously within the neoliberal hegemony of market economies (Uğurlu, 2011, p. 248). This dual function renders surveillance an institutionalized technology of power on both horizontal and vertical levels.

Questions of "how, by whom, and for what interests" surveillance is exercised directly point to the nature of power relations. The history of surveillance practices has been shaped not only by technological advancements but also by the body politics and disciplinary strategies inherent to the power structures of different periods. From the intelligence-gathering activities of warrior groups and the moral regulation functions of religious institutions to the surveillance, registration, and intelligence operations of nation-states, and extending to the global electronic surveillance practices of contemporary developed states, surveillance techniques have served both instrumental and ideological functions across a broad historical spectrum (Dolgun, 2008, pp. 23–24).

The Transformation of Surveillance Today and the Impact of Technology

Surveillance practices today show significant changes compared to the past. Whereas control mechanisms were previously more visible and took place in physical and localized spaces, contemporary surveillance tools have become less visible or even entirely invisible. This shift leads to a greater disregard for the will of those being observed. The cost of surveillance has decreased, enabling the widespread use of remote control and monitoring systems. Coercive and oppressive methods have been largely replaced by softer techniques based on persuasion and guidance.

The data collection process has become integrated into routine activities and, once conducted by human or animal agents, is now partially or fully carried out by automated machines. The collected data transcends local boundaries and enters large-scale circulation. The timing of control has also changed; intermittent surveillance has become continuous and real-time, resulting in individuals being constantly monitored with regard to their past, present, and future.

While surveillance technologies were once limited to a select elite, they have now acquired a more democratic character and are accessible to broad segments of society. The focus of data collection has shifted from individuals to categories and groups based on interests. Traditional surveillance was superficial and based on direct representation, whereas new systems also utilize simulations of reality.

Identification processes have evolved; anonymous crowds and personalities now play an important role in surveillance. Surveillance is no longer solely the domain of experts—individuals have begun to surveil themselves. The organization, storage, and analysis of data have become easier, with visual, auditory, textual, and numerical data in various formats and locations being integrated and assessed collectively. Finally, information acquisition and transmission have become much faster and more accessible than before (Marx, 2002, p. 11).

Technological advancements during the new imperialism have fundamentally transformed the form and operation of power. The design of power as omniscient, omnipresent, and omnipotent is no longer a mere theoretical assumption but a practice penetrating social reality with the support of technological infrastructures. In this process, power has been constructed around axes of coercion and consent, embedding an invisible yet continuous control mechanism within the fabric of social life that directs individual behaviors (Çoban, 2008, pp. 116–117).

Throughout its historical development, especially intertwined with capitalism, surveillance has become a key tool for social control and economic restructuring (Dolgun, 2008, p. 114). The globalization process, enabled by new communication technologies, has made surveillance more widespread and effective, transforming it into a power mechanism operating not only locally or nationally but also on a global scale. This globalized form of surveillance plays a role not only in controlling individuals but also in reshaping the economic and political structures of societies (Çoban, 2008, p. 117).

Information technologies have a decisive role in this transformation. Digital tools permeating every aspect of daily life have eroded the concept of privacy, making individuals continuously observable. These technologies create a new regime of control that sustains the continuity of panoptic power, a foundational philosophical element of surveillance. Through the mass collection and analysis of personal data, individuals can be monitored invisibly and without their awareness (Dolgun, 2008, p. 144). This development reveals a form of control more flexible than the classical panoptic model but equally effective.

Today's understanding of surveillance evolves not only through technological advancements but also through individuals' everyday interactions with these tools. The internalization of surveillance devices as indispensable components of daily life, rather than as instruments of a culture of fear, is one of the most notable aspects of this transformation. Individuals are subjected, often unknowingly, to subtle and continuous forms of surveillance through digital information channels they use daily alongside modern surveillance technologies (Güven, 2011, p. 173). This indicates that surveillance has shifted from an external control mechanism to an internalized process functioning with voluntary participation.

Discipline, Pleasure, and Consumption: The Functioning of Postmodern Power in the Surveillance Society

Panoptic structures, characterized by forms of surveillance in which the watcher remains unseen, evoking a god-like gaze and continuous monitoring, inherently possess an asymmetric nature. This asymmetry creates submission through the ambiguity of the gaze. Panoptic surveillance aims not only to control individuals by making them visible but also, as part of modernity, to eliminate alternative power structures that are remnants of traditional societies. Therefore, panoptic principles hold a central significance for the continuity of modern power. Surveillance practice, however, only gains meaning under conditions such as power imbalance or lack of trust (Lyon, 2009; cited in Güven, 2011, p. 175).

The impact of this form of control on individuals extends beyond external regulation; it also operates on a deeper level by nurturing a sense of helplessness within the individual's inner world. The continuity and invisibility of panoptic surveillance trigger an expectation of uncontrollability in individuals, producing learned helplessness. This condition weakens individuals' motivation to respond effectively to their environment and diminishes their cognitive appraisal capacities, while emotionally it results in diminished self-esteem and feelings of sadness. Consequently, individuals become passive in the face of the system's surveillance and position themselves as objects of power (cited in Güven, 2011, p. 187, from Nolen-Hoeksema & Girgus).

From a Foucauldian perspective, surveillance is not merely a practice of discipline and control but is also intertwined with bodily pleasure and voyeurism. The subject who watches derives pleasure from the act of watching, while every object of surveillance gains erotic meaning. Thus, surveillance functions not only as a mechanism of power but also as an aesthetic and libidinal mode of relation. As Foucault emphasizes, the pleasurable nature of watching enables power to operate in more subtle and intrinsic ways (cited in Çoban, 2008, p. 122).

In the postmodern era, this dimension of pleasure is further reinforced through the culture industry and consumer society. As Bauman points out, in an age where identities are polished and lifestyles commodified, the condition of existence for the individual becomes intertwined with consumption. Continuous renewal, dissatisfaction, and the pursuit of pleasure become necessary for individuals to remain in harmony with the system. Within this context, surveillance shapes not only the control of the individual but also their desires integrated with consumption. Those who cannot consume, feel pleasure, or perform adequately are marginalized and excluded from the system (cited in Uğurlu, 2011, p. 254, from Bauman).

These dynamics of the culture industry facilitate the voluntary participation of individuals in a surveillance order shaped through pleasure. The entertainment and pleasure-providing aspects of surveillance tools not only legitimize surveillance but also transform individuals into active components of the surveillance regime. As Adorno notes, the culture industry turns the individual into a subject who chases pleasure, wanders, and may even resort to violence, while simultaneously hollowing out subjectivity and reducing the individual to a passive member of the masses (cited in Güven, 2011, p. 192). This process of subjectivity loss results in the individual becoming an object of the culture industry.

In light of all these transformations, for contemporary subjects, loss of privacy ceases to be one of the primary issues of surveillance; instead, the pleasure offered by surveillance tools becomes paramount. Individuals who voluntarily engage in micro-surveillance techniques perceive these technologies not only as instruments of control but also as platforms that provide entertainment and a sense of belonging. Thus, individuals become active participants in the surveillance society and do not resist the internalized forms of power (Güven, 2011, p. 192).

From Panopticon to Cyber Surveillance Society: Body, Power, and Control

Although the body is often perceived as the "measure of all things," in reality it is the intersection of multiple forces and processes that shape it. Like the self and identity, it is an internal relationship; porous, permeable, and open to the world. Therefore, the body is not merely a biological entity but also a practical field situated at the center of political, cultural, and social conflicts (Harvey, 2011, p. 164).

This multilayered structure of the body transforms it into a site of contradictions. Metabolic processes are directly linked to the environment; any change in these processes causes the body either to transform or cease to exist. Simultaneously, the social and spatial use of the body depends on the contextual determinants—technological, economic, physical, and cultural—of the setting it inhabits. The representation of the body in society is shaped through clothing, postures, and symbols, while any critique or attack on the representation of the body—such as those by feminist or queer theorists—directly targets bodily practices. As a result, these resistances lead to the production of differentiated bodies on both material and symbolic levels (Harvey, 2011, p. 127).

According to sociologist Anthony Giddens (2010, p. 132), the body is not only a biological basis for individual identity formation but also a practical system shaped through everyday interactions. The active participation of the body in daily life interactions is fundamental for maintaining a coherent sense of self. In this framework, the body is the carrier of both pleasure and comfort as well as tension and pain.

From the perspective of power, the body has historically been the focus of discipline and control, constructed by knowledge and power (Appadurai, 1996; Özarslan, 2008, citing McLaren and Leonardo, p. 140). Since the 18th century, with population growth, new disciplines analyzing individual characteristics have emerged; institutions (schools, prisons, hospitals) disciplining bodies based on this data were established. Thus, the individual transformed into an object of knowledge whose personal life history is tracked through documents (Özarslan, 2008, citing Featherstone, p. 140).

Social control includes directing, monitoring, and, if necessary, correcting individuals' behaviors. Modern society exists within a surveillance order based on Bentham's panoptic system. This system is not limited to monitoring prisoners but aims for the continuous observability of the entire society, becoming the utopia of power (Dolgun, 2008, p. 22). In this order, everyone is both the watcher and the watched. This leads to a fragmented society where individuals distrust each other, facilitating the domination of power—a structure also described as a "society of fear" (Çoban, 2008, p. 122).

One striking aspect of the panoptic system is the invisibility of power. Surveillance mystifies and sacralizes power. The state's decentralized, invisible, and inaccessible nature naturalizes and internalizes power within society (Çoban, 2008, p. 119).

This surveillance order has deepened with the development of technology. Thanks to information technologies, panoptic surveillance now extends beyond physical spaces to digital realms. All interactions occurring in public and private spheres are continuously monitored by "electronic eyes." In summary, Bentham's theoretical model, Foucault's analyses, and George Orwell's dystopian imagination have practically come to fruition today through digital surveillance technologies (Dolgun, 2008, pp. 105–106).

Entertainment, Domination, and Surveillance Practices: The Allegorical Layers of *Squid Game*

Netflix's 2021 release *Squid Game* is not merely a captivating survival drama; it functions as an allegorical representation of neoliberalism, class divisions, and surveillance society. The games that the characters face throughout the series serve as metaphors for their entrenched class constraints. Each character represents an individual marginalized by the system, and their forced competition underscores the structure of modern society that renders interpersonal solidarity nearly impossible.

Almost all participants in *Squid Game* are individuals burdened by debt, unemployment, illness, or the inability to provide for their families. In this regard, the series explicitly portrays the economic inequalities produced by South Korea and global capitalism. The background stories humanize the characters not only as "players" but also as victims of the system. For instance, the Pakistani immigrant worker Ali's experience of wage theft by his employer, culminating in a violent escape, highlights issues of job insecurity and racial discrimination.

The premise that only one contestant survives while others perish can be read as an extreme depiction of the neoliberal system where individuals compete solely for their self-interest. The capitalist reality in which a person's value is defined by economic success is mirrored in the game's reward system—each death corresponds to an incremental increase in prize money for the winner.

One of the series' most striking aspects is that the games are conducted before a secret audience. The so-called "VIPs" are wealthy, masked elites who watch the contestants' survival struggles live. These scenes critique both the commodification of violence by media and the class-based culture of spectatorship.

The VIPs' indulgence—drinking and betting in luxury lounges while human lives hang in the balance—symbolizes how the struggles of the poor become entertainment for the rich. In this respect, *Squid Game* aligns with texts like *The Hunger Games* (Gary Ross, 2012), interrogating the ethical implications of spectatorship: Are those who watch complicit in the system?

According to Debord (1994), in modern society, reality is replaced by representation, and lived experience by spectacle. The games in *Squid Game* exist not for the players but for the elite spectators. The spectacle becomes both a tool of torture and entertainment for the system.

Surveillance in the series operates not only on physical but also regulatory and psychological levels. Those who fail are instantly executed, symbolizing not merely exclusion but direct elimination from the system. This represents contemporary capitalism's logic of "disposability" towards non-producers.

Squid Game exposes the invisible pressures of modern societies. This allegory constructed through class, surveillance, violence, and spectatorship critiques not only South Korea but the global order. The helplessness of individuals, the system's ruthlessness, and the audience's passivity converge, transforming the series into a powerful systemic critique in both narrative and symbolic terms.

Moreover, *Squid Game* is a multilayered narrative that questions not only the tragedy experienced by the contestants but also the manner in which this tragedy is observed. The survival struggles are transformed into a "show," consumed not only by the series' viewers but also by the ultra-wealthy VIP characters within the story. Thus, the series emphasizes that violence is not merely a means but a consumable content, even a product of entertainment.

Reinterpretation of Symbols in Popular Game Culture: Triangle, Square, Circle

Squid Game employs the aesthetics and iconography of gaming culture to depict a gamified manifestation of the capitalist system. The contestants are individuals burdened by debt who are ostensibly given a chance at salvation through "playing games." However, these games are far from fair and frequently culminate in death. Through its use of symbols referencing the PlayStation controller, *Squid Game* simultaneously draws upon modern gaming culture and subverts it, offering a critique of the gamified violence intrinsic to capitalist structures. Symbols traditionally associated with entertainment here become part of a life-threatening competition, demonstrating that popular culture can serve not only as an object of amusement but also as a platform for systemic critique.

The geometric shapes of triangle, square, and circle in the series represent a layered social hierarchy. Moreover, these shapes function metaphorically to symbolize the roles and power relations among participants and enforcers, reflecting class divisions and mechanisms of control within capitalist society. Visually, these symbols articulate dynamics of oppression, order, and power relations.

The triangle, which denotes middle-tier personnel in the series, visually presents a sharp and dynamic form through its pointed vertices, evoking notions of aggression and mobility. Within the narrative, triangle-masked guards operate centrally within the game structure, directly overseeing participants and enforcing rules akin to military personnel. They typically carry weapons and project a threatening presence. Thus, the triangle symbolizes the intermediary managerial class within the system: those who execute pressures imposed from above but lack control over the highest levels. The sharp points of the triangle further signify conflict, coercion, and forces that instigate discord.

The square, utilized by the highest-ranking guards representing order and authority, conveys solidity, stability, and permanence through its visual form. Its sharp corners also denote rigidity and immutability. Square-masked individuals govern the game, make decisions, and occupy the apex of the system's hierarchical structure. Although less visible, they possess ultimate control and power. This group represents the summit of capitalist hierarchy: capital owners, the elite class, and bureaucracy. They are the "rule-makers" and "game organizers." As the locus of power and control, squares embody the often-invisible forces that establish and maintain the system. They set the rules of the game (and life) but delegate accountability elsewhere.

At the lowest tier are the obedient circle-masked workers, who occupy subordinate roles, performing basic tasks and generally remaining removed from direct violence. They do not interfere directly with players but serve as the system's foundational operatives. Visually, the circle signifies softness, infinity, and wholeness. Lacking sharp edges, it is a less threatening, more "compliant" form. Within the series, circles symbolize the lower working class—those obedient yet situated at the system's base. They are system-dependent, aware of their position, and often compelled to act under duress. Circles function as the unseen "motors" of the system; without them, order could not persist, though they lack power and are easily replaced.

These shapes—commonly recognized in popular culture as emblems of gaming, entertainment, and competition—are transposed into a deadly context in *Squid Game*. Consequently, games intended for enjoyment transform into brutal competitions for survival. This phenomenon offers a potent critique of the concept of "gamification" in contemporary society and the notion that human lives are increasingly managed as if they were games. Fundamentally, *Squid Game* dramatizes lethal childhood games. In this regard, the incorporation of triangle, square, and circle symbols from the PlayStation controller functions as a visual allusion to control over games and competitions, immediately evoking a "gaming" atmosphere for the viewer and reinforcing the idea that human beings are manipulated like game pieces.

The symbolic congruence between the shapes in *Squid Game* and those on the PlayStation controller establishes a strong visual and thematic connection, reinforcing the series' central motif: life as a game governed by rules with death as the penalty for losing. This interplay also fosters a powerful link with the audience through shared popular culture references.

Gaming consoles such as PlayStation have become major consumer commodities within popular culture. The deployment of these symbols in *Squid Game* conveys how individuals are controlled, commodified, and exploited under the guise of "entertainment" within capitalist systems. Game and entertainment culture thus functions as a mechanism entrapping individuals within the system. These symbols serve not only as signifiers within gaming culture but also as universal communicative tools in global popular culture. For a South Korean production like *Squid Game*, this facilitates bridging Eastern and Western cultural spheres. Audiences readily recognize these symbols both visually and conceptually, thereby enhancing intercultural communication.

Normalization of Violence, Ethical Decay, and Death as Entertainment

In the series, the VIP characters are positioned as privileged elites who wear masks, lounge in luxurious seats, sip their drinks, and place bets on the contestants' struggle for survival. Their mere physical presence reveals how violence and death are transformed into "spectacles." From the perspective of the VIPs, the harsh reality faced by the contestants is reduced to a mere game simulation; here, individuals do not die but merely perform a "script."

The presence of the VIPs extends beyond passive spectatorship; they occasionally intervene directly in the structure of the games, select their "favorite" players, and even attempt physical contact with contestants to satisfy their personal desires. A striking example is the male VIP's sexual advance toward contestant number 240. This scene not only highlights the instrumentalization of violence and the human body but also points to their sexualized and objectified representations.

In this context, violence is presented not merely as a physical act but as a systematically produced and staged object of pleasure. While the games signify an existential struggle for life and death for the contestants, for the VIPs, they function merely as a temporary thrill and an aesthetic experience. This distortion reveals how class injustice is institutionalized not only economically but also emotionally and morally. While the "watched" perish, the elite "watchers" continue their lives in luxury.

Squid Game invites not only the fictional VIP characters but also its real-world viewers to engage in an ethical reflection. Audiences often witness the characters' pain, deaths, and helplessness with emotional distance and curiosity. This situation inevitably raises questions about the limits of ethical spectatorship: under what conditions does the representation of violence produce critical awareness, and at what point does it serve exploitation, emotional manipulation, or desensitization?

At this juncture, Laura Mulvey's "gaze" theory offers a significant explanatory framework. According to Mulvey, the spectator's position in media is predominantly shaped by a dominant gaze—often gendered or class-based (Mulvey, 1975, pp. 6–18). In the case of *Squid Game*, while viewers watch violence as an aesthetic experience, they may unwittingly become complicit in its reproduction. This dynamic calls for a critical reconsideration not only of the fictional narrative but also of viewing practices themselves.

Significance and Purpose of the Study

Methodology

In this study, symbolic analysis was employed to reveal the layers of meaning embedded within selected visual and spatial elements. Symbolic analysis is an interpretative approach that seeks to move beyond the surface narratives of cultural and artistic texts, aiming to examine the ideological, social, and cultural significances of these elements.

During the analytical process, recurring visual codes in the series—such as color usage, spatial arrangements, costumes, and architectural structures—were examined alongside the positioning of characters and the function and composition of spaces. These elements were evaluated in relation to themes including class differentiation, power relations, surveillance practices, and ethical decay.

The visual components were treated as symbolic representations of social realities. Accordingly, the study situates the analysis within theoretical frameworks including Laura Mulvey's gaze theory, Michel Foucault's concepts of the panopticon and disciplinary society, and Guy Debord's theory of the society of the spectacle. This method aims to make visible how the series functions not only aesthetically but also within political, ideological, and cultural contexts.

Determination of the Research Sample: The Thematic and Theoretical Significance of *Squid Game*

The international impact of a South Korean television series like *Squid Game* presents a compelling subject for analysis, as it exemplifies the transformation of a local narrative into a global phenomenon while simultaneously offering a powerful critique of contemporary society through its cinematographic language. Moreover, elements such as the game scenes, spatial usage, color palette, and character development provide rich material for in-depth visual analysis. Therefore, it is considered that the examination of *Squid Game* offers a timely, critical, and aesthetically productive field of study.

Furthermore, *Squid Game* serves as an ideal case for investigating the transformation of violence into entertainment as a cultural consumption object and for exploring the relationship between the audience and the text. The series provides valuable data on how social inequalities, economic crises, and individual/collective psychology under capitalist conditions are represented, making its analysis both meaningful and relevant in contemporary discourse.

CONCLUSION

Squid Game operates violence not only as a narrative strategy but also as a cultural commodity. The aesthetic presentation and transformation of violence into entertainment extend beyond the VIP characters within the narrative, opening a multilayered critical space that also implicates the series' real-world viewers. The moral collapse of viewing culture, the absence of empathy, and the normalization of class-based injustices are powerfully and disturbingly exposed through the series. Thus, *Squid Game* becomes not merely a fictional text but a self-reflexive structure that reverses the viewer's position through the act of watching, placing them within the spectator role.

The series employs architectural arrangements not simply as visual backdrops but as symbolic spaces where ideological representations are visualized. The cramped, gray-toned dormitories emphasize the de-individualized collectivity and surveillability of the contestants; pastel-colored spiral staircases, labyrinthine corridors, and game arenas grotesquely juxtapose childhood-associated innocence with lethal competition. The VIP rooms, meanwhile, are decorated with colors such as deep red, gold, and emerald green—codes signifying "luxury," "power," and "privilege." These colors are not merely aesthetic choices but visual metaphors for class distinctions. In this way, the spaces transform into ideological apparatuses that reveal how capitalist systems are constructed materially and symbolically.

At this juncture, Guy Debord's concept of the *Society of the Spectacle* (1994) proves illuminating. According to Debord, contemporary life is increasingly mediated through representations rather than direct experience; reality is consumed as spectacle. The way VIPs watch the games—and how the audience observes these scenes—becomes a layered example of this "representative reality." For the VIPs, the games are not struggles for survival but ephemeral moments of pleasure and aesthetic excitement. This aestheticization banalizes violence and creates an empathic rupture between the viewer and the narrative.

The physical design of the game arenas—closed, isolated from the outside world, equipped with numerous cameras—directly evokes Michel Foucault's notion of the panopticon. Bentham's panopticon prison model exemplifies a disciplinary regime in which individuals internalize the possibility of constant surveillance and modify their behavior accordingly. Contestants in *Squid Game* act under the constant awareness of being watched, deprived of privacy, autonomy, and resistance. This surveillance imposes not only physical but also psychological and ontological confinement.

The symbols used in *Squid Game*'s visual narrative are not merely aesthetic but carry profound social critiques through images embedded in cultural memory. The triangle, square, and circle buttons on the PlayStation controller—widely recognized worldwide—evoke the subconscious sense of "playing a game" while simultaneously delivering a social critique of violence and competition gamified within society. These shapes, accepted in popular culture as emblems of play, entertainment, and rivalry, take on deadly dimensions in *Squid Game*. Thus, games that are supposed to be entertaining transform into ruthless competitions and struggles for survival. This strongly critiques the modern concept of "gamification" and the notion of life being managed like a game. These symbols illustrate how the modern individual is controlled and transformed into a "player" within the system. Thus, the positive associations of entertainment and technology signified by the PlayStation iconography are inverted in *Squid Game*; innocent game images become part of a lethal competitive struggle.

This striking contrast demonstrates how popular culture can function both as a tool and as a subject of critique. By using symbols familiar to a global audience, *Squid Game* constructs a universal visual language and highlights the entanglement of game-entertainment worlds with social structures. In this respect, the series is not only a thriller but also a dystopian allegory of the intertwining of play, entertainment, and survival struggles within the capitalist system.

In conclusion, *Squid Game* processes violence not simply as representational content but as a systemic ideology through the language of colors, visual aesthetics, and spatial coding. The narrative woven around the concepts of surveillance, class, pleasure, and visual representation calls its viewers to a disturbing ethical inquiry: What is our share in watching this visuality, and what does this mode of viewing produce?

REFERENCES

Agamben, G. (2005). *State of exception* (Attell, K., Trans.). University of Chicago Press.

Arkan, G.N. (2024). Örneklerle sinemada renk ve anlam yaratma. [E-Book]. Literatürk academia.

Arnheim, R. (1977). *Film as art*. University of California Press.

Bauman, Z. (2000). *Liquid modernity*. Polity Press.

Bazin, A. (1967). *What is cinema?* University of California Press.

Bellantoni, P. (2005). *If it's purple, someone's gonna die: The power of color in visual storytelling*. Elsevier.

Bong, J. (Director). (2019). *Parasite* [Film]. CJ Entertainment.

Bordwell, D., & Thompson, K. (2010). *Film art: An introduction* (10th ed.). McGraw-Hill.

Bourdieu, P. (1984). *Distinction: A social critique of the judgement of taste*. Harvard University Press.

Brown, W. (2015). *Undoing the demos: Neoliberalism's stealth revolution*. Zone Books. DOI: 10.2307/j.ctt17kk9p8

Çoban, B. (2008). "Gözün iktidarı üzerine." B. Çoban & Z. Özarslan (Ed.), *Panoptikon gözün iktidarı* içinde (pp. 111-137). Su Yayınları.

Debord, G. (1994). *The society of the spectacle* (Nicholson-Smith, D., Trans.). Zone Books. (Original work published 1967)

Demirer, H. (2020). Sinemada renk kullanımı: Aşçı, hırsız, karısı ve aşığı filmi. *Social Mentality And Researcher Thinkers Journal*, *6*(31), 570–580. DOI: 10.31576/smryj.497

Dolgun, U. (2008). *Şeffaf hapishane yahut gözetim toplumu*. Ötüken Yayınları.

Dong, H. (Director). (2021). *Squid Game*. [TV series]. Netflix.

Foucault, M. (1977). *Discipline and punish: The birth of the prison*. Pantheon Books.

Gage, J. (1999). *Color and meaning: Art, science, and symbolism*. University of California Press.

Giddens, A. (2010). *Modernite ve bireysel-kimlik* (Tatlıcan, Ü., Trans.). Say Yayınları.

Greenaway, P. (Director). (1989). *The Cook, The Thief, His Wife & Her Lover* [Film]. Allarts Production.

Gücüyener, M. (2001). Panoptikonik gözetimden synoptisizme gözetim toplumu. [Unpublished master's thesis]. Kocatepe Üniversitesi.

Güven, S. K. (2011). "Gözetimin toplumsal meşruiyeti." H. Köse (Ed.), *Medya mahrem* (pp.173-199). Ayrıntı Yayınları.

Harvey, D. (2011). *Umut mekânları* (Gambetti, Z., Trans.). Metis Yayınları.

Hirokazu, K. (Director). (2018). *Shoplifters* [Film]. Aoi Pro, Inc.

Kırık, A. M. (2013). Sinemada renk öğesinin kullanımı: Renk ve anlatım ilişkisi. *21. Yüzyılda Eğitim ve Toplum*, *2*(6), 71–83.

Koca, Ş. E. (2019). Sinemada anlam yaratma sürecinde rengin metaforik kullanımı. *Sanat Tasarım Dergisi*, *23*, 223–239.

Laughey, D. (2010). *Medya çalışmaları teoriler ve yaklaşımlar* (Toprak, A., Trans.). Kalkedon Yayınları.

Lefebvre, H. (1991). *The production of space* (Nicholson-Smith, D., Trans.). Blackwell.

Lyon, D. (1997). *Elektronik göz gözetim toplumunun yükselişi* (Hattatoğlu, D., Trans.). Sarmal Yayınevi.

Marx, G. T. (2002). What's new about the "new surveillance?" Classifying for change and continuity. *Surveillance & Society*, *1*(1), 9–29. www.surveillance-and-society.org. DOI: 10.24908/ss.v1i1.3391

Metz, C. (1974). *Film language: A semiotics of the cinema*. University of Chicago Press. DOI: 10.1515/9783110816044

Monaco, J. (2009). *How to read afFilm: Movies, media, and beyond* (4th ed.). University Press.

Mulvey, L. (1975). Visual Pleasure and Narrative Cinema. *Screen*, *16*(3), 6–18. DOI: 10.1093/screen/16.3.6

Mulvey, L. (1989). *Visual and other pleasures*. Palgrave Macmillan. DOI: 10.1007/978-1-349-19798-9

Özarslan, Z. (2008). "Gözün iktidarı: Elektronik gözetim sistemleri." *Panoptikon gözün İktidarı*.

Pastoureau, M. (2017). *Renklerin tarihi* (Toprak, A., Trans.). Yapı Kredi Yayınları.

Sözen, M. (2003). *Sinemada renk – sembolik anlamlar*. Detay Yayıncılık.

Todd, H. (Director). (2002). *Far from Heaven* [Film]. Focus Features.

Uğurlu, Ö. (2011). "Postmodern pazarlama stratejisi olarak dönüşen mahremiyet: 'Fulya'nın intikamı' viral kampanyası örneği." Hüseyin Köse (Ed.), *Medya mahrem* (pp. 173-199). Ayrıntı Yayınları.

ADDITIONAL READING

Bordwell, D. (2008). *The way hollywood tells it: Story and style in modern movies*. University of California Press.

Dursun, A. (2021). *Sinemada mekân, renk ve sınıf: Görsel kodların ideolojik işlevleri*. Ayrıntı Yayınları.

Foucault, M. (2006). *Hapishanenin Doğuşu* (Kılıçbay, M. A., Trans.). İmge Kitabevi.

Lefebvre, H. (2007). *Modern Dünyada Gündelik Hayat* (Gürbüz, I., Trans.). Metis Yayınları.

Rotha, P. (2000). *Sinemanın Öyküsü* (Şener, İ., Trans.). İzdüşüm Yayınları.

Sigmund, F. (1955). "The Uncanny". The Standard Edition of the Complete Work of Sigmund Freud. The Hogarth Press.

Sigmund, F. (2001). *Haz İlkesinin Ötesinde, Ben ve İd* (Babaoğlu, A., Trans.). Metis Yayınları.

Şişman, A. (2021). *Sanat ve Sanat Kavramlarına Giriş*. Literatür Yayınları.

Süalp, Z. T. A. (2004). *Zaman Mekan Kuram ve Sinema*. Bağlam Yayınları.

Sunal, G., & Erutku, B. Ü. (Eds.). (2024). *Görüntüleri izinde*. Nobel Yayınları.

Woolen, P. (2008). *Sinemada Göstergeler ve Anlam* (Alacakök, Z., & Doğan, B., Trans.). Metis Yayınları.

Žižek, S. (2005). *Yamuk Bakmak: Popüler Kültürden Jaques Lacan'a Giriş* (Birkan, T., Trans.). Metis Yayınları.

KEY TERMS AND DEFINITIONS

Aestheticized Violence: This refers to the stylized visual presentation of violence, often rendered as attractive, intriguing, or pleasurable. Such representations transform violence from a mere narrative device into a commodified, spectacle-like form, contributing to the blurring of moral boundaries. Aestheticized violence can create emotional distance for the viewer or facilitate the normalization of violence.

Class Distinction: This term refers to the structural hierarchy that positions individuals differently based on their levels of economic, cultural, and social capital. Class distinction encompasses not only income levels but also living environments, cultural preferences, access to education, and even modes of representation. Media products can either expose these distinctions or reproduce and normalize them.

Color and Visual Codes: Colors are powerful symbolic tools that shape the meaning and emotional resonance of visual narratives. Color palettes are employed to code characters' psychological states, class differences, and thematic subtexts. For instance, bright and vibrant hues may suggest hope or false security, while dark and cold tones emphasize danger and despair. Through visual codes, subconscious messages are conveyed to the viewer, enhancing the narrative depth.

Ethical Spectatorship: This approach questions the moral position adopted by viewers in response to media content. Particularly in representations of violence, poverty, or trauma, it becomes crucial to consider how viewers respond, the extent to which they empathize, and how their role as consumers transforms the experience. Ethical spectatorship requires not only an analysis of media content but also a critical reflection on viewing practices.

Panopticism: A concept developed by Michel Foucault, panopticism describes modern forms of power in which individuals regulate their own behavior under the constant possibility of surveillance. Beyond physical prisons, the panoptic model operates visibly or invisibly in various domains such as schools, factories, media, and digital platforms. Surveillance instills an internalized discipline in the subject, ensuring the continuity of power.

Spatial Ideology: A concept that explores how spatial arrangements represent and reproduce social relations and power structures. Space is not merely physical but also the product of an ideological construction. Architectural configurations determine individuals' positions, movements, and interactions within a system, presenting an invisible yet palpable form of power.

The Society of the Spectacle: Coined by Guy Debord, this concept refers to a cultural transformation in capitalist societies where lived experience is replaced by representations and images. The spectacle is not merely a passive viewing experience; it functions as a mechanism of power that shapes individuals' identities, desires, and social relations. Reality is constructed and consumed through visualized and commodified representations.

Chapter 11 The Effect of Digital Streaming Platforms on Visual Media Consumption Within the Framework of Binge-Watching Concept

Dilar Diken Yücel
https://orcid.org/0000-0003-3139-2038
İnönü University, Turkey

Ömer Faruk Yücel
https://orcid.org/0000-0002-3035-004X
İnönü University, Turkey

ABSTRACT

Increasing spread of digitization has transformed many fields, one of which has been in the consumption of visual media products. The growth in the use of digital streaming platforms, which emerged from the development in Internet technologies, has shifted the viewing practices of the mainstream media viewers significantly. The scope of this study is a discussion on the viewing experiences during the activity named Binge-watching, which has been shaped by the new-normal of the digital era. The paper first presents the definition of the concept of Binge-watching. Next, an analysis of the interviews with 16 participants is presented. The analysis of these interviews indicated that Binge-watching promises freedom to individuals as they manage their time and taste accompanied by a considerable number of negative

DOI: 10.4018/979-8-3373-1967-4.ch011

consequences. The interviewees reported such physical effects as loneliness, sleep deprivation, headaches, bad posture, and nausea as well as fear of exclusion from social communities and exposure to a spoiler as related to Binge-watching.

INTRODUCTION

Digital streaming platforms are gradually occupying a more central place in our lives, transforming media product consumption and purchasing processes as such platforms promise to their viewers what may be coined as *plastic* viewing processes. Viewers who has the comfort of accessing any content in any location through any device they wish have been further liberated to manage and stretch their viewing experiences at any rate. Particularly the Covid-19 pandemic acted as a catalyzer in the adoption of digital viewing culture. For the viewers who were locked down at home due to the pandemic, the content presented by digital platforms has presented an alternative visual media consumption method to viewers to prefer over the mainstream television channels and the cinema in time.

Digital platforms pioneered great changes in widespread viewing practices besides viewing experiences at home. It has reached a point where viewers have the power over deciding on the kind of content to consume as well as when, how much, and with what device to consume it. This has proven to be very attractive to the mainstream media viewers who had to wait a week to watch the next episode of the content that they followed. Moreover, such digital platforms as Netflix share with their viewers all the episodes of a season at the same time to strengthen their subscribers' loyalty. This promised viewers liberty, leading to a novel type of viewing called *Binge-watching*. The Oxford English Dictionary defines *binge-watching* as *"The practice of watching several episodes of a TV Show on one occasion, usually by means of DVDs or digital streaming"* (Oxford English Dictionary, 2014). We can understand from this definition that waiting for the next part of the content that one follows and consuming the content by watching the content episode by episode is gradually becoming old-fashioned. This new type of viewing culture is termed in the literature as *"watching without breaks, watching non-stop, excessive watching, serial watching, binge-watching, watching like binge-eating, watching spree, series marathon, over-watching, serial viewing"* (Terzi, 2021, p. 82), and it has shifted the traditional television viewing behavior, bringing about a serious transformation in the means of production, distribution and consumption of media products.

This new viewing culture has ignited discussion in a wide range of topics from health to communication sciences. The common motive in these disciplines is the attempts to define *Binge-watching* and understand/discuss its physiological and psychological impact on humans. The main aim of this study is to discuss the effects of this new viewing culture called *Binge- watching* on individuals who reside in Türkiye aged between 20-40, who can be coined as digital natives, thereby shedding light on the transformation that digital viewing facilities have created. With this aim, interviews were held with 20 participants between the ages of 20 and 40, however, of these, the data of 16 interviews were found to be suitable to report. Open-ended questions were used in the interviews conducted between 10.03.2025-05.04.2025 in the offices of the researchers within about half an hour each.

Introduction of Digital Streaming Platforms in Türkiye

With the advancement of Internet technologies, digital platforms have started to be viewed as an alternative to mainstream television channels in Türkiye. Bilge Şenyüz views digital platforms as an "escape platform". This escape is one that aims to evade censorship, advertisements, and time restrictions that exist in mainstream media. The Internet was used as a resort to catch up on the episodes of television series that were missed, which entered our lives as a new viewing practice to watch series (Şenyüz, 2023, p. 44).

The history of viewing methods that came as an alternative to mainstream television channels is not as recent as that of digital platforms. The service that Netflix provided in 1997 to homes by distributing DVDs pioneered the use of Internet-based digital platforms. The introduction of Netflix to Türkiye is an important stepping stone on the path of digitization. Netflix started to provide services as of January 2016 in Türkiye, leading to the emergence of many digital platforms (Sarı & Sancaklı, 2020, p. 248). Today such digital platforms as Netflix, Max (Blu TV), Disney+, MUBI, Amazon Prime, beIN CONNECT, Gain, and Exxen are active service providers in Türkiye. According to the data of JustWach, the platform with the highest number of subscribers in Türkiye in the first quarter of 2025 was Max with 30%, followed by Netflix with 25 percent (Cumhuriyet, 2025).

The New-Normal of the Digital World: Binge-Watching

With the advancement of Internet technologies, the access rates to digital content are growing by the day. This digitization has led to distinctive changes in many areas both in production and consumption processes. The focus of this study is the transformation experienced in the consumption of media products. One of the fundamental factors that shifted viewers' status is that it has become possible today to view as much as one wants anywhere with the use of such mobile devices as mobile phones and tablets. The is mainly because viewers are no longer passive about making choices as active viewers who can select what content to consume. Meanwhile, today, the number of individuals who follow weekly series on mainstream television channels is falling by the day. The data collected by Exeter University in their study called "*The Future of TV Distribution*" (2024, p. 8) is noteworthy in this respect. According to this study, up to the year 2040, 95% of the households in the United Kingdom will have gained access to TV content through the Internet. In the year 2023, when this study was conducted, this percentage was 87 percent. The number of Netflix subscribers, one of the digital platforms with the highest number of subscribers globally, even in the last quarter of 2024, was 4.82 million in the USA and Canada, 4.15 million in Latin America, 5 million in the Middle East, Europe and Africa regions, 4.94 million in Asia-Pacific region, and the total number of subscribers reached 301.63 million globally (Maas, 2025), which indicates that the popularity of following content via digital platforms is continually increasing. .

The results of an online survey in the USA for Netflix on 25-27 November 2013 administered by a company called Harris Interactive are significant as they indicated that these subscriptions signaled the transformation of the viewing practices. The study on 3,078 adults at the age of 18 and above asked the participants to define watching a series, and in response, 73% of the participants defined watching a series as viewing 2 to 6 episodes of a TV content consecutively in one sitting (Deadline, 2013). Swati Panda and Satyendra C. Pandey define the concept as watching a minimum of 2-3 episodes of the same series or the same content in one sitting within a minimum of one hour (Panda & Pandey 2017, p. 426). This new viewing practice, which is termed in the literature as *Binge -watching* or *Marathon watching,* is seen as feasting by Grant McCracken. McCracken stated in his essay entitled "*TV Got Better*" that digital platform smade a positive impact on both visual media content and the viewers in the following words:

> *"Quite suddenly, a series of virtuous cycles unfolded. Good TV created smarter viewers. Smarter viewers in turn made for better TV. Good TV created smarter writers who in turn made for better TV. Good TV created better critics who in turn made for better TV. Opposers threatened us with a descent into bad. TV somehow managed an ascent into better. Even Hollywood says so. Look at the migration of talent. Bona fide movie stars are prepared to make TV because that's where real creative opportunities now lie." (McCracken, 2015).*

A statement made in 2017 by Netflix reminded readers of McCracken's reference to a feast: "*There's much to say about what we watched, but how we watched was also intriguing. Though bingeing was standard in 2017, not all series were enjoyed in the same way - there were those we devoured, those we savored, those we cheated on, and those that brought us together*" (Netflix, 2017). With the focus on the way we view rather than what we view, the state of *savoring what is viewed* is emphasized, which is associated with viewing more. In addition, the statement informed that more than 140 million hours of content is being viewed daily by the subscribers globally, and a positive association was established between much viewing and savoring what is viewed.

McCracken called consecutive viewing of the same TV content a feast, associating this novel viewing practice with a positive meaning, while there exist numerous academic studies focusing on the negative consequences of *Binge-watching*, in contrast to McCracken's hypothesis. In the literature, this concept is linked with lack of control, being anti-social, depression and loneliness, and poor physical health (Wallenstein, 2015). Mareike Jenner (2015, p. 306) defines the concept of *Binge-watching* as excessive viewing, linking the concept with such unwanted behaviors as excessive alcohol consumption and over-eating. According to this researcher, over-viewing suggests an action in which one can harm oneself due to the negative connotation of the prefix "over-". Another study on the negative impact of *Binge-watching* links over-viewing behavior with excessive smoking, excessive alcohol consumption, over-eating and gambling, highlighting that the individuals with such behavior can be said to be socially stigmatized (Da Costa, 2019, p. 3).

The study titled "*Binge-Watching and Psychological Well-Being: Media Use between Lack of Control and Perceived Autonomy*" (2018), conducted by Viola C. Granow, Leonard Reinecke, and Marc Ziegele collaboratively, is noteworthy for its discussion of both the positive and negative impact of this novel viewing practice on individuals. The questionnaire that was administered on 499 participants indicated that 58% watched series online, and it was found that over-viewing could have positive and negative effects on individuals' welfare. The independence provided by digital platforms had a positive impact on welfare, while in some cases, such independence led to a conflict of goals, which was found to be causing feelings of guilt. The participants who had difficulty in exercising self-control reported feeling guilty because of over-viewing (Granow et al., 2018, pp. 395-398). Bridget Rubenking and Cheryl Campanella defined *Binge-watching* as a "stressful emotional habit", dwelling on such negative effects that may gradually develop in individuals as decrease in physical activities, problems in sleeping and diet, mental procrastination or evading tasks (Rubenking & Bracken, 2018). In the same study, it was concluded that the tendency of over-viewing behavior was observed among the young, who were prone to over-view to regulate their emotions, and were in search of suspense and excitement, while being old was associated with slow-viewing which is also described as viewing on appointment (Rubenking & Bracken, 2018, p. 388).

Motivations for Binge-Watching

In academic studies conducted on individuals who consume digital content through *Binge-watching*, more than one motivation factor identified to direct individuals towards this new viewing experience. These studies, which refer to the approach of uses and satisfaction, the underlying motivations for *Binge-watching* were listed as entertainment, relaxation, spending free time, the desire for momentary enjoyment, and satisfaction (Starosta & Izydorczyk, 2020; Flayelle et al., 2019; Pittman & Sheehan, 2015; Panda & Pandey,2017; Rubenking & Bracken, 2018).

The role of social inclusion should be emphasized among the basic motivation factors for excessive viewing because individuals may choose to watch content in response to the advice of their social circles, and they may consume the said content fast to have an idea about it, or what we may call catching up with the hot topic or trend, as it is the topic of conversation among their social circles. In the literature, this situation is termed Fear of Missing Out (FoMO). The study by Lindsey Conlin et al. entitled "*Time-shifting vs. appointment viewing: the role of fear of missing out within TV consumption behaviors*" is noteworthy in this regard. The concept of Fear of Missing Out (FoMO) can be defined as constantly following up on the popular content and making extra effort to avoid falling behind due to the desire to keep up with what others are doing, which is increasing by the day with the advancement of technology (Conlin et al. 2016, p. 152).

The motivations of the university students related to excessive viewing were examined by Swati Panda and Satyendra C. Pandey, who reported six fundamental motivations related to *Binge-watching*. These can be listed as social involvement, entertainment, unwinding to cope with negativities, escaping from reality and problems, retreating to spend time with fictional characters, and seeking consolation. In addition, the study reported further motivations besides these six factors, listed as the attraction of subscription fees and the possibility of consuming a high amount of content for a low fee, the effect of promotion campaigns of digital platform content, and the ease of access to content (Panda & Pandey, 2017, pp. 429, 433). The study conducted by Jolanta Starosta and Bernadetta Izydorczyk Sebastian Lizińczyk where they discuss the characteristics of excessive viewing behaviors in early-adulthood, present similar findings that were reported by Panda and Pandey. The researchers stated that viewers had five fundamental motivations: the need to access information and education, identification with the characters, entertainment and strengthening social ties by excessive viewing. The findings of the same study related to gender are noteworthy. In the results of the study, it was reported that females watched digital series more compared to males, and that their reasons for such preference were coping with loneliness, entertainment and experiencing positive emotions (Starosta et al. 2019, pp. 150, 161).

METHOD[1]

Within the scope of this study, phenomenological research design was selected as a qualitative research method. The starting point of the phenomenological research studies is to discuss humans' responses and perceptions towards a selected phenomenon. The present study aimed to discuss the responses of the viewers to the novel viewing experiences brought about by digitalization and the ways in which the viewers access the content specifically related to the concept of *Binge-watching*. It was decided that the most suitable method that would allow such a discussion was the phenomenological research method. The main aim in phenomenological research studies is to understand the meaning of individuals' experiences through their own eyes (Creswell, 2018, p. 77). In phenomenological research studies, the focus areas are individuals' experiences of daily lives and their interpretations of these, after which these experiences and interpretations are utilized as qualitative descriptive data. The preferred data collection method in phenomenological studies is usually the interview method to "relay the phenomenon as the phenomenon feels, tells and describes it" (Erdoğan, 2012, pp. 157-158).

In the scope of this study, the participants were asked 20 open-ended questions, and the interviews were conducted in line with these questions. The main research questions for which the study sought answers are the following:

1. In what ways have *Binge-watching* processes changed individuals' viewing habits?
2. What are the physiological and psychological changes that *Binge-watching* processes caused on individuals?
3. What are the differences between *Binge-watching* and traditional viewing?
4. How does the presence of artificial intelligence-based algorithms affect viewing experiences?
5. What kinds of changes will *Binge-watching* lead to in the future viewing culture?

As can be understood from the questions above, the study focuses on the viewer's experiences in *Binge-watching* processes and aims to facilitate comments on physiological and psychological changes in the viewers in this process.

A Ritual for Self-Coping: Binge-Watching

In this study, the participants were asked when they had started *Binge-watching* and whether they enjoyed being alone or having company while they followed the content. The participants' responses are noteworthy because they are the pre-existing motivations that pushed them to *Binge-watch.* Most of the participants stated that they started *Binge-watching* in their university years. The second more frequently mentioned period emerged as the Covid-19 pandemic. The participants added that they were involved in *Binge-watching* by obtaining subscription to a digital platform. One of the participants, B. D. (student aged 22), said the following: *"I started the first year of university. I can say that I started by opening a Netflix account."*, summarizing the linear relationship between *Binge-watching* process and digital platforms.

Most of the participants stated that they preferred to be alone while *Binge-watching*. One participant, T. Y. (student aged 30) said, *"I prefer to watch alone because I enjoy speeding it up and watching the way I want when I want."*, emphasizing that being alone in this new viewing experience opened free space for this individual. Another participant, B. A. (student aged 24) explained, *"I prefer to watch by myself because our taste with the people around me are generally different"*, indicating that *Binge-watching* process helped in reaching personal state and likes. Yet another participant, Y. D. (teacher aged 25) said, *"I prefer to watch the productions that I watch my myself with others afterwards. I find the comments of my friends about the productions whose stories I know interesting."*, mentioning that watching the content alone positively contributed to socialization.

The participants also emphasized that watching the content by themselves helped to focus on the content more effectively, and that chatting with friends while watching overshadowed the viewing experience. One of the participants, B. D., replied by stating *"I prefer watching by myself because when I have company I cannot focus. I focus on the conversation"*. Another participant, H. F. (a student aged 24) said, *"I watch by myself because I lose concentration when I watch with others and I find myself in a conversation"*, explaining that watching alone helps to concentrate better. The responses indicate that *Binge-watching*, which can be said to be the new-normal of viewing experiences, points to a process in which watching together has lost its importance as opposed to the mainstream media with which individuals tend to spend time on their own. Only two of the participants mentioned viewing with people they are close with or friends as they referred to their experiences of *Binge-watching*. Of these participants, D. A. (student aged 22) said *"Actually, it does not make a difference but I prefer to watch with a friend, in this way we can talk about the content that we watch and discuss"* and another participant M. E. (student aged 25) *"It does not make a difference but consulting with someone makes it more fruitful"*, mentioning the idea of including other individuals along with themselves into the process of *Binge-watching*. Yet, both participants emphasized that this is not essential by starting their response by saying "makes no difference". The participants' tendency to prefer to be alone while *Binge-watching* reminds us of Panda and Pandey's findings of their study. According to the researchers, individuals tend to what can be called lost among the fictional characters while *Binge-watching*, thus retreating from and escaping the challenges of daily life (Panda & Pandey 2017,

p. 429). The findings of the present study are consistent with those of the above-mentioned study. The participants preferred to view alone most of the time rather than watch with company.

The Attraction of the Screen: Screen Time İncreases

The findings from the interviews conducted in this study indicate that the amount of screen time is consistent with the previously reported negative aspects of *Binge-watching*. Time spent in front of the screen means that individuals miss out on physical activities and adopt accompanying rituals (eating and drinking habits), which have repercussions on body health. As can be remembered, Rubenking and Bracken stated that *Binge-watching* has such physical effects as a decrease in physical activity and problems with sleep and diet among individuals. In addition, these negative physical effects are accompanied by negative effects on mental health such as procrastination or task avoidance (Rubenking & Bracken, 2018). The findings of this study are consistent with these reported effects.

Of the participants, F. Ş. (academician aged 35) stated, "*I watched 10 or 12 episodes of the series Lost consecutively. Each episode is about 1 hour, so this means 10 or 12 hours of viewing. Naturally, such a series as Lost, which is long and captivating, had a negative effect on my sleep schedule. As I was following that series (I was doing my undergraduate studies at the time), I would watch it at night and sleep during the day. Binge-watching forces one to be active at night, so I can say that it limits social relationships to a degree*", mentioning the negative impact of staying up in front of the screen for a long time on both one's sleeping patterns and social relationships. Another participant, T. Y., also talked about the negative effects of spending a long time viewing the screen on social life, adding that they suffered from sleep deprivation, headache and sore eyes. Of the participants, B. D., said "*I experience negative effects due to looking at the screen all the time. I understand that my posture is not right anymore. Also, I feel tired during the day as if the screen has sucked up all my energy*", commenting that staying in front of the screen for long hours may lead to such negative physical effects as bad posture and drained energy. Of the participants, M. T. said, "*The last series I watched was "Round of Applause" directed by Berkun Oya. One episode lasted on average 30 minutes. It was a mini-series, and I completed the entire series. It had a total of 6 episodes, so I was tied to the screen for about 3 hours. Besides this, I finished the series called "Stranger Things" by watching one season per day. The series consists of 4 seasons. Each episode is on average for 40 minutes. I would spend 4-5 hours sometimes. They were a captivating series for me. When the series ended, I felt hallow. I had difficulty moving onto a new series. So, if I finish a season in one day, I need some time to go onto a new series. I cannot get out of the characters quickly. I think I enter that world."*, emphasizing the psychological dimension of the time spent viewing the screen. M.T., different from the other participants, indicated that the time spent across the screen builds an emotional bond between the content and the viewer. Such emotional bond is described by the words "I enter their world" by the participant by drawing a conclusion, indicating that the future viewing culture is affected by screen time. There were participants who linked screen time to *nausea*. B. A. said *"I watched for about 8 hours. I was sometimes happy that I watched because my curiosity was satisfied. Sometimes I felt that I got seriously nauseous and that I was drawn into the movie for no good reason."*, stating that on the one hand *Binge-watching* was satisfying while, on the other hand, it had negative physical and psychological effects. Of the participants, D. E. (teacher aged 22) said, "*There have been times when I finished two seasons of a series that captured me strongly. For instance, I felt excited to observe how the personality of my favorite characters evolved from one episode to the other in a fantastic series called Game of Thrones. The striking transformation of the characters in the series called Breaking Bad and its captivating story made me watch the series breathlessly,* emphasizing the emotional satisfaction that *Binge-watching* provided.

Among the participants interviewed in this study was H.F., whose explanation of the effect of *Binge-watching* on personality is worth noting. This participant stated "*I sometimes speed up the series and sometimes I skip certain parts while watching it. When I try to exercise the same liberty in my social life, I get accused of being impatient"*, saying that this process transformed them into impatient individuals. Another participant, just like H.F, stated that their *Binge-watching* habit encouraged them to speed up. N.Ş. (student aged 22) said, "*I fast forward as I watch the content, so I rewind some parts and watch them again. I want to be fast in my daily life, too, but this speed causes me to revise many things to redo them just like I do as I watch series.*", explaining that *Binge-watching* encouraged them to be fast, but such speed disrupted both the process of viewing series and their social lives.

As can be understood from the statements of the participants, the screen is rather attractive in the process of *Binge-watching*. However, this attraction also brings about negative results such as sleep deprivation, headache and sore eyes, nausea, posture problems, impatience, and getting caught up. Besides these negative effects, many of the participants talked about the undesirable impact of spending long hours in front of the screen on their daily tasks and responsibilities. M. T. summarized this by saying, "*I cannot not realize how time passes just like that when I take my phone in my hands before I start whatever I need to do. The number of hours that passes by the time I raise my head can affect my tasks I need to complete. Sometimes I can be late." Binge-watching* brings about problems in managing time for many participants.

Fear of Missing Out (FoMO)

What Conlin and colleagues defined as Fear of Missing Out (FoMO) (Conlin et al. 2016), which involves the tendency of following popular content for fear of falling behind one's social circle, emerged in our interviews in this study. One of the participants, M. T., explained the contentment due to the similarity of series viewing habits among their circle of friends and the undesirable scenarios that may occur in an opposite situation in the following statement: "*When I talk to my friends, I hear that they too have similar habits of watching consecutively. In such a case, there is no chance to get a spoiler because all have seen the series. If one of us finished an episode of a series in one week and one completed the entire series at the same time, then we would have a consistency problem, and we could only talk about the parts that we both watched but in such a case there would the risk of giving away a spoiler." According* M.T., having a friend who follows a series weekly would mean such negative results as "getting a spoiler" or having trouble talking about the content that each viewed.

Of the participants, H. F. also pointed out that they had such trouble discussing with friends the digital content of what they viewed that they cannot keep up with their speed. This participant said "*I enjoy talking about what I watched with my friends, but sometimes I watch all the episodes at once. Then they do not want to talk to me for fear of a spoiler.*", mentioning a negative impact of *Binge-watching* on socialization. As can be seen, the process that is termed "getting a spoiler" is a significant driving force that affects *Binge-watching*.

Of the participants, F. Ş. said "*Having a friend or a group of friends who watch a series almost at the same time opens the way to longer and better conversations about the said series because humans want to talk about what they like. The joy is even more so when we converse about a series of a film which was prepared wisely and shot aesthetically.*", highlighting the importance of watching the same series at the same time frame to engage in social relations with their circle of friends. This explains the need to follow up on the popular content that is on the current agenda in order to keep up with the social circle.

The participant B. U. said, "*Because I love recommending the series that I like and talking about this topic, I prefer watching the series that my friends recommend and the subject matter of those that I like by Binge-watching quite fast.*", emphasizing that making recommendations on content among friends encourages *Binge-watching*. Of the participants, A, D., said that now many people around their circle consume content through *Binge-watching,* that they adopted this style of consumption in time, and that in this way they did not have trouble socializing with their close friends. This explanation indicates that keeping up with one's social circle's viewing rhythms is no longer considered as a need in today's digital viewing practices, but it is rather seen as a must.

Will Algorithms Decide for Us What to Watch?

Stacks of data that increased through digitalization are collected and analyzed with the use of algorithms. Today, when algorithms are used in the field of viewing experiences, they act as a means that presents new alternative viewing suggestions based on observations of viewers' tastes and likes. Digital platforms market personalized content tuned to viewers' tastes and likes via algorithms that digital platforms utilize. They direct their viewers to watching new content with such following statements: "Series we thought you might like; Exclusively for you; We thought you would like this, too." Are such statements that sound like friendly advice innocent? This is open to being seen as a commercial and marketing strategy for digital platforms that charge subscription fees. As viewers' demands and viewing rates increase, the platform becomes more well-known, and the number of subscribers also increases by the day. Osman Sümer called this process *Algorithm Tyranny*, stating that algorithms almost direct individuals' lives, impede independent choice and decision-making skills, and push them towards undesirable behaviors (Sümer, 2024, p. 105).

In the scope of this study, the common thread of that emerged from the interviews did not quite match how Sümer used the term *tyranny* to refer to algorithms but indicated that algorithms were quite effective in adopting the habit of *Binge-watching*. Of the participants, B. U., "*I did not have such habits of Binge-watching much before digital platforms and algorithms. My habit grew the moment when I subscribed to different digital applications*", stating that the presence of algorithms directed them towards *Binge-watching*. Of the participants, M. T., said, "*I observed that the number of mini-series increased on digital platforms lately, and I notice that I also started to prefer such mini-series. The idea of starting series that lasts many seasons scares me a little, so instead I can reach satisfaction that I desire by completing the 5–6-episode mini-series. However, long series intimidate me, so I try to choose series with fewer seasons,* indicating that the content on digital platforms give direction to viewers' demands. Similarly, H.T. (academician aged 40) said, *"To me, content produced like this are designed with the idea of Binge-watching. I consider it as a long film. Now they produce a season to be consumed in one sitting. Because 'Adolescence' was produced as a 4-episode format to be viewed in one sitting, it was marketed with this feature and stayed on the agenda for weeks as the most viewed content on Netflix",* highlighting the influence of algorithms on content production. Digital platforms introduce viewers to mini-series by promising "Do not spend about three hours watching only one episode on mainstream television channels anymore; Do not spend one hour to watch one episode only; Avoid exposure to ads and waiting for the next episode for one week", and they make use of algorithms to produce such content and promote them to viewers. Of the participants, A. D., explained in a similar way by saying, "*Digital platforms facilitate the habit of Binge-watching through the features that they offer. As the series with fewer and shorter episodes are watched, this habit starts for form.*" Another feature that encourages viewers to follow content without leaving the platform is the availability to play automatically on digital platforms. Of the participants, U. Y. (academician aged 30), emphasized this with the following statements: "*The automatic suggestions of the platforms and going onto the next episode automatically can make the viewing time longer. This strengthens the habit to watch in this format.*" The automatic start of the next episode is seen as a motivating element to continue watching the next part.

The statements of another participant, H. F., confirm that the algorithms that digital platforms use encourage the users to view more content: "*I constantly encounter parts from series and films on digital platforms, and I want to finish them. In time, you become a consumption maniac. You constantly want more, like an endless bottomless well.*", stating that algorithms directed individuals' will-power. Likewise, of the participants, Y. D. stated: "*When you complete a series in one platform, different productions similar to that series are suggested. They do not want you to leave that platform.*", indicating that algorithms offer attractive suggestions to stay loyal to their platform subscriptions.

It is clear that algorithms will give directions to our viewing experiences in the future just as they do today. The news published on a site on the Internet called Webtekno dated April 12, 2025, is noteworthy with respect to presenting a glimpse of the future viewing practices. The news piece talks about Netflix's OpenAI-supported novel search feature. According to this news article, with the help of artificial intelligence technology, individuals can be presented with content that matches their mental state when they search with such prompts as "I am demoralized; things to blow off some steam" (Ögenler, 2025). Based on the news content, one can predict that artificial intelligence will be indispensable in the future viewing experiences.

Are Mainstream Television Content Turning to Nostalgia?

Today the effect of digitalization is strong, so perhaps the most effective way to understand the viewing transformation experiences of today by focusing on the past viewing experiences. The first transformation that took place in the viewing experiences in Türkiye in the 1970s. Prior to the 70s, viewers could watch films only in movie theatres. Films entered homes when television sets entered the stage in the 70s, when the number of viewers of movie theatres fell significantly. With the introduction of televisions into households, the experience of viewing has become an activity to do at home. The next change to the viewing experiences that had changed with television was the availability of digital platforms as of 2010, when the necessary technical infrastructure was built (Diken Yücel, 2021, p. 52). Now the viewers are no longer tied to the television screens or homes. They can enjoy viewing experiences anywhere any time and via any device they wish via an Internet connection. In fact, they no longer need high-level media literacy for this digital viewing experience. With a subscription, individuals at any age can access digital viewing platforms. What we previously called a plastic viewing opportunity that digital platforms offer led to a decrease in the number of individuals that follows mainstream television channels and the nostalgic to be coined to the mainstream television content.

As a result of the interviews in this study, it was concluded that majority of the participants preferred to follow the content of digital platforms rather than viewing mainstream media content. Of the participants, H. T. said, *"Digital platforms promise more comfortable viewing experiences. Because viewers got used to such comfort, we can say that the viewing experiences of watching TV series has become history. The clearest indicator of this is the YouTube accounts that TV series broadcast their episodes. Now, YouYube streaming ratings are the topic of discussion more than ratings. This shows that traditional viewing has decreased."*, stating that following content on mainstream television channels has become nostalgic.

Of the participants, M. T., stated, "*I lose my curiosity in weekly series, I lose contact with the series. When I'm Binge-watching, I can satisfy all my curiosity on the same day. I get more engaged with the story in this way. When I complete one season in one sitting, I remember more. When I watch week by week, there occurs a discontinuity, and I get disengaged from the series. When I watch it in one sitting, I do not have full control of the story and I do not forget it in this way."*, explaining that having to wait for the next episode on mainstream media takes away the feeling of curiosity, preventing them to get into the story.

Moreover, the participants mentioned the effect of these new viewing experiences brought about by digital platforms on perceiving the content. A. D. said, "*Today we can access all series and films whenever we want, so we do not watch them as carefully as we did in the past. This encourages us to view less carefully. Before digital platforms became so widespread, we knew that we could not access the weekly series through other media, so we watched them more intently,* emphasizing that the ease of access to content in this novel viewing experiences transformed the approach to the content, as well.

Of the participant, H. F., said "I *can remember the series that I viewed in one sitting because there was no time between the episodes, and I can establish the connections between two episodes. Moreover, the emotional load in each episode accumulates one after another when I watch them in one sitting, making the effect last longer. However, in weekly series, this emotional accumulation does not occur as there is time between episodes*, highlighting the differences between weekly viewing and consecutive viewing with regard to creating emotional intensity and perception of content. K.S., another participant, said *"To me, weekly viewing brought about a different excitement. The habit of viewing series weekly is over. Because of digital platforms, the habit of viewing TV series has also decreased because when watched one after another, emotional scenes are more effective. For instance, a character dies in the last scene of a series, and you can start the next episode and continue watching, thus the emotion goes on."*, explaining the effect of the presence of digital platforms on television series. In a similar way, B. A. said, *"I do not remember the last time I watched a weekly series."*, talking about the practice of watching series weekly as a nostalgic activity.

CONCLUSION

Parallel to the fast digitalization of today, significant transformation has been experienced in the visual media product consumption practices, just as in many other fields. In the period when Internet technologies were not widespread, the viewing experiences offered by mainstream television channels or movie theatres have been taken over by viewing in the comfort of homes. With Internet-based digital streaming platforms, the viewing experiences of today's viewers have become what may be called plastic. Now, viewers determine the content to watch, where to watch and how long to watch. This may seem to be a positive aspect at first, but in time some negative effects of this new viewing experience have emerged.

The practice of *Binge-watching,* which is the focus of this discussion, is the most prominent one among such viewing experiences as this viewing experience which is also referred to viewing like devouring or binge-eating, has certain negative effects on viewers. The conclusion to draw from the interviews conducted within the scope of this study is that the negative impact of Binge-watching is significantly higher than its positive impact. The participants stated that they usually prefer to be alone when *Binge-watching*, and that this process had such physical effects on their body as sleep deprivation, headaches, bad posture, and nausea, and that they felt obligated to *Binge-watch* to be able to keep up with the media content so that they could talk about them with their friends and avoid exposure to spoiler. The participants emphasized the length of time that they spent in front of the screen was high, stating that this time may be as much as 10-12 hours and noting that this made an impact even on their perception of night and day. Even if there are negative effects of Binge-watching, it seems that it has gained acceptance into our lives together with such effects. The new normal of the digitalizing world in terms of viewing is now *Binge-watching*. The participants stressed that following content from mainstream media is now nostalgic, and that the presence of digital platforms distances them from mainstream media. It seems that not so far in the future the present study and similar ones will be conducted on a limited number of viewers as the sample who follow mainstream media, thus the experiences of the past will be examined through lived experiences.

REFERENCES

Conlin, L., Billings, A. C., & Averset, L. (2016). Time-shifting vs. appointment viewing: The role of fear of missing out within TV consumption behaviours. *Comunicar: Media Education Research Journal, 29*(4), 151–164.

Creswell, J. W. (2018). *Nitel araştırma yöntemleri: Beş yaklaşıma göre nitel araştırma ve araştırma deseni* (S. B. Demir & M. Bütün, Çev.). Ankara: Siyasal Kitabevi.

Cumhuriyet. (2025, April 24). *Türkiye'de en çok abonesi bulunan içerik platformları.* https://www.cumhuriyet.com.tr/bilim-teknoloji/turkiyede-en-cok-abonesi-bulunan-icerik-platformlari-2332056. [Erişim tarihi: 15.04.2025].

Da Costa, J. C. R. (2019). Binge-watching: A life course perspective. *Journal for Social Thought, 3*(1), 1–10.

Deadline. (2013). *Netflix: Binge-watching is the new normal for TV streamers.* https://deadline.com/2013/12/netflix-binge-watching-tv-streaming-study-651962/. [Erişim tarihi: 18.04.2025].

Diken Yücel, D. (2021, Kasım 11–12). Seyir kültüründe dijital dönüşüm: Sinema salonlarından dijital platform üyeliklerine. *3rd International Cinema and Youth Congress*, Eskişehir, Türkiye.

Erdoğan, İ. (2012). *Pozitivist metodoloji ve ötesi.* Erk Yayınları.

Flayelle, M., Canale, N., Vögele, C., Karila, L., Maurage, P., & Billieux, J. (2019). Assessing binge-watching behaviours: Development and validation of the "Watching TV Series Motives" and "Binge-Watching Engagement and Symptoms" questionnaires. *Computers in Human Behavior, 90*, 26–36. DOI. 10.1016/j.chb.2018.08.022

Future of TV Distribution. (2024). A report prepared by the University of Exeter with the University of Leeds, MTM, 3 Reasons and Real Wireless. https://assets.publishing.service.gov.uk/media/672cafe262831268b0b1a2f4/Future_of_TV_Distribution_FINAL__7_Nov_2024_-accessible.pdf. [Erişim tarihi: 20.04.2025].

Granow, V. C., Reinecke, L., & Ziegele, M. (2018). Binge-watching and psychological well-being: Media use between lack of control and perceived autonomy. *Communication Research Reports, 35*(5), 392–401. DOI: 10.1080/08824096.2018.1525347

Jenner, M. (2015). Binge-watching: Video-on-demand, quality TV and mainstreaming fandom. *International Journal of Cultural Studies, 20*(3), 304–320. DOI: 10.1177/1367877915606485

Maas, J. (2025, January 21). Netflix adds nearly 19 million subscribers to end 2024 with more than 300 million globally. Variety. https://variety.com/2025/tv/news/netflix-subscribers-300-million-q4-2024-1236280419/. [Erişim tarihi: 22.04.2025].

McCracken, G. (2015). *TV got better: How we got from bingeing to feasting.* https://grant27.medium.com/tv-got-better-how-we-got-from-bingeing-to-feasting-782a67ee0a1. [Erişim tarihi: 24.04.2025].

Netflix. (2017). *2017 on Netflix – A year in bingeing.* https://media.netflix.com/en/press-releases/2017-on-netflix-a-year-in-bingeing. [Erişim tarihi: 18.04.2025].

Ögenler, Y. (2025, Mayıs 3). Netflix, OpenAI ile iş birliğine mi gidiyor? *Webtekno.* https://www.webtekno.com/netflix-openai-arama-h157844.html. [Erişim tarihi: 26.04.2025].

Oxford English Dictionary. (2014). Oxford University Press.

Panda, S., & Pandey, S. C. (2017). Binge-watching and college students: Motivations and outcomes. *Young Consumers, 18*(4), 425–438. DOI: 10.1108/YC-07-2017-00707

Pittman, M., & Sheehan, K. (2015). Sprinting a media marathon: Uses and gratifications of binge-watching television through Netflix. *First Monday,* •••, 20. DOI: 10.5210/fm.v20i10.6138

Rubenking, B., & Bracken, C. C. (2018). Binge-watching: A suspenseful, emotional habit. *Communication Research Reports, 35*(5), 381–391. DOI: 10.1080/08824096.2018.1525346

Sarı, Ü., & Sancaklı, P. (2020). Küyerelleşmenin Dijital Platformların İçerik Tanıtımına Etkisi [The effect of glocalization on the content promotion of digital platforms]. *Erciyes İletişim Dergisi, 7*(1), 243–260. DOI: 10.17680/erciyesiletisim.647463

Şenyüz, F. B. (2023). Türkiye'de dizi endüstrisi ve dijital dönüşüm: Yapım, dağıtım, içerik ve izleme pratiklerinin dijitalleşmesi. In *Kültürel Değişim ve Endüstrileşme Sürecinde Türk Dizileri 2010-2020* (pp. 41–53). İstanbul Ticaret Odası.

Solis, L. (2014). Expert analyzes students' Netflix usage. *The Daily Toreador.* https://www.dailytoreador.com/lavida/expert-analyzes-students-netflix-usage/article_0dfc194a-5412-11e4-9415-001a4bcf6878.html. [Erişim tarihi: 26.04.2025].

Starosta, J., Izydorczyk, B., & Lizińczyk, S. (2019). Characteristics of people's binge-watching behavior in the "entering into early adulthood" period of life. *Health Psychology Report, 7*(2), 149–164. DOI: 10.5114/hpr.2019.83025

Sümer, O. (2024). Algoritma tiranlığı: Dijital dünyada bireyin savunmasızlığına yönelik bir sosyo-teknik analiz [Algorithm tyranny: A socio-technical analysis of vulnerability of the individual in the digital world]. *İmgelem*, 15, 87–114.

Terzi, A. (2021). Binge-watching kavramı ve Türkçe karşılıkları üzerine. *Türk Dili, 70*(833).

Wallenstein, A. (2015, January 29). Binge-watching TV linked to depression, loneliness. *Variety*.

ADDITIONAL READING

Caldwell, J. T. (2008). *Production culture: Industrial reflexivity and critical practice in film and television*. Duke University Press.

Ellis, J. (2000). *Seeing things: Television in the age of uncertainty*. I.B. Tauris. DOI: 10.5040/9780755604784

Kubey, R., & Csikszentmihalyi, M. (1990). *Television and the quality of life: How viewing shapes everyday experience*. Routledge.

Lotz, A. D. (2014). *The television will be revolutionized* (2nd ed.). NYU Press.

Lotz, A. D. (2017). *Portals: A treatise on internet-distributed television*. Michigan Publishing. DOI: 10.3998/mpub.9699689

Mittel, J. (2015). *Complex TV: The poetics of contemporary television storytelling*. NYU Press.

Skovmand, M., & Schrøder, K. C. (Eds.). (2016). *Media cultures: Reappraising transnational media*. Routledge. DOI: 10.4324/9781315511931

Spigel, L., & Olsson, J. (Eds.). (2004). *Television after TV: Essays on a medium in transition*. Duke University Press. DOI: 10.2307/j.ctv11qdz4m

Tryon, C. (2013). *On-demand culture: Digital delivery and the future of movies*. Rutgers University Press.

Uricchio, W. (Ed.). (2004). *Media Cultures: Reappraising Transnational Media*. Routledge.

KEY TERMS AND DEFINITIONS

Binge-watching: Binge-watching is the viewing practice that gained prominence with the spread of Internet technologies. It means viewers watching more than one episode of a series or seasons consecutively or consuming more than one film consecutively via digital platforms. For this reason, it is often likened to fast-food consumption in literature.

Digital Streaming Platforms: Digital streaming platforms are defined as the services that facilitate digital content distribution via the Internet. With the help of these services, which are used commonly today, fundamental changes have emerged in viewing experiences today. Viewers can determine the content, time and location that they wish as they consume the media content via such digital streaming platforms as Netflix, Max (Blu TV), Disney+, MUBI, Amazon Prime, beIN CONNECT, Gain, Exxen and Hulu, and thus they are more independent compared to mainstream media.

Digitalization: Digitalization, which signifies a new order in the process of generating, consuming and sharing knowledge, is defined as restructuring of individual, public and institutional processes through digital technologies.

Fear of Missing Out (FoMO): Fear of Missing Out is described as the tendency to constantly follow up on the popular content with the fear of falling behind of one's social circles, and it is one of the most significant motivations that direct individuals to Binge-watching. Individuals remain in front of the screen for fear of exposure to a spoiler or the anxiety of falling behind their social circles.

Mainstream media: This term refers to organizations which address a wide proportion of the public, widespread with a wide-ranging influence. To follow these media organizations, specific skills such as digital media literacy are not necessary. In mainstream media, content is formed by considering viewer surveys and ratings, still it is rather passive compared to digital platforms with respect to planning the viewing process.

New media: All the media content which is produced and distributed with the use of technologies can be regarded as new media. Different from traditional media, in new media, users are in the position of being both the content producer and the consumer. For this reason, the users are more active compared to traditional media. A few of the prominent examples of new media are the Internet, social media platforms, blogs, and digital streaming platforms.

Viewers: Viewers are those who follow, experience and interact with the artistic and performance products along with the media products. Viewers not only consume the content with such interaction, but they also become part of such processes as perception, critique, sharing and reproduction.

ENDNOTE

1 Ethics committee approval was received for the study from the Inonu University Social and Human Sciences Scientific Research Ethics Committee on 02-05-2025.

Compilation of References

Aarseth, E. (1997). *Cybertext: Perspectives on Ergodic Literature*. The Johns Hopkins University Press. DOI: 10.56021/9780801855788

Acar, A. (2017). *Sessiz Sinema Tarihi*. Doruk.

Adams, S. (24 September 2010). "Interview: Gaspar Noé". https://web.archive.org/web/20110731142928/http://www.avclub.com/articles/gaspar-noe,45554/ (Access 24.04.2025)

Agamben, G. (2005). *State of exception* (Attell, K., Trans.). University of Chicago Press.

Ahlers, K. H., Kramer, A., Breen, D. E., Chevalier, P. Y., Crampton, C., Rose, E., Tuceryan, M., Whitaker, R. T. ve Greer, D. (1995). Distributed Augmented Reality for Collaborative Design Applications, *European Computer-Industry Research Centre Gmbh, Proceedings of Computer Graphics International*, Maastricht, Munich, 1-21.

Ahlers, D. (2006). News Consumption and the New Electronic Media. *The Harvard International Journal of Press/Politics*, *11*(1), 29–52. DOI: 10.1177/1081180X05284317

Akıncı, S., & Başer, E. (2020). Reklamdan Kaçınma Bağlamında Geleneksel ve Modern Film İzleme Ortamlarının Genç İzleyiciler Üzerinden Karşılaştırılması: Netflix ve Sinema Salonları Örneği. *Erciyes İletişim Dergisi*, *7*(1), 1. Advance online publication. DOI: 10.17680/erciyesiletisim.622176

Akmeşe, Z. (2020). Televizyon Ana Haber Bültenlerinde Çerçeveleme: Türk Telekom Reklam Filmi Örneği. *Middle Black Sea Journal of Communication Studies*, *5*(1), 37–49.

Akmeşe, Z., & Deniz, K. (2015). *Dijital Aktivizm Olarak Video Aktivizm: Redhack Belgeselleri. Yeni Medya Çalışmaları II. Ulusal Kongresi*. Kadir Has Universty.

Aksoy, Ü. (2025). *Müzeverse'nin ikinci içeriği "Son Kale - Orta Çağ'ın Keşfi" ziyaretçilerle buluştu,* https://www.aa.com.tr/tr/kultur/muzeversenin-ikinci-icerigi-son-kale-orta-cagin-kesfi-ziyaretcilerle-bulustu, [Erişim tarihi: 02.06.2025].

Aldemir, A. (2022). Dijital Çağda Fotoğrafik İmge ve Sözün Düşüşü. Intermedia. *International Journal (Toronto, Ont.), 9*(17), 192–211. DOI: 10.56133/intermedia.1109495

Allam, R., & Dinana, H. (2021). The Future of TV and Online Video Platforms: A Study on Predictors of Use and Interaction with Content in the Egyptian Evolving Telecomm, Media & Entertainment Industries. *SAGE Open, 11*(3), 215824402110408. DOI: 10.1177/21582440211040804

Altınbay, R., & Gümüş, N. (2020). Sosyal bilgiler öğretmenlerinin sanal tur uygulamalarıyla ilgili görüşleri. *Journal of Innovative Research in Teacher Education, 1*(1), 60–71. DOI: 10.29329/jirte.2020.321.5

Altun, F. (2022) Geleneksel Medyadan Yeni Medyaya Propaganda: Süreklilik ve Değişim. Mevzu: Sosyal Bilimler Dergisi, 7 (Mart 2022): 1-22.)

Andrejevic, M. (2009). The twenty-first-century telescreen. In Turner, G., & Tay, J. (Eds.), *Television Studies After TV: Understanding Television in the Post-Broadcast Era* (pp. 31–40). Routledge.

Antmen, A. (2012). *20. Yüzyılın batı Sanatında Akımlar: Sanatçılardan Yazılar ve Açıklamalarla.* Sel Publication.

Apaydın, C. (2024). *Film üretiminde gerçek zamanlı görsel efekt kullanımına ilişkin bir araştırma: Genişletilmiş gerçeklik (XR) stüdyoları örneği. (Doktora Tezi).* Selçuk Üniversitesi, Sosyal Bilimler Enstitüsü.

Arapoğlu, F. (2017). *Video Sanatının Öyküsü.* [Belgesel / Kısa Film / Video Sanatı]. (N. Ç. Bikiç ve F. Özgür, Ed.). İstanbul: Doruk Publication.

Arda, Ö. (2015). Belgesel Filmde Yeni Yaklaşımların Idfa Doclab 2010-2015 Belgesel Film Örneklemi Üzerinden Yapısal Ve Kullanım Özelliklerine Göre İçerik Analizi. İstanbul Üniversitesi İletişim Fakültesi Dergisi| Istanbul University Faculty of Communication *Journal*, (48), 1-23.

Arda, Ö., & Şentürk, O. (2021). Dijital Çağda Belgesel Film: Dijital Dönüşüme Türk Belgesel Sinemacılarının Yaklaşımları Üzerine Bir İnceleme. *Bildiri Kitabı Proceedings Book*, 100.

Arditi, D. (2021). *Streaming culture: Subscription platforms and the unending consumption of culture*. Emerald Publishing. DOI: 10.1108/9781839827686

Arıkan, B. A. (2023). Sanal Gerçeklik ve Medya. *TRT Akademi*, *8*(17), 5–7. DOI: 10.37679/trta.1245474

Aristoteles (1987). Poetika, (Çev: İsmail Tunalı), Remzi Kitabevi, İstanbul

Arkan, G.N. (2024). Örneklerle sinemada renk ve anlam yaratma. [E-Book]. Literatürk academia.

Armes, R. (2006). *Sinema ve Gerçeklik Tarihsel Bir İnceleme, Çev*. ZeynepÖzen Barkot, Doruk Yayımcılık.

Arnheim, R. (1977). *Film as art*. University of California Press.

Arnheım, R. (1985). *Film ve Gerçeklik-Sinema Kuramları, Derleyenler: Seçil Büker Oğuz Onaran*. Dost Kitabevi.

As, E. ve Parsa, A. F. (2023). Teknolojik Gelişmeler Bağlamında Belgesel Film Yaklaşımları. *Pamukkale Üniversitesi Sosyal Bilimler Enstitüsü Dergisi*, (59), 423–444.

Ashley, J. (2015). Marshall McLuhan and Understanding Digital Reality, https://www.imaginativeuniversal.com/blog/2015/06/23/marshall-mcluhan-and-understanding-digital-reality, [Erişim Tarihi: 17.04.2025].

Aslan, Ö., & Öztekin, H. (2023). İnternet Tabanlı Televizyona Geçiş Sürecinde İzleme Pratiklerinde Yaşanan Değişim veslie Dönüşüm: Erciyes Üniversitesi Öğrencileri Örneği. *Erciyes Akademi*, *37*(4), 4. Advance online publication. DOI: 10.48070/erciyesakademi.1394517

Ayaz, F. (2024). Instagram'da Büyümek: Takipçi Etkinliklerine Yönelik Karma Bir Analiz. *Hitit Sosyal Bilimler Dergisi*, *17*(2), 303–330. DOI: 10.17218/hititsbd.1437098

Aydın, A. A. (2021). *Yeni Televizyon Teknolojileri ve Değişen Habercilik*. Kriter Yayınevi.

Aydoğan, A., & Başaran, F. (2012). Yeni Medyayı Alternatif Medya Bağlamında Anlamak. İn Ö. Özer (Ed.), *Alternatif Medya Alternatif Gazetecilik*. Literatürk.

Aytaş, M., & Yavuz, G. (2024). Netflix Çağı: Dijital Seyir Platformları Ve Belgesel Sinemada Anlatının Dönüşümü. *Idil: Journal Of Art & Language, 13*(113).

Bajura, M. ve Neumann, U. (1995). Dynamic Registration Correction in Augmented Reality Systems, *Virtual Reality Annual International Symposium*, March 11-15, 1995 Research Triangle Park, North Carolina, 189-196. DOI: 10.1109/VRAIS.1995.512495

Bal, H. (2016). *Nitel Araştırma Yöntem ve Teknikleri*. Sentez Yayınları.

Bal, M. (2009). *Narratology: Introduction to the theory of narrative* (3rd ed.). University of Toronto Press.

Barlas, M. (2024). Hiçlik Ve Ölümün Gölgesinde Bir Hayaleti Hatırlatmak: Kayıp Ev/ La Maison Manquante. *SDÜ ART-E Güzel Sanatlar Fakültesi Sanat Dergisi*. 17/33, 190-207. ISSN 1308-2698

Barnard, M. (2002). *Sanat Tasarım Ve Görsel Kültür* (Korkmaz, G., Trans.). Ütopya Publication.

Barnhart, A. (2022). *The primetimer guide to streaming TV*. Primetimer.

Başaran Alagöz, S., & Ekici, N. (2016). Tartışmalı bir konu olarak sanat pazarlama: Kavramsal bir değerlendirme. *Kahramanmaraş Sütçü İmam Üniversitesi İktisadi ve İdari Bilimler Fakültesi Dergisi*, *6*(2), 189–202.

Başer, A. (2010). Yeni İletişim Teknolojilerinin Evriminde Televizyonun Yer. İn F. Aydoğan & A. Akyüz (Eds.), *İkinci Medya Çağında İnternet*. Alfa Yayınları.

Basu, A. (2019). A brief chronology of Virtual Reality, Cornell University arxiv, 1, 2-18.https://doi.org//arXiv.1911.09605.DOI: 10.48550

Baudrillard, J. (1981). *Simülakrlar ve Simülasyon, 1981* (O. Adanır & A. K. Mollaoglu, Çev.). İstanbul: Yüksekistan Yayınları. (Orijinal eser yayımlanma yılı: 1981).

Baudrillard, J. (2003). *Simülakrlar ve Simülasyon*. Doğu Batı Yayınları.

Bauman, Z. (2000). *Liquid modernity*. Polity Press.

Bayer, H. ve Bulut Özek, M. (2021). Yeni Medya Eğitimi ve Önemi. *Journal of Communication Science Researchs*, *1*(2), 127–138. DOI: 10.7456/100102100/004

Bazin, A. (1967). *What is cinema?* University of California Press.

Bellantoni, P. (2005). *If it's purple, someone's gonna die: The power of color in visual storytelling*. Elsevier.

Benjamin, W. (1995). *Estetize Edilmiş Yaşam*. Der Publication.

Benjamin, W. (2001). *Tekniğin Olanaklarıyla Yeniden Üretildiği Çağda Sanat Yapıtı, Pasajlar* (Cemal, A., Trans.). Yapı Kredi Publication.

Bergson, H. (2017). *Yaratıcı Tekamül. (Çev. M. Tunç)*. Dergah Yayınları.

Bernaschina, D. (2023). Screen Video Art: A New Experimental Proposal for Creative Study. *Art Vision*, *29*(51), 170–177. DOI: 10.5152/ArtVis.2023.23202

Birchall, D. (2008). Online Documentary. In Austin, T., & de Jong, W. (Eds.), *Rethinking Documentary New Perspectives, New Practices* (pp. 278–283). Open University Press.

Blaustein, J. (2016) Jonathan Blaustein, Penelope Umbrico Röportajı, aphotoeditor, https://www.aphotoeditor.com/2016/04/27/penelope-umbrico-interview/

Bong, J. (Director). (2019). *Parasite* [Film]. CJ Entertainment.

Bordwell, D. (2008). *Poetics of Cinema*. Routledge.

Bordwell, D. (2008). *The Way Hollywood Tells It: Story and Style in Modern Movies*. University of California Press.

Bordwell, D., & Thompson, K. (2010). *Film art: An introduction* (10th ed.). McGraw-Hill.

Bordwell, D., & Thompson, K. (2013). *Film Art: An Introduction* (10th ed.). McGraw-Hill.

Bourdieu, P. (1984). *Distinction: A social critique of the judgement of taste*. Harvard University Press.

Brecht, B. (2011). *Epik Tiyatro*. Agora Kitaplığı.

Brown, W. (2013). *Supercinema: Film-philosophy for the digital age*. Berghahn Books. DOI: 10.3167/9780857459497

Brown, W. (2015). *Undoing the demos: Neoliberalism's stealth revolution*. Zone Books. DOI: 10.2307/j.ctt17kk9p8

Bruner, J. (1991). The Narrative Construction of Reality. *Critical Inquiry, 18*(1), 1–21. DOI: 10.1086/448619

Buck, A. M., & Plothe, T. (2019). Introduction. In Plothe, T., & Buck, A. M. (Eds.), *Netflix at the Nexus Content, Practice, and Production in the Age of Streaming Television* (pp. 1–9). Peter Lang Publishing.

Burnett, R. (2012). *İmgeler Nasıl Düşünür* (Pusar, G., Trans.). Metis Publication.

Busselle, R., & Van den Bulck, J. (2019). Cultivation theory, media, stories, processes, and reality. *Media effects*, 69-82.

Butler, J. G. (2018). *Television visual storytelling and screen culture*. Routledge. DOI: 10.4324/9781315181295

Çakın, Ö. ve Yaman, D. (2020). Fijital (Phygital) Pazarlama Uygulamaları Üzerine Bir İnceleme: Amazon Go Örneği. *Middle Black Sea Journal of Communication Studies, 5*(1), 1–10.

Çakın, Ö., & Bolat, N. (2018). *Sosyal Medya Reklamcılığı ve Dijital Medya Etkileşimleri. Dijital Kültür ve İletişim* (Çakmak, V., & Çavuş, S., Eds.). Literatürk Publication.

Çakın, Ö., & Yaman, D. (2022). *Dijital Oyunlar- Fijital Deneyim Alanları, Gamevertising Dijital Çağda Oyunlaştırma ve Reklam İçinde* (Nur Erdem, M., Ed.). Eğitim Yayınları.

Çalışkan, Ö. (2024). Yapay Zekâ ile Retorik Farklılıklarla Üretilen Türkçe ve İngilizce Film Diyalogları. *Yedi*, (Sanatta Dijitalizm Özel Sayısı), 357-374.

Canán, A. C. (2008). McLuhan, Flusser, and the Mediatic Approach to Mind. Flusser Studies 6.

Castells, M. (2010). *The rise of the network society* (2nd ed.). Wiley-Blackwell.

Castells, M. (2013). *Ağ Toplumunun Yükselişi*. İstanbul Bilgi University Publication.

Çelik, A. G., & Bolat, N. (2025). The Recreation of Space in Digital Broadcasting: Virtual Production. In Bolat, N. (Ed.), *Innovative Applications and Changing Framework in Digital Broadcasting* (pp. 513–539). IGI Global. DOI: 10.4018/979-8-3373-1957-5.ch019

Çelikel, A., & Akın, M. B. (2017). Yenilebilir filmler ve peynir teknolojisinde kullanımı. *Batman Üniversitesi Yaşam Bilimleri Dergisi, 7*(2/2), 50-58.

Cereci, S. (1992). *Belgesel Filmde Senaryo. İstanbul Üniversitesi Sosyal Bilimler Enstitüsü Radyo Televizyon Bölümü*. Doktora Tezi.

Cereci, S. (2022). *Belgesel Yapımı*. İksad Yayınevi.

Chatman, S. (2009). *Öykü ve Söylem: Filmde ve Kurmacada Anlatı Yapısı, Çev. Özgür Yaren, 1.b.* De Ki Basım Yayın.

Çiftçi, A., & Karakaş, Y. (2019). Dijitalleşen Zamanın İzdüşümünde: Kimliğin, Bedenin ve İletişimin Dönüşümü. AJIT-E. *Academic Journal of Information Technology, 10*(37), 7–30. DOI: 10.5824/1309-1581.2019.2.001.x

Çil, S., & Yurdigül, Y. (2018). Sosyal Paylaşım Siteleri ve Demokratikleşme İlişkisi: Facebook Örneği. *Uluslararası Kültürel ve Sosyal Araştırmalar Dergisi, 4*(2), Article 2.

Cincinatus, R. B., & Sheffet, M. (2016). “With Percentages the 100 Is Always in the Denominator”: From the Field to Pre-Service Teachers. *International Journal of Research in Education and Science*, *2*(1), 143–155. DOI: 10.21890/ijres.75271

C-Istudios. (2024). 360-Degree Video: Immersive Experiences and Virtual Reality, https://c-istudios.com/360-degree-video-immersive-experiences-and-virtual-reality, [Erişim tarihi: 11.04.2025].

Clarke, J. (2012). *Sinema Akımları: Sinema Dünyasını Değiştiren Filmler. Şahan Yatarkalkmaz (Çev.)*. Kalkedon.

Çoban, B. (2008). “Gözün iktidarı üzerine.” B. Çoban & Z. Özarslan (Ed.), *Panoptikon gözün iktidarı* içinde (pp. 111-137). Su Yayınları.

Çokokumuş, B. (2012). Dijital Ortamda Kültür ve Sanat. *İnternational Journal of New Trends in Arts, Sports &. Science Education, 1*(3).

Çolak, Y., & Zinderen, A. (2023). YouTube Haberciliği Üzerine Dijital Etnografik Bir Araştırma ve Gazeteci Görüşleri: Cüneyt Özdemir ve Nevşin Mengü YouTube Haber Kanalları Örneği. *Gümüşhane Üniversitesi İletişim Fakültesi Elektronik Dergisi, 11*(1), 1. Advance online publication. DOI: 10.19145/e-gifder.1194597

Comolli, J. L. (1974), Teknik ve İdeoloji, Çev. Yakup Barokas Çağdaş Sinema Dergisi, Sayı:1,İstanbul

Conlin, L., Billings, A. C., & Averset, L. (2016). Time-shifting vs. appointment viewing: The role of fear of missing out within TV consumption behaviours. *Comunicar: Media Education Research Journal, 29*(4), 151–164.

Coşkun, M. (2015). Sayılsal İletişim Teknolojileri Bağlamında Tv Haberciliği. İn S. Özel (Ed.), *Yeni Medya Çağında Televizyon*. Derin Yayınları.

Creswell, J. W. (2018). *Nitel araştırma yöntemleri: Beş yaklaşıma göre nitel araştırma ve araştırma deseni* (S. B. Demir & M. Bütün, Çev.). Ankara: Siyasal Kitabevi.

Cumhuriyet. (2025, April 24). *Türkiye'de en çok abonesi bulunan içerik platformları.* https://www.cumhuriyet.com.tr/bilim-teknoloji/turkiyede-en-cok-abonesi-bulunan-icerik-platformlari-2332056. [Erişim tarihi: 15.04.2025].

Da Costa, J. C. R. (2019). Binge-watching: A life course perspective. *Journal for Social Thought, 3*(1), 1–10.

Deadline. (2013). *Netflix: Binge-watching is the new normal for TV streamers.* https://deadline.com/2013/12/netflix-binge-watching-tv-streaming-study-651962/. [Erişim tarihi: 18.04.2025].

Debord, G. (1994). *The society of the spectacle* (Nicholson-Smith, D., Trans.). Zone Books. (Original work published 1967)

Deleuze, G. (1986). *Cinema I: The Movement Image, Sinema I: Hareket İmge* (Özdemir, S., Trans.). Norgunk Publication.

Demetriades, G. C. (2003). *Streaming Media building and ımplementing a complete streaming system*. Wiley.

Demirer, H. (2020). Sinemada renk kullanımı: Aşçı, hırsız, karısı ve aşığı filmi. *Social Mentality And Researcher Thinkers Journal*, *6*(31), 570–580. DOI: 10.31576/smryj.497

Demirkıran, C. (2010). Geleneksel Televizyonun Sanal Modeli Olarak Web Tv Ve Livestream Portalında Tv Yönetimi. *Akdeniz Üniversitesi İletişim Fakültesi Dergisi, 14*, Article 14.

Digitalreality, (2022). *Common uses for AI in virtual reality today,* https://digitalreality.ieee.org/publications/ai-in-virtual-reality, [Erişim tarihi: 17.01.2025].

Dijk, J. V. (2016). *Ağ Toplumu*. Kafka Publication.

Diken Yücel, D. (2021, Kasım 11–12). Seyir kültüründe dijital dönüşüm: Sinema salonlarından dijital platform üyeliklerine. *3rd International Cinema and Youth Congress*, Eskişehir, Türkiye.

Dimaggio, P., & Useem, M. (1978). Social Class and Arts Consumption. *Theory and Society*, *5*(2), 141–161. DOI: 10.1007/BF01702159

Doane, M. A. (2002). *The Emergence of Cinematic Time*. Harvard University Press.

Dolgun, U. (2008). *Şeffaf hapishane yahut gözetim toplumu*. Ötüken Yayınları.

Dong, H. (Director). (2021). *Squid Game*. [TV series]. Netflix.

Dove, M., Balasubramanian, A., & Narayanan, B. G. (2020). Transparency as a way of attaining quality, safety and optimal food purchases.

Eisenstein, S. (1977). *Film Form-Essays in Film Theory* (Leda, J., Ed. & Tranş.). Harcourt Brace&Company.

Ekim, B. (2022). The Museum Of Me: A Visual Presentation Of Your Online Social Life On Facebook. Visualist 2012, Uluslar Arası Görsel Kültür Kongresi, İletişimde, Sanatta ve Tasarımda Yeni Yaklaşımlar. In Zeybek, I., & Yengin, D. (Eds.), *Diijitalleşme* (pp. 273–280).

Ellul, J. (2004). *Sözün Düşüşü* (Arslan, H., Trans.). Paradigma Publication.

Elsaesser, T. (2015). *Digital Cinema: The Transformation of the Film Industry*. Routledge.

Ensslin, A., & Goggin, J. (Eds.). (2020). *Digital fiction and the unnatural: Transmedial narrative theory, method, and analysis*. The Ohio State University Press.

Ercansungur, D. O. & Çetin, E. (2023). Popüler Kültür ve Dijitalleşme: Yeni Medya Ekosisteminde Yakınsama ve Kitle Üzerine Sosyolojik Bir Analiz . İnsan ve Toplum Bilimleri Araştırmaları Dergisi, 12 (2), 565-584 . https://doi.org/ DOI: 10.15869/itobiad.1215377

Erdem, M. N. (2017). Dijital Çağda Kültür, Yeni Tüketici Kapitalizmi ve Reklam Anlatısı. *1st International Conference on New Trends in Communication*. İstanbul: Ticaret Universty.

Erdem, S. (2022). Görsel İletişim Tasarımında Dijital Dönüşümü Nft Üzerinden Tartışmak.

Erdem, S. (2024). Çapraz medya stratejileri: Markaların çoklu platformlarda etkileşim sanatı. In M. Nur ERDEM & Cansu MAYADAĞLI (Eds.), *Marka İletişimi Tüm Boyutlarıyla Marka ve Tüketici Etkileşimi* (pp. 131-142). Palet Yayınları.

Erdem, S. (2023). Görsel Okuryazarlığının İşaret Ettiğine Dair: Görsel İletişim Çağında Bilinçli Medya Tüketimi. In Bayraktar, R. (Ed.), *Medya Okuryazarlığı Analizleri*. Eğitim Yayınevi.

Erdoğan, İ. (2012). *Pozitivist metodoloji ve ötesi*. Erk Yayınları.

Erkılıç, H., & Dönmez, S. C. (2016). Sanal Gerçeklik Anlatısının İzini Sürmek: Trinity VR ve Selyatağı VR Örnekleri. *SineFilozofi*318-344. DOI: 10.31122/sinefilozofi.674107

Erkılıç, H., & Toprak, A. G. (2012). Belgesel Sinemanin Alternatif Dağitim Ve Gösterim Olanaği Olarak İnternet. The Turkish Online Journal of Design, Art and Communication – TOJDAC, 2(2), 10-16.

Erzurum, F. (2019). Transmedya Anlatım Sinema Filmleriyle Mi Başladı? *Dördüncü Kuvvet Uluslararası Hakemli Dergi*, *2*(1), 32–42. DOI: 10.33464/dorduncukuvvet.542962

Esslin, M. (2001). *Televizyon Çağı*. Pınar Yayınları.

Ettema, J. S. (2005). Crafting cultural resonance: Imaginative power in everyday journalism. *Journalism*, *6*(2), 131–152. DOI: 10.1177/1464884905051005

Ferhat, S. (2016). Dijital Dünyanın Gerçekliği, Gerçek Dünyanın Sanallığı Bir Dijital Medya Ürünü Olarak Sanal Gerçeklik. *TRT Akademi, 1*(2), 724–746.

Field, S. (2005). *Screenplay: The foundations of screenwriting* (Rev. ed.). Delta.

FILMD. (2024). Emerging Filmmaking Technologies: Shaping the Future of Cinema, https://www.filmd.co.uk/articles/emerging-filmmaking-technologies-shaping-the-future-of-cinema, [Erişim tarihi: 15.04.2025].

Filmmaker, (2016). *The Best VR of 2016: Arnaud Colinart and Amaury La Burthe on Notes on Blindness: Into Darkness,* https://filmmakermagazine.com/100779-the-best-vr-of-2016-arnaud-colinart-and-amaury-la-burthe-on-notes-on-blindness-into-darkness, [Erişim tarihi: 09.04.2025].

Fitzek, F. H. (2006, June). The medium is the message. In *2006* [). IEEE.]. *IEEE International Conference on Communications : [proceedings]. IEEE International Conference on Communications, 11,* 5016–5021.

Flayelle, M., Canale, N., Vögele, C., Karila, L., Maurage, P., & Billieux, J. (2019). Assessing binge-watching behaviours: Development and validation of the "Watching TV Series Motives" and "Binge-Watching Engagement and Symptoms" questionnaires. *Computers in Human Behavior, 90,* 26–36. DOI: 10.1016/j.chb.2018.08.022

Foss, B. (2012). *Sinema ve Televizyonda Anlatım Teknikleri ve Dramaturji*. Hayalperest Kitap.

Foucault, M. (1977). *Discipline and punish: The birth of the prison.* Pantheon Books.

Franck Atalay, O. (2002). *Düşünce İçin Mimarlık: Sanallığın Gerçekliği, Mimarlık ve Sanallık. Çağdaş Mimarlık Sorunları Dizisi.* Boyut Yayın Grubu Publication.

Frank, A. W. (1995). *The Wounded Storyteller: Body, Illness, and Ethics.* University of Chicago Press. DOI: 10.7208/chicago/9780226260037.001.0001

Freund, G. (2008). *Fotoğraf ve Toplum.* Sel Yayıncılık.

Future of TV Distribution. (2024). A report prepared by the University of Exeter with the University of Leeds, MTM, 3 Reasons and Real Wireless. https://assets.publishing.service.gov.uk/media/672cafe262831268b0b1a2f4/Future_of_TV_Distribution_FINAL__7_Nov_2024_-accessible.pdf. [Erişim tarihi: 20.04.2025].

Gage, J. (1999). *Color and meaning: Art, science, and symbolism.* University of California Press.

Gençcelep, B., & Özkemahlı, K. K. (2023). "Gerçek Olandan Simülasyona: Fotoğraf Stüdyosu Eğitiminde "Set.A.Light 3D" Örneği", International Academic Social Resources Journal, (e-ISSN: 2636-7637), Vol:8, Issue:51; pp:3062-3071. DOI: 10.29228/ASRJOURNAL.70969

Genel, M. (2018). *Online Medya Mecrasında Video İçerik Üretimi ve Reklam Geliri Elde Etme: "Youtube Örneği.* 491–500.

Genette, G. (1980). *Narrative discourse: An essay in method.* Cornell University Press.

Geraghty, C., & Lusted, D. (1998). *The Television Studies Book.* Arnold Publication.

Gerbner, G., & Morgan, M. (2002). *Against the mainstream: The selected works of George Gerbner.* No Title.

Giddens, A. (2010). *Modernite ve bireysel-kimlik* (Tatlıcan, Ü., Trans.). Say Yayınları.

Giorgi, S. (2017). The mind and heart of resonance: The role of cognition and emotions in frame effectiveness. *Journal of Management Studies, 54*(5), 711–738. DOI: 10.1111/joms.12278

Girgin, F. (2018). *Çağdaş Sanat ve Yeniden Üretim Alıntı, Öykünme, Kolaj, Taklit.* Hayalperest.

Granow, V. C., Reinecke, L., & Ziegele, M. (2018). Binge-watching and psychological well-being: Media use between lack of control and perceived autonomy. *Communication Research Reports, 35*(5), 392–401. DOI: 10.1080/08824096.2018.1525347

Greenaway, P. (Director). (1989). *The Cook, The Thief, His Wife & Her Lover* [Film]. Allarts Production.

Grotticelli, M. (2021). *BT Sport's Live VR 360 Coverage Of Premier League Brings Fans Closer To The Action,* https://www.thebroadcastbridge.com/content/entry/17429/bt-sports-live-vr-360-coverage-of-premier-league-brings-fans-closer-to-the, [Erişim tarihi: 17.04.2025].

Gücüyener, M. (2001). Panoptikonik gözetimden synoptisizme gözetim toplumu. [Unpublished master's thesis]. Kocatepe Üniversitesi.

Gültekin, G. (2019). Cınematıc Use Of Tıme And Space Through Example Fılms, *V Human and Civilization Congress From Past To Future Full Text Book,* Antalya.

Gündeş Öngören, S. (1991). *Belgesel Filmin Yapısal Gelişimi ve Türkiye'ye Yansıması.* Der Yayıncılık.

Gündeş, S. (1998). *Belgesel Filmin Yapısal Gelişimi Türkiye'ye Yansıması.* Alfa Basım Yayım Dağıtım.

Güney, E. (2014). Dijital Görsel Kültür ve Yeni Medya Ekseninde Sanatın Değişen Rolü. *PhD Thesis*. İzmir: Ege Universty.

Güngör, N. (2018). *İletişim Kuramlar-Yaklaşımlar*. Siyasal Kitabevi.

Gunning, T. (2004). The Cinema of Attractions: Early Film, Its Spectator, and the Avant-Garde. *Wide Angle*, *3*(4), 63–70.

Guo, W. (2022)... *The Comparative Analysis of the Utilization of Traditional Broadcast TV in the UK and China, 65–69*. Advance online publication. DOI: 10.2991/assehr.k.220502.015

Güven, S. K. (2011). "Gözetimin toplumsal meşruiyeti." H. Köse (Ed.), *Medya mahrem* (pp.173-199). Ayrıntı Yayınları.

Ha, A. (2019). *'The Key,' a VR story about dreams and refugees, wins Storyscapes prize at Tribeca*, https://techcrunch.com/2019/05/03/the-key-tribeca-immersive-storyscapes, [Erişim tarihi: 20.04.2025].

Hacking, J. (2015). *Photography The Whole Story*. Thames&Hudson.

Ha, L. (2021). The changing audience. In Mahoney, M., & Tang, T. (Eds.), *Handbook of media management and business* (pp. 21–38). Rowman & Littlefield.

Harvey, D. (2011). *Umut mekânları* (Gambetti, Z., Trans.). Metis Yayınları.

Hasançebi Ayaz, E., & Çelik Yılmaz, N. (2025). Dijital Yerliler ve Dijital Göçmenlerin Sosyal Medyadaki Haberlere Karşı Güven Algıları. İn E. Küçük Durur, A. Yılmaz, & Z. Cinisli (Eds.), *İletişim Bilimlerinde Yenilikçi Araştırmalar I* (pp. 59–74). Gazi Kitabevi.

Hirokazu, K. (Director). (2018). *Shoplifters* [Film]. Aoi Pro, Inc.

Hughes, P. (1990). Today's television, tomorrow's world understanding television. In Goodwin, A., & Whannel, G. (Eds.), *Understanding Television* (pp. 170–187). Routledge.

Hutcheon, L. (2006). *A theory of adaptation*. Routledge. DOI: 10.4324/9780203957721

Interaction Design Foundation - IxDF. (2023). *What is Presence in Virtual Reality (VR)?* Interaction Design Foundation - IxDF. https://www.interaction-design.org/literature/topics/presence

İspir, N., & Kaya, Z. (2011). Sinemada Postmodern Arayışlar. *Atatürk İletişim Dergisi*, (2), 81–99.

Jenkins, H. (2006). *Convergence Culture: Where Old and New Media Collide*. New York University Press.

Jenkins, H. (2016). *Cesur Yeni Medya: Teknolojiler ve Hayran Kültürü. (N. Yeğengil, Dü.)*. İletişim.

Jenkins, H. (2016). *Cesur Yeni Medya" Teknolojiler ve Hayran Kültürü* (Yeğengil, N., Trans.). İletişim Publication.

Jenkins, H., & Thorburn, D. (2003). *The New Media Reader*. MIT Press.

Jenner, M. (2017). Binge-watching: Video-on-demand, quality TV and mainstreaming fandom. *International Journal of Cultural Studies, 20*(3), 304–320. DOI: 10.1177/1367877915606485

Jenner, M. (2021). Introduction. In Jenner, M. (Ed.), *Binge-Watching and Contemporary Television Research* (pp. 1–20). Edinburgh University Press.

Joomuck, N. (2024). Exploring AI in Movies: Present, Future, and Implications. [IJSR]. *International Journal of Scientific Research, 13*, 1066–1069.

Kalaman, S. (2019). Yeni Medya ve Dijital Gözetim: Türkiye'deki Sosyal Medya Kullanıcıları Üzerine Bir Araştırma, Yönetim ve Ekonomi, Yıl:2019, Cilt:26. *Sayı, 2*, 575–594.

Kandemir, C. (2013). *IPTV Yayıncılığının Sorunları ve Geleceği: Sektörün Aktörleriyle Derinlemesine Mülakatlar*. Derin Yayınları.

Kaplan, A. M., & Haenlein, M. (2016). Dünyanın Bütün Kullanıcıları Birleşin- Sosyal Medya Zorlukları ve Sağladığı FırsatlarDünyanın Bütün Kullanıcıları Birleşin-Sosyal Medya Zorlukları ve Sağladığı Fırsatlar (D. Işık, Trans.). İn H. Hülür & C. Yaşın (Eds.), *Yeni Medya Kullanıcının Yükselişi* (2016th ed.). Ütopya Yayınevi.

Kara, K. (2022). *Cumhurbaşkanlığı İletişim Başkanı Altun: Türkiye gibi bir güç, "metaverse"nin inşa sürecinin parçası olmalı*. Anadolu Ajansı (AA) Kültür Haberleri. https://www.aa.com.tr/tr/gundem/cumhurbaskanligi-iletisim-baskani-altun-turkiye-gibi-bir-guc-metaversenin-insa-surecinin-parcasi-olmali. [Erişim tarihi: 19.05.2025].

Kara, O. Ç. (2019). Vilém Flusser'in Soyutlama Modeli ve Sinematografik İmge. *SineFilozofi*221-240. DOI: 10.31122/sinefilozofi.514984

Karadoğan, S. (2001). *Belgesel Film Ve Propaganda. Marmara Üniversitesi İletişim Fakültesi Radyo Tv Ve Sinema Bölümü*. Araştırma Projesi.

Karaduman S., & Ünlü Karataş N. (2022). Dijital Yayıncılık ve İzleme Pratikleri: "Gain" Üzerine Bir Analiz. *İstanbul Ticaret Üniversitesi Sosyal Bilimler Dergisi*, 21(45), 982-1007. DOI: 10.46928/iticusbe.1001254

Kaynarca, F. (2024). Dijitalleşme İle Değişen Belgesel Filmlerdeki Anlatı Estetiğinin Dijital Yayın Platformları Üzerinden İncelenmesi. *Journal Of Media & Cultural Studies*, *6*(2).

Kazan, F. (2023). İlk Yerli Dijital İnsan Örneği: Aypera. *RumeliDe Dil ve Edebiyat Arastirmalari Dergisi*, (34), 819–838. DOI: 10.29000/rumelide.1316233

Kılıç, L. (2000). *Görüntü Estetiği*. İnkılap Publication.

Kılıç, L. (2012). *Fotoğraf ve Sinemanın Toplumsal Tarihi*. Dost Kitapevi.

Kinder, M. (1999). Rewiring the spectator: The reception of narrative in digital cinema. In Sturken, M., Thomas, D., & Ball-Rokeach, S. (Eds.), *Technological visions: The hopes and fears that shape new technologies* (pp. 183–204). Temple University Press.

King, C. (2017). In Farthing, S. (Ed.), *Dijital Sanat, Sanatın Tüm Öyküsü* (pp. 548–549).

Kırık, A. M. (2015). İnternet Teknolojisi ve Sayısal yayıncılık Bileşkesinde Gelişen IPTV'nin Günümüzdeki Durumu. İn S. Özel (Ed.), *Yeni Medya Çağında Televizyon* (pp. 129–169). Derin Yayınları.

Kırık, A. M. (2010). *Etkileşimli Televizyon*. Anahtar Kitaplar Yayınevi.

Kırık, A. M. (2013). Sinemada renk öğesinin kullanımı: Renk ve anlatım ilişkisi. *21. Yüzyılda Eğitim ve Toplum*, *2*(6), 71–83.

Kırık, A. M., & Karakuş, M. K. (2013). Sosyal Medya ve İnternet Teknolojisi İle Yöndeşen Televizyon Yayıncılğı: Sosyal TV. *AJIT-E: Academic Journal of Information Technology*, *4*(11), 61–73. DOI: 10.5824/1309-1581.2013.2.003.x

Kizirian, S. (2025). 1927: *October* (Sergei Eisenstein) https://www.sensesofcinema.com/2017/soviet-cinema/1927-october-sergei- eisenstein/

Kobylinski, P., & Pochwatko, G. (2020). Detection of strong ´ and weak moments in cinematic virtual reality narration with the use of 3D eye tracking. *Proceedings of the 13th International Conference on Advances in Computer–Human Interaction*, pp. 280–284.

Koca, Ş. E. (2019). Sinemada anlam yaratma sürecinde rengin metaforik kullanımı. *Sanat Tasarım Dergisi*, *23*, 223–239.

Koçyiğit, A., & Darı, A. (2023). *"Yapay Zekâ İletişiminde ChatGPT: İnsanlaşan Dijitalleşmenin Geleceği", Stratejik ve Sosyal Araştırmalar Dergisi, C.7.* Temmuz.

Kotan, S. (2022). *Korku filmlerinin ideolojisi ben'in içindeki öteki.* Eğitim Yayinevi.

Kuipers, G. (2012). *Medium is the Message.* The Wiley-Blackwell Encyclopedia of Globalization. DOI: 10.1002/9780470670590.wbeog381

Kuleshov, L. (1971). *Kuleshov on Film: Writings by Lev Kuleshov.* University of California Press.

Kümbetoğlu, B. (2020). *Niteliksel Araştırmalarda Analiz.* Bağlam Yayıncılık. https://www.baglam.com/home/book/niteliksel-arastirmalarda-analiz

Kuyucu, M. (2019, August 11). *The Differences between Television Watching and Digital Media Using Habits of Y and Z Generations.*

Kuyucu, M. (2021). Geleneksel Televizyon Mecrasının Yaşadığı Dijital Dönüşüm ve Bunun Mecraya Olan Etkileri. *Avrasya Sosyal ve Ekonomi Araştırmaları Dergisi, 8*(3), Article 3.

Kuyucu, M., & Karahisar, T. (2013). *Yeni İletişim Teknolojileri ve Yeni Medya.* Zinde Reklam Yayıncılık.

Landow, G. P. (1997). *Hypertext 2.0: The Convergence of Contemporary Critical Theory and Technology, Johns Hopkins University Press, Manovich, L. (2001). The Language of New Media.* MIT Press.

LaRose, R., & Atkin, D. (1991). Attributes of movie distribution channels and consumer choice. *Journal of Media Economics, 4*(1), 3–17. Advance online publication. DOI: 10.1080/08997769109358200

Laughey, D. (2010). *Medya çalışmaları teoriler ve yaklaşımlar* (Toprak, A., Trans.). Kalkedon Yayınları.

Lauristin, M., & Vihalemm, P. (2014). *Reading in Changing Society. Studies in Reading and Book Culture.* University of Tartu press. DOI: 10.26530/OAPEN_496790

Lefebvre, H. (1991). *The production of space* (Nicholson-Smith, D., Trans.). Blackwell.

Lehman, K., & Wickham, M. (2014). Marketing Orientation and Activities in The Arts Marketing Context: Introducing A Visual Artists' Marketing Trajectory Model. *Journal of Marketing Management, 30*(7–8), 664–696. DOI: 10.1080/0267257X.2013.838987

Levy, D. A. L. (1999). *Europe's Digital Revolution: Broadcasting Regulation, the EU, and the Nation State*. Psychology Press. DOI: 10.4324/9780203278284

Lewis, E. (2018). *İzmler Fotoğrafı Anlamak*. Hayalperest.

Li, B., & Yin, H. (2007). Peer-to-peer live video streaming on the internet: Issues, existing approaches, and challenges [Peer-to-Peer Multimedia Streaming]. *IEEE Communications Magazine*, *45*(6), 94–99. DOI: 10.1109/MCOM.2007.374425

Limon, Ö. (2025). *Türkiye'nin ilk sanal gerçeklik müzesi "Müzeverse", tarihi alışılmışın dışında keşfetme imkanı sunuyor.* Anadolu Ajansı (AA) Kültür Haberleri. https://www.aa.com.tr/tr/kultur/turkiyenin-ilk-sanal-gerceklik-muzesi-muzeverse-tarihi-alisilmisin-disinda-kesfetme-imkani-sunuyor. [Erişim tarihi: 21.05.2025].

Li, Z., & Sun, Y. (2019). OTT Industry Development in China. *Journalism and Mass Communication*, *9*(4), 164–175. DOI: 10.17265/2160-6579/2019.04.002

Lobato, R. (2019). *Netflix Nations: The geography of digital distribution*. New York University Press. DOI: 10.18574/nyu/9781479895120.001.0001

Lotz, A. D. (2014). *The television will be revolutionized*. New York University Press.

Lotz, A. D. (2014). *The Television Will be Revolutionized*. NYU Press.

Lotz, A. D. (2017). *Portals: A treatise on ınternet-distributed television*. Michigan Publishing. DOI: 10.3998/mpub.9699689

Lu, X. (2019). *"Clouds over Sidra": Empathy and Immersion in VR Storytelling*, https://medium.com/@irenelu728/clouds-over-sidra-empathy-and-immersion-in-vr-storytelling, [Erişim tarihi: 06.02.2025].

Lyon, D. (1997). *Elektronik göz gözetim toplumunun yükselişi* (Hattatoğlu, D., Trans.). Sarmal Yayınevi.

Maas, J. (2025, January 21). Netflix adds nearly 19 million subscribers to end 2024 with more than 300 million globally. Variety. https://variety.com/2025/tv/news/netflix-subscribers-300-million-q4-2024-1236280419/. [Erişim tarihi: 22.04.2025].

Manovich, L. (2001). Reality Media, https://manovich.net/index.php/projects/reality-media, [Erişim Tarihi: 13.04.2025].

Manovich, L. (2001). *The Language of New Media*. MIT Press.

Marr, B. (2015). *Big Data: Using SMART big data; analytics and metrics to make better decisions and improve performance*. Wiley.

Marshall, L. (Summer, 2014). "The Budapest Hotel: Wes Anderson's Fabulous Fancy". *Queen's Quarterly*. Vol. 121. Kingston, Ontario. pp. 242–251.

Marx, G. T. (2002). What's new about the "new surveillance?" Classifying for change and continuity. *Surveillance & Society*, *1*(1), 9–29. www.surveillance-and-society.org. DOI: 10.24908/ss.v1i1.3391

McCracken, G. (2015). *TV got better: How we got from bingeing to feasting*. https://grant27.medium.com/tv-got-better-how-we-got-from-bingeing-to-feasting-782a67ee0a1. [Erişim tarihi: 24.04.2025].

McKee, R. (1997). *Story: Substance, structure, style and the principles of screenwriting*. HarperCollins.

McKee, R. (2011). *Öykü, (Çev. Nuray Yılmaz, Ertan Yılmaz ve Fatih Kınalı)*. Plato.

McLane, B. A. (2022). *A new history of documentary film*. Bloomsbury Publishing USA.

McLuhan, M. (2005). The medium is the message.

McLuhan, M. (1964). *Understanding Media: The Extensions of Man*. McGraw-Hill.

Meehan, E. R. (1986). Conceptualizing culture as commodity: The problem of television. *Critical Studies in Mass Communication*, *3*(4), 448–457. DOI: 10.1080/15295038609366675

Metz, C. (1974). *Film Language: A Semiotics of the Cinema*. University of Chicago Press. DOI: 10.1515/9783110816044

Mikos, L. (2016). Digital Media Platforms and the Use of TV Content: Binge Watching and Video-Ondemand in Germany. *Media and Communication*, *4*(3), 154–161. DOI: 10.17645/mac.v4i3.542

Mitchell, W. J. T. (2005). *What Do Pictures Want?* University of Chicago Press. DOI: 10.7208/chicago/9780226245904.001.0001

Mohamed, E., Abbasher, M. H., & Mohamed, M. A. (2024). Navigating the Ethical Horizon: Artificial Intelligence-Generated Content and the Imperative for Transparency and Ethics.

Monaco, J. (2009). *How to read afFilm: Movies, media, and beyond* (4th ed.). University Press.

Morgan, M. (2011).George Gerbner: A Critical Introduction to Media and Communication Theory.

Mulvey, L. (1975). Visual Pleasure and Narrative Cinema. *Screen, 16*(3), 6–18. DOI: 10.1093/screen/16.3.6

Mulvey, L. (1989). *Visual and other pleasures*. Palgrave Macmillan. DOI: 10.1007/978-1-349-19798-9

Mulvey, L. (2006). *Death 24x a Second: Stillness and the Moving Image*. Reaktion Books.

Murray, J. H. (2016). *Not a Film and Not an Empathy Machine*, https://immerse.news/not-a-film-and-not-an-empathy-machine, [Erişim tarihi: 19.04.2025].

Mutlu, E. (1991). *Televizyonu Anlamak*. Gündoğan Publication.

Mutlu, E. (2005). *Globalleşme Popüler Kültür ve Medya*. Ütopya Yayınevi.

Nalven, J. ve Jarvis J.D. (2005). Going Digital: The Practice and Vision of Digital Artists. *Thomson Course Technology*. USA.

Napoli, P. M. (2011). *Audience Evolution: New Technologies and the Transformation of Media Audiences*. Columbia University Press.

Netflix. (2017). *2017 on Netflix – A year in bingeing*. https://media.netflix.com/en/press-releases/2017-on-netflix-a-year-in-bingeing. [Erişim tarihi: 18.04.2025].

Nick, C. (2012). *Media, Society, World: Social Theory and Digital Media Practice* (Vol. 30). Polity.

O'Driscoll, G. (2008). *Next generation IPTV services and technologies*. Wiley.

Ögenler, Y. (2025, Mayıs 3). Netflix, OpenAI ile iş birliğine mi gidiyor? *Webtekno*. https://www.webtekno.com/netflix-openai-arama-h157844.html. [Erişim tarihi: 26.04.2025].

Ong, W. (2007). *Sözlü Kültür, Yazılı Kültür, Sözün Teknolojileşmesi* (Postacıoğlu Banon, S., Trans.). Metis Publication.

Oscars (2017). The academy's board of governors awards an Oscar to alejandro g. Iñárrıtu's "carne y arena" vırtual realıty ınstallatıon, https://www.oscars.org/news/academys-board-governors-awards-oscarr-alejandro-g-inarritus-carne-y-arena-virtual-reality, [Erişim tarihi: 26.04.2025].

Oxford English Dictionary. (2014). Oxford University Press.

Özarslan, Z. (2008). "Gözün iktidarı: Elektronik gözetim sistemleri." *Panoptikon gözün İktidarı*.

Özdemir, E. (2005). *Anlatıbilim: Kuramsal bir giriş*. Phoenix Yayınevi.

Özdemir, M., & Şeyma, Y. (2020, January). Medya ve Gerçeklik İlişkisini Küresel Biyopolitik Dijital Çağda Yeniden Düşünmek. *Erciyes İletişim Dergisi*, *7*(1), 507–532. DOI: 10.17680/erciyesiletisim.541652

Özdil, N. M. (2021). Dijital Platformlarda Belgesel Film Yapim ve Gösterim Pratikleri: Blutv Örneği. İstanbul Üniversitesi Sosyal Bilimler Enstitüsü Radyo, Televizyon Ve Sinema Anabilim Dali, (Yüksek lisan tezi), İstanbul, Türkiye.

Özdoğlar, E. (2015). Fantezi ve Korku Kurguda Kurgusal Karakter ve Kurgusal Mekân İlişkisi Yayınlanamış sanata yeterlik tezi.

Özdoyran, G. (2019). Yeni Medya'ya Yönelik 'Yeni' Yaklaşımlar: Yeni Medya ve Post-Yapısalcılık/Postmodernizm. *Etkileşim.*, *4*(4), 186–218. DOI: 10.32739/etkilesim.2019.4.70

Özel Sağlamtimur, Z. (2017). Yeni Medya Sanatı ve Fotoğraf. *Sanat Ve Tasarım Dergisi*, *7*(2), 82–100. DOI: 10.20488/www-std-anadolu-edu-tr.393543

Özel, S. (2015). Televizyon Ekseninde İzleyici ve İnternet Üzerindeki Video Hizmetlerinin İzleyiciye Etkisi. İn S. Özel (Ed.), *Yeni Medya Çağında Televizyon* (pp. 1–31). Derin Yayınları.

Özgür, F., & Bigiç, N. Ç. (2017). *Belgesel/Kısa Film/Video Sanatı*. Doruk Yayımcılık.

Özkemahlı, K. K., & Er, M. H. (2024). "Yapay Zekâ Destekli Fotoğrafik Yaklaşım: Portre ve Mekân Temelinde Mustafa Kemal Atatürk Fotoğraflarının Yeniden Üretilmesi". (Ed.), Yağmur, Ö. & Şengünalp, C., Disiplinlerarası Sanat Pratikleri 2. Ankara: Detay Yayıncılık.

Özkemahlı, K. K., & Gençcelep, B. (2024). Konstrüktivizmin Fotoğraf Sanatına Etkisi. In Bulat, M., & Daşkesen, H. (Eds.), *Sanatsal Araştırmalar II.* Detay Yayıncılık.

Özkoçak, Y. (2014). *Ses'li Sinema*. Der Yayınları.

Özsoy, A. (2011). *Televizyon ve İzleyici: Türkiye'de Dönüşen Televizyon Kültürü ve İzleyici*. Ütopya Yayınevi.

Öztopçu, H. A. (1989) Kurgusal Mekân, https://www.avnioztopcu.com/avni/kurgusalmekan.htm (Erişim, 20.04.2025)

Öztürk, G., & Dayanç Kıyat, G. B., Yılmaz Altuntaş, E. (2024) Halkla İlişkiler ve Reklam Perspektifinden Yaratıcılığın Geleceği, Turkish Review of Communication Studies, Year: 2024, Issue: Special Issue, pp. e148-171 • ISSN: 2630-6220 • DOI: DOI: 10.17829/turcom.1360999

Özturk, A. (2024). Sinemada Çok Katmanlı Anlatı: Inception (Başlangıç) Filmi Üzerinden Bir İnceleme. *Ahi Evran Üniversitesi Sosyal Bilimler Enstitüsü Dergisi, 10*(3), 579–598. DOI: 10.31592/aeusbed.1505072

Panda, S., & Pandey, S. C. (2017). Binge-watching and college students: Motivations and outcomes. *Young Consumers, 18*(4), 425–438. DOI: 10.1108/YC-07-2017-00707

Papathanasopoulos, A., & Varoutas, D. (2022). On the competition between Video OTTs Platforms vs Traditional TV: A Niche Case Study in Greece. *31st European Regional ITS Conference, Gothenburg 2022: Reining in Digital Platforms? Challenging Monopolies, Promoting Competition and Developing Regulatory Regimes*, Article 265664. https://ideas.repec.org//p/zbw/itse22/265664.html

Parikka, J. (2017). *Medya Arkeolojisi Nedir?* (Kılıç, E., Trans.). Koç Üniversitesi Yayınları.

Parsa, S. (1994). *Televizyon Estetiği*. Ege Üniversitesi İletişim Fakültesi Publication.

Pastoureau, M. (2017). *Renklerin tarihi* (Toprak, A., Trans.). Yapı Kredi Yayınları.

Pavlik, J. (1998). *New Media Technology: Cultural and Commercial Perspectives* (2nd ed.). Ally and Bacon.

Pavlik, J. V. (2020). Drones, Augmented Reality and Virtual Reality Journalism: MappingTheir Role in Immersive News Content [Erişim Tarihi: 28.03.2025]. *Media and Communication*, *8*(3), 137–146. DOI: 10.17645/mac.v8i3.3031

Pisters, P. (2012). *The neuro-image: A Deleuzian film-philosophy of digital screen culture*. Stanford University Press.

Pittman, M., & Sheehan, K. (2015). Sprinting a media marathon: Uses and gratifications of binge-watching television through Netflix. *First Monday*, •••, 20. DOI: 10.5210/fm.v20i10.6138

Playful, (2021). *Creating Empathy in Notes on Blindness VR,* https://www.playfulrealities.com/2021/01/creating-empathy-in-notes-on-blindness.html, [Erişim tarihi: 03.04.2025].

Polat, A. (2022). Nitel Araştırmalarda Yarı-Yapılandırılmış Görüşme Soruları: Soru Form ve Türleri, Nitelikler ve Sıralama. *Anadolu Üniversitesi Sosyal Bilimler Dergisi, 22*(Özel Sayı 2). *Article Özel Sayı*, *2*(Özel Sayı 2), 161–182. Advance online publication. DOI: 10.18037/ausbd.1227335

Postman, N. (2020). *Televizyon Öldüren Eğlence* (Akınhay, O., Trans.). Ayrıntı Yayınları.

Punch, K. F. (2011). *Sosyal Araştırmalara Giriş Nicel ve Nitel Yaklaşımlar (Z. Etöz* (Bayrak, D., Arslan, H. B., & Akyüz, Z. Eds. & Trans.). Siyasal Kitabevi.

Randall, W. (1999). *Bizi Biz Yapan Hikâyeler. Çev. Şen Süer Kaya.* Ayrıntı Yayınları.

Ricoeur, P. (1984). *Time and Narrative* (Vol. 1). University of Chicago Press.

Ritchin, F. (2012). *Fotoğraftan Sonra.* Espas Yayınları.

Robertson, A. (2016). *The UN wants to see how far VR empathy will go*, https://www.theverge.com/2016/9/19/12933874/unvr-clouds-over-sidra-film-app-launch, [Erişim tarihi: 19.05.2025].

Ross, M., & Munt, A. (2018). Cinematic virtual reality: Towards the spatialized screenplay. *Journal of Screenwriting, 9*(2), 191–209. DOI: 10.1386/josc.9.2.191_1

Roth, P. C. H. (2016). *Experiencing Interactive Storytelling.* [PhD-Thesis – Research external, graduation internal, Vrije Universiteit Amsterdam].

RTÜK. (2022). *Gençlerin Medya Kullanımı ve Dijital Okuryazarlık Araştırması 2022* (p. 216). Radyo ve Televizyon Üst Kurulu. https://www.rtuk.gov.tr/genclerin-medya-kullanimi-ve-dijital-okuryazarlik-arastirmasi-2022/4402

Rubenking, B., & Bracken, C. C. (2018). Binge-watching: A suspenseful, emotional habit. *Communication Research Reports, 35*(5), 381–391. DOI: 10.1080/08824096.2018.1525346

Ryan, M.-L. (2006). *Avatars of Story.* University of Minnesota Press.

Sağlamtimur, Z. Ö. (2010). Dijital Kültür. *Anadolu Üniversitesi Sosyal Bilimler Dergisi, 10*(3), 213–238.

Sarı, Ü. (2021). Televizyon Yayıncılığının Dijital Dönüşümü. İn H. S. Türkoğlu (Ed.), *Dijitalleşme Çağında Değişen İletişim Dinamikleri* (pp. 171–197). Beta Basım Yayım.

Sarıberberoğlu, M. T. (2019). Sinematik Kurgunun Bilinçaltı Mekânları-Tekinsiz Mekânlar. *Journal of Architecture and Life* 5(1), 27-39. ISSN: 2564-6109 DOI: DOI: 10.26835/my.645548

Sarı, Ü., & Sancaklı, P. (2020). Küyerelleşmenin Dijital Platformların İçerik Tanıtımına Etkisi [The effect of glocalization on the content promotion of digital platforms]. *Erciyes İletişim Dergisi, 7*(1), 243–260. DOI: 10.17680/erciyesiletisim.647463

Sarı, Ü., & Türker, H. (2021). Dijital Platform Kullanıcılarının İzleme Alışkanlıklarına Yönelik Bir Araştırma: Netflix Örneği. *Anadolu Üniversitesi Sosyal Bilimler Dergisi, 21*(1), 1. Advance online publication. DOI: 10.18037/ausbd.902558

Scott, G. (2020). *New Ways of Seeing, The Democratic Language of Photography*. Routledge. DOI: 10.4324/9781003103639

Seel, P. B. (2018). Digital television & video. In Grant, A. E., & Meadows, J. H. (Eds.), *Communication Technology Update and Fundamentals* (pp. 67–76). Routledge. DOI: 10.4324/9780203702871-6

Şenyüz, F. B. (2023). Türkiye'de dizi endüstrisi ve dijital dönüşüm: Yapım, dağıtım, içerik ve izleme pratiklerinin dijitalleşmesi. In *Kültürel Değişim ve Endüstrileşme Sürecinde Türk Dizileri 2010-2020* (pp. 41–53). İstanbul Ticaret Odası.

Shahaby, A., Setiawan, S. A., & Handoko, D. (2024). *Dampak Penerapan Etika Periklanan Terhadap Efektivitas Iklan Layanan Masyarakat*. Harmoni.

Shattuc, J. (2020). Netflix, Inc. and online television. In Wasko, J., & Meehan, E. R. (Eds.), *A Companion to Television* (pp. 145–164). John Wiley & Sons. DOI: 10.1002/9781119269465.ch7

Shaviro, S. (1993). *"The Cinematic Body" (1993)*. University of Minnesota Press.

Sheets, P., Rowling, C. M., Gilmore, J., & Melcher, N. (2022). Us and Them: The Role of Group Identity in Explaining Cultural Resonance and Framing Effects. *Mass Communication & Society*.

Shropshire, J., Gowan, A., & Guo, C. (2016). Defining Audience Awareness for Information Systems Research.*ACM SIGMIS Conference on Computers and People Research*. DOI: 10.1145/2890602.2890614

Shrum, L. J. (2017). Cultivation Theory: Effects and Underlying Processes. In Rössler, P., Hoffner, C. A., & Zoonen, L. (Eds.), *The International Encyclopedia of Media Effects*. DOI: 10.1002/9781118783764.wbieme0040

Silverstone, R. (1994). *Television and Everyday Life* (Routledge). Routledge.

Simons, N. (2009). "Me TV": Towards changing TV viewing practices? *Proceedings of the 7th European Conference on Interactive TV and Video*, 219–222. DOI: 10.1145/1542084.1542128

Şimşek, H., & Yıldırım, A. (2016). *Sosyal Bilimlerde Nitel Araştırma Yöntemleri (Hasan Şimşek-Ali Yıldırım)*. Seçkin Yayıncılık.

Sirer, E. (2020). Televizyon Yayıncılığında Büyük Veri. *Atatürk Üniversitesi Sosyal Bilimler Enstitüsü Dergisi, 24*(4), Article 4.

Sirer, E. (2020). *Zamanın Ruhu: Eşzamansızlık. Yeni İzlence Yeni İzlerkitle* (Sirer, E., Ed.). Literatürk Publication.

Snyder, H. (2019). Literature review as a research methodology: An overview and guidelines. *Journal of Business Research, 104*, 333–339. DOI: 10.1016/j.jbusres.2019.07.039

Söğüt, F. (2020, December 11). *Dizilerin Yeni Mecrası Olarak İnternet: Televizyon Dizileri İle İnternet Dizileri Arasında Karşılaştırmalı Bir Analiz.*

Solis, L. (2014). Expert analyzes students' Netflix usage. *The Daily Toreador.* https://www.dailytoreador.com/lavida/expert-analyzes-students-netflix-usage/article_0dfc194a-5412-11e4-9415-001a4bcf6878.html. [Erişim tarihi: 26.04.2025].

Sönmezer, Z., & Büyükbaykal Ilgaz, A. C. (2024). Medya ve Teknolojinin Kolektif Kullanmı. *The Turkish Online Journal of Design Art and Communication, 14*(3), 753–762. DOI: 10.7456/tojdac.1464676

Sözen, M. (2003). *Sinemada renk – sembolik anlamlar.* Detay Yayıncılık.

Spampinato, F. M., & Catricalà, V. (2021). Contemporary Art and Virtual Reality: New Conditions of Viewership. *Cinergie – Il Cinema E Le Altre Arti, 10*(19), 121–133. DOI: 10.6092/issn.2280-9481/12322

Stallabras, J. (2016). *Sanat A.Ş. Çağdaş Sanat ve Bienaller* (Soğancılar, E., Trans.). İletişim Publication.

Stam, R. (2004). *Film Theory: An Introduction.* Wiley-Blackwell.

Starosta, J., Izydorczyk, B., & Lizińczyk, S. (2019). Characteristics of people's binge-watching behavior in the "entering into early adulthood" period of life. *Health Psychology Report*, *7*(2), 149–164. DOI: 10.5114/hpr.2019.83025

Strangelove, M. (2015). *Post-TV: Piracy, cord-cutting, and the future of television.* University of Toronto Press. DOI: 10.3138/9781442666184

Sümer, O. (2024). Algoritma tiranlığı: Dijital dünyada bireyin savunmasızlığına yönelik bir sosyo-teknik analiz [Algorithm tyranny: A socio-technical analysis of vulnerability of the individual in the digital world]. *İmgelem*, 15, 87–114.

Sun, P. (2024). A Study of Artificial Intelligence in the Production of Film.SHS Web of Conferences. DOI: 10.1051/shsconf/202418303004

Sunal, G. (2016). Sanal Gerçeklik ve Dijital Sinemanın Olanakları. İnönü Üniversitesi İletişim Fakültesi Elektronik Dergisi (İNİF E-Dergi), 1(2), 294-309.

Suner, A. (2006). *Hayalet ev: Yeni Türk sinemasında aidiyet, kimlik ve bellek.* Metis Yayınları.

Sütçü, C. S. (2012). *Sosyal Medyaya Girmeden Önce Bilinmesi Gerekenler. Yeni Medya ve...* (Yengi, D., Ed.). E Publication.

Tağ, Ş. (2003). *Belgesel Sinema ve Türleri* (Master's thesis, Anadolu University (Turkey)).

Tang, W., & Wei, M. (2023). Streaming media business strategies and audience-centered practices: A comparative study of Netflix and Tencent Video. *Online Media and Global Communication*, *2*(1), 3–24. DOI: 10.1515/omgc-2022-0061

Tang, Y., Luo, J. G., Zhang, Q., Zhang, M., & Yang, S. Q. (2007). Deploying P2P networks for large- scale live video-streaming service [Peer-to-Peer Multimedia Streaming]. *IEEE Communications Magazine*, *45*(6), 100–106. DOI: 10.1109/MCOM.2007.374426

Taşçı, G., & Çelebi, M. (2020).Eğitimde Yeni Bir Paradigma: "Yükseköğretimde Yapay Zekâ."

Taşdelen, B., & Kesim, M. (2020). User central content design in the age of interactive television. *Erciyes İletişim Dergisi*, *7*(1), 703–722. DOI: 10.17680/erciyesiletisim.612545

Taylor, S., & Choi, M. (2022). An Initial Conceptualization of Algorithm Responsiveness: Comparing Perceptions of Algorithms Across Social Media Platforms. *Social Media + Society*, *8*(4), 205630512211443. DOI: 10.1177/20563051221144322

Tekbey, O., & Alaybeyoğlu, A. (2023). Yapay Zekanın Kullanım Alanları ve Etik Sorunlar. *International Journal of Advanced Natural Sciences and Engineering Researches*, *7*(9).

Temel, F., & Önürmen, O. (2017, March).. . *Gelenekselin Dönüşümünde İletişim Akademisyenleri ve Sosyal Medya, E-Gifder, Volume*, *5*(1), 292–313.

Terzi, A. (2021). Binge-watching kavramı ve Türkçe karşılıkları üzerine. *Türk Dili, 70*(833).

Thakur, G. B. (2008). *Of Sture and Signifier in Michael Haneke's Cache. Psychoanalysys, Culture & Society. Sayı: 13*. Palgrave Macmillan.

The Guardian. (2016). *Can tearjerker virtual reality movies tempt donors to give more aid?* https://www.theguardian.com/global-development/2015/dec/31/virtual-reality-movies-aid-humanitarian-assistance-united-nations, [Erişim tarihi: 09.03.2025].

Thompson, J. B. (2019). *Medya ve Modernite* (Öztürk, S., Trans.). Kırmızı Yayınları.

Thompson, K. (2003). *Storytelling in the new Hollywood: Understanding classical narrative technique*. Harvard University Press.

Todd, H. (Director). (2002). *Far from Heaven* [Film]. Focus Features.

Tokdil, E. (2018) Yeni Medya, Temel Bileşenleri ve Sanatın Değişen Estetik Dili, İletişim Kuram ve Araştırma Dergisi - Sayı 47 /2018, pp.167-190

Toprak Ökmen, A. G. (2022). Dijital dönemde belgesel film yapımındaki yenilikler üzerine bir değerlendirme. *Medya ve Kültür.*, *2*(2), 184–207.

Torun, H. (2019). Belgesel Sinema Dili Ve Savaş. *Karadeniz Uluslararası Bilimsel Dergisi*, (41), 144–158.

Tosun, M. (2020). *15 Temmuz anma programlarında dijital çalışmalar öne çıktı*, https://www.aa.com.tr/tr/15-temmuz-darbe-girisimi/15-temmuz-anma-programlarinda-dijital-calismalar-one-cikti, [07.04.2025].

TRT. (2017). *"360 derece, karanlık gece" tırı Türkiye turunda,* https://www.trthaber.com/haber/gundem/360-derece-karanlik-gece-tiri-turkiye-turunda, [Erişim tarihi: 13.03.2025].

Tuğal, S. A. (2018). *Oluşum Süreci İçinde Dijital Sanat*. Hayalperest.

Turner, G., & Tay, J. (2009). Introduction. In Turner, G., & Tay, J. (Eds.), *Television Studies After TV Understanding Television in the Post-Broadcast Era* (pp. 1–6). Routledge. DOI: 10.4324/9780203878316

Türnüklü, A. (2000). Eğitimbilim Araştırmalarında Etkin Olarak Kullanılabilecek Nitel BirAraştırma Tekniği: Görüşme. *Kuram ve Uygulamada Eğitim Yönetimi, 24*(24), Article 24.

Tutar, C. (2020). Postmodern Kentsel Mekanda Kültürel Belleğin İnşa Süreci. *Akdeniz Üniversitesi İletişim Fakültesi Dergisi*, *34*(34), 102–121. DOI: 10.31123/akil.781252

Uğurlu, Ö. (2011). "Postmodern pazarlama stratejisi olarak dönüşen mahremiyet: 'Fulya'nın intikamı' viral kampanyası örneği." Hüseyin Köse (Ed.), *Medya mahrem* (pp. 173-199). Ayrıntı Yayınları.

Ünlü, T. T., & Akyol, O. (2025). Türkiye'de Dijital Platformlarda Yayınlanan İnteraktif Yapımların İncelenmesi. *Communicata*, *29*(29), 29. Advance online publication. DOI: 10.32952/communicata.1619972

Ürper, O. (2012). *Dijital Teknoloji Çağında Reklam Fotoğrafçılığı*. Say Publication.

Uyanık, G. (2023). Dijital medya platformlarını Henry Jenkins'in kavramları üzerinden okumak. *Etkileşim*, *11*(11), 432–444. DOI: 10.32739/etkilesim.2023.6.11.204

Van Esler, M. (2020). Reproducing television canons: Streaming services and the legacy of linear TV. *Journal of Popular Culture, 53*(4), 946–966. DOI: 10.1111/jpcu.12940

Vaughan, W. (2005). *Digital Art History. (Eds: Anna-Bentkowska-Kafel, Trish Canhen, Hazel Gardiner) History of Art in the Digital Age: Problems and Possibilities.* USA: Intellent Ltd.

Vives, V. V. (2024). *El presente de la IA en la actividad cinematográfica.* Revista Policía y Seguridad Pública. DOI: 10.26439/vent.indiscreta2024.n032.7474

Wallenstein, A. (2015, January 29). Binge-watching TV linked to depression, loneliness. *Variety.*

Wands, B. (2006). *Dijital Çağın Sanatı* (Akınhay, O., Trans.). Akbank Sanat Publication.

We are Social. (2024). *Digital 2024 Global.* https://wearesocial.com/uk/blog/2024/10/digital-2024-october-global-statshot-report/

We are Social. (2024). *Digital 2024 Turkey.* https://datareportal.com/reports/digital-2024-turkey

Wiesing, M., Comadran, G., & Slater, M. (2025). Confusing Virtual Reality with Reality – an Experimental Study. *iScience, 28*(6), 112655. DOI: 10.1016/j.isci.2025.112655 PMID: 40520086

Wilkinson, J. S. (2010). IPTV: Streaming media. In Grant, A. E., & Jennifer, H. M. (Eds.), *Communication technology update and fundamentals* (pp. 117–131). Focal Press. DOI: 10.1016/B978-0-240-81475-9.50008-6

Williams, R. (2003). *Televizyon, Teknoloji ve Kültürel Biçim* (Türkbağ, A. U., Trans.). Dost Kitabevi.

Wolfenstein, K. (2025). *VR gelişiminde AI dönüşümü: Meta, 3D dünyaları tek görüntülerden nasıl çağırıyor,* https://xpert.digital/tr/vr-gelisiminde-ki-donusumu, [Erişim tarihi: 17.04.2025].

Wu, Z, (2024). "Automated Cinematographer for VR Viewing Experiences", *Dartmouth College Master's Theses.* 162.

Wulf, T., Schneider, F., & Beckert, S. (2020). Watching Players: An Exploration of Media Enjoyment on Twitch. *Games and Culture, 15*(3), 328–346. DOI: 10.1177/1555412018788161

Yakışkan, N. N. (2024). Sergileme Mekânlarının Sanal ve Fiziksel Ortam Arasındaki Farkının İncelenmesi. *İstanbul Modern örneği. yedi: Sanat, Tasarım ve Bilim Dergisi*, Sanatta Dijitalizm Özel Sayısı, 327-337. DOI: 10.17484/yedi.1458408

Yaşar, İ. H. (2020). s.192). İletişim tarihine kısa bir bakış: Analogdan dijitale iletişim araçlarının gelişimi ve bireye yansıması. *Ufkun Ötesi Bilim Dergisi, 20*(1), 183–207.

Yavuz, S. (2024). Belge Filmden İnteraktif Belgesele: Teknolojik Gelişmeler Bağlamında Belgeselin Değişen Estetiği. *İnönü Üniversitesi İletişim Fakültesi Elektronik Dergisi (İnif E-Dergi), 9*(1), 111-133.

Yaykın, M. (2010). *Sanat, Teknoloji, Bilim ve Fotoğraf.* Kalkedon Publication.

Yetişkin, E. (2012). *Sular Artık Tersine Akıyor: Yeni Medya Sanatı Nasıl Okunabilir? Art Unlimited.* Temmuz.

Yıldırım, E. (2019). Televizyon Belgesel Filmleri Anlatı Yapısının İzler Kitle Beklentilerine Etkileri Ve Örnek Belgesel Film Çözümlemesi. *Dördüncü Kuvvet.*, *2*(2), 134–149.

Yılmaz, R. (2017). Türkiye'de Reklam Anlatısının Dijital Çağdaki Yapısal Dönüşümü. *1st International Conference on New Trends in Communication.* İstanbul Ticaret Universty.

Yılmaz, R. (2017). Türkiye'de Reklam Anlatısının Dijital Çağdaki Yapısal Dönüşümü. *1st International Conference on New Trends in Communication.* İstanbul: Ticaret Universty.

Yücel, D. (2012). *Yeni Medya Sanatı ve Yeni Müze.* İstanbul Kültür Üniversitesi Publication.

Yu, R. (2023). *Algorithmic audience in the age of artificial intelligence: Tailored communication, information cocoons, algorithmic literacy, and news literacy.* Peter Lang Publishing.

Yurdigül, Y., & Yüksel, H. (2012). Gazeteciliğin Dönüşümü: Yeni Medyaya Entegrasyon Sürecinde Değişen Habercilik Pratikleri. *Akdeniz Üniversitesi İletişim Fakültesi Dergisi, 18*, Article 18.

Yurdigül, Y., & Yurdigül, A. (2014). Tv Haberlerinin Anlatı Yapısının Oluşturulması Sürecinde Özel Efekt Teknolojileri: Ntv ve Cnn Türk Ana Haber Bültenleri Üzerinden Bir İnceleme. *İstanbul Üniversitesi İletişim Fakültesi Dergisi | Istanbul University Faculty of Communication Journal*(*46*), 121-146

Yurdigül, Y., Bayraktar, R., & Çil, S. (2024). A study on Kahramanmaraş Earthquake survivors: "Social media platforms were more effectively used than the traditional media.". *Connectist: Istanbul University Journal of Communication Sciences, 66*(66), 66. Advance online publication. DOI: 10.26650/CONNECTIST2024-1460282

About the Contributors

Serhat Erdem, born in 1982 in Kırıkkale, completed his high school education at Samsun Ondokuz Mayıs High School. In 2004, he earned his bachelor's degree from Atatürk University, Faculty of Communication, Department of Radio, Television, and Cinema. In 2005, he began his academic career as a Research Assistant in the Department of Cinema and Television at Atatürk University's Faculty of Fine Arts. Erdem completed his master's degree in 2009 at Atatürk University, Institute of Social Sciences, Department of Journalism, with a thesis titled "An Analysis of the Transformations Caused by Matte Painting Applications in Cinema." In 2010, he began his doctoral studies at Istanbul University, Institute of Social Sciences, Department of Radio, Television, and Cinema. He received the title of Doctor in 2015 after successfully completing his dissertation titled "The Use of Visual Effects in the Reconstruction of Reality: Spatial Construction in the Spartacus Television Series." Between 2005 and 2022, Erdem served as an Assistant Professor in the Department of Cinema and Television at Atatürk University's Faculty of Fine Arts. In 2022, he earned the title of Associate Professor in the field of Visual Communication Design. Since February 2022, he has continued his academic work in the Department of Radio, Television, and Cinema at Atatürk University's Faculty of Communication.

Zuhal Akmeşe completed her first undergraduate studies in Social Studies Teaching at Afyon Kocatepe University Faculty of Education (2005). She then completed her second undergraduate degree with honors in the Department of Radio, Television, and Cinema at Ege University Faculty of Communication (2010). She completed her Master's degree in Radio, Television, and Cinema at the Institute of Social Sciences at Ege University with a thesis titled "Representation of Youth in Post-2000 Turkish Cinema" (2013), and her PhD at Istanbul University Institute of Social Sciences in the Department of Radio, Television, and Cinema with a thesis titled "Framing in Television and the Construction of Societal Perspective" (2017). In 2020, she earned the title of Associate Professor in the field of Communication

Studies. Dr. Akmeşe's research interests include documentary cinema, new communication technologies, communication theories, social media, media literacy, visual communication design, production and direction in television and cinema, and the sociology of communication. She has authored several co-written studies in the field of communication and presented papers at international conferences and symposiums. She has also been involved as a consultant in film projects and has participated in numerous film and documentary film works as a director. Her documentary works have been shown at various national and international festivals, and her photographs have been exhibited in art galleries. Since 2013, Dr. Akmeşe has been conducting academic work at the Department of Radio, Television, and Cinema at Dicle University Faculty of Communication. In addition to her academic work, she has served as a member of the main jury at numerous national and international film festivals, as well as being part of the selection committees for many other events. She is a member of the editorial board of TRT Academy Journal, a board member of the Dicle University Cultural and Art Practice and Research Center, a member of the Dicle University Quality Board, and has held the position of Chair of the Department of Radio, Television, and Cinema at Dicle University Faculty of Communication. Dr. Akmeşe has authored several books in the fields of cinema and communication. She is also the founding president and current board chair of the Cinema Researchers' Culture and Arts Association, where she continues her work in cinema, arts, and communication. Within the association, she holds the roles of Festival President and Festival Director of the International Dicle Documentary and Short Film Festival. Additionally, she is the head editor of the SineAkademi cinema studies journal, which was initiated within the association.

***

Gülgin Altuğ Born in 1976 in Erzurum, Gülgin ALTUĞ earned her bachelor's degree from Atatürk University, Faculty of Fine Arts, Department of Painting in 2003. Having completed her master's degree in 2015 at Atatürk University, Institute of Social Sciences, Department of Painting with her thesis titled "Surface Inquiry in Modern Painting from Two Dimensions to Three Dimensions", Altuğ has been continuing her academic studies as a lecturer at Erzincan Binali Yıldırım University, Faculty of Fine Arts, Department of Painting since February 2020.

Gözde Sunal completed her undergraduate studies in Visual Communication Design at Istanbul Commerce University between 2004 and 2009. She earned her master's degree in Radio, Television and Cinema from Marmara University between 2009 and 2011, and completed her PhD in Communication Sciences at Maltepe University between 2011 and 2015. Since 2009, Sunal has been working at the university from which she graduated, and she currently serves as the Vice Dean of the Faculty of Communication at Istanbul Commerce University. In addition to her academic publications, she serves as a jury member in various artistic events and has received several awards. Sunal is the author of the books "In Pursuit of Images: Photography-Cinema-Time" and "The Unconscious in Cinema and New Surveillance." Her areas of expertise include communication studies, film and media research, film production and management, cinema and psychology, film analysis, and aesthetics.

Elif Çanğa Bayer completed her undergraduate and graduate studies in Fine Arts, and earned her Ph.D. in Communication. She is currently a faculty member at the Faculty of Communication, Tirebolu, Giresun University, where she continues her academic career and conducts research on cinema and artificial intelligence.

Dilar Diken Yücel graduated from Atatürk University, Faculty of Communication, Department of Radio Television and Cinema in 2012. Afterwards, she completed her master's degree at Gazi University Institute of Social Sciences, Department of Radio, Television and Cinema (2016) and her doctorate at Ankara Hacı Bayram Veli Graduate Education Institute (2021). In 2023, she was awarded the title of associate professor in the field of cinema. She is currently working as Assoc. Prof. Dr. at İnönü University, Faculty of Communication, Department of Radio, Television and Cinema. Diken Yücel, who is interested in the fields of surveillance in films, semiotics, sociology and philosophy of cinema, Turkish cinema, audience in cinema, art cinema, has various studies in these fields, as well as a book titled "The Adventure of Social Realism in Turkish Cinema" and a book titled "Popular Themes of Recent Turkish Cinema".

Murat Han Er, born in Erzurum in 1981, completed his undergraduate studies in the Graphic Design Department at Atatürk University. He earned his Master's degree from the Photography Department of the Faculty of Fine Arts at Atatürk University and his Doctorate from the Photography Department of the Faculty of Fine Arts at Mimar Sinan Fine Arts University. In 2005, he began working as a Research Assistant in the Photography Department at the Faculty of Fine Arts at Atatürk University. Between 2010 and 2015, he worked in the Photography Department at Mimar Sinan Fine Arts University. He currently serves as an Associate Professor in the Photography Department of the Faculty of Fine Arts at Atatürk University. The artist has participated in numerous exhibitions both nationally and internationally and has received various awards.

Mustafa Gülsün Born in Muş in 1986. After completing his primary and secondary education here, he completed his bachelor's degree in Business

Administration at the Faculty of Economics and Administrative Sciences at Karadeniz Technical University in 2006, and his master's degree in Basic Communication at the Faculty of Communication at Malatya İnönü University. As of 2025, he is pursuing his doctorate in Basic Communication at Atatürk University's Faculty of Communication. In 2015, he began working as a lecturer at Muş Alparslan University's Malazgirt Vocational School of Higher Education. He currently continues to perform this role. He conducts academic research in the fields of cinema and communication.

Gül Gürsoy holds a B.Sc. in Chemical Engineering from Istanbul Technical University (1993), an M.Sc. in Environmental Technologies from Bogazici University (1998), and an MBA in Marketing from Istanbul Bilgi University (2008). Rooted in a long-standing interest in photography, her academic focus gradually shifted toward visual culture, cinema history, and film analysis. After years of professional experience in international companies, she completed an MA in Film Studies at Istanbul Commerce University in 2025. Her thesis examined the historical representation and transformation of the femme fatale archetype in film noir. Her research interests include film theory, representation, gender, and contemporary cinema.

Semra Kotan is a faculty member at the Department of Radio, Television and Cinema of the Faculty of Communication at Atatürk University. She completed her BA (2011), MA (2015) and PhD (2020) degrees at Atatürk University in the field of Radio, Television and Cinema. In her PhD, she focused on the dynamic ties between cinema and social structure by examining the relationship between the individual and ideology in the context of horror cinema. Her academic interests include cinema theories, semiotics, representation theories, cultural studies and ideology criticism; in this context, she conducts theoretical and analytical studies on Turkish cinema, popular culture products and digital media content. Her studies, which focus especially on film genres, national cinema narratives, visual culture and media representations in the digital age, contribute to communication research by adopting an interdisciplinary approach. Kotan, who has numerous national and international articles, book chapters and papers published within the scope of her research, has taken part in various scientific research projects as a director or researcher. She also serves as a thesis advisor at the MA level and teaches courses requiring expertise in the fields of cinema and media studies. Kotan, who has served as program coordinator, Erasmus institutional coordinator, department vice president and various commission members within the scope of his academic-administrative duties, continues his academic studies in the fields of social sciences and humanities and actively participates in national/international scientific activities.

Yelda Özkoçak started her education at Anatolian Communication High School and completed the associate degree program in R-TV Broadcasting. She received her bachelor's degree from Istanbul University Radio, Television and Cinema program in 2007, then her master's degree in 2009 and her doctorate in the same field from Istanbul University Institute of Social Sciences in 2013. She played important roles in the cinema and television industry between 2001-2009 and gained extensive experience during this time. Özkoçak started her academic career in 2009 and received the title of Associate Professor in 2018 and Professor in 2023. She has actively participated in numerous national and international conferences, symposiums and seminars and has shared her expertise with four books and numerous academic publications. Her research interests include a wide range of topics such as television programs, gender studies, Turkish cinema and audiovisual editing, and she makes significant contributions to these fields with her in-depth research and academic studies. Özkoçak, who worked at different universities, held positions such as center director, assistant dean, department head. Özkoçak served as the Rector's Advisor at Doğuş University between December 2018 and July 2024. He has been serving as the Vice Rector at the same University since July 2024.

Gözde Sunal completed her undergraduate studies in Visual Communication Design at Istanbul Commerce University between 2004 and 2009. She earned her master's degree in Radio, Television and Cinema from Marmara University between 2009 and 2011, and completed her PhD in Communication Sciences at Maltepe University between 2011 and 2015. Since 2009, Sunal has been working at the university from which she graduated, and she currently serves as the Vice Dean of the Faculty of Communication at Istanbul Commerce University. In addition to her academic publications, she serves as a jury member in various artistic events and has received several awards. Sunal is the author of the books "In Pursuit of Images: Photography-Cinema-Time" and "The Unconscious in Cinema and New Surveillance." Her areas of expertise include communication studies, film and media research, film production and management, cinema and psychology, film analysis, and aesthetics.

Gökhan Toptepe completed his undergraduate studies in Philology, and received his Master's and Ph.D. degrees in Communication. He is currently a faculty member at the Faculty of Communication, Tirebolu, Giresun University, where he continues his academic career and conducts research on photography and communication.

Ömer Yücel was born in Ankara. He completed his primary and secondary education in Ankara. He graduated from Marmara University, Faculty of

Communication, Department of Journalism in 2012. In 2013, he was appointed as a research assistant at İnönü University, Faculty of Communication, Department of Journalism. He completed his master's degree in Gazi University, Institute of Social Sciences, Department of Journalism. He graduated from Ankara Hacı Bayram Veli University, Institute of Graduate Education, Department of Journalism in 2022. He continues to work as an assistant professor at İnönü University, Faculty of Communication, Department of Journalism. He continues his studies in the fields of journalism and media studies, communication sciences, and press history.

Yusuf Yurdigül was born in Erzurum. After completing his primary, secondary and high school education in various cities of Turkey, he completed his education in Erzurum. He completed his BA in Journalism at the Istanbul University Faculty of Communication in 1999 and his MA in Radio, Television and Cinema at the Istanbul University Institute of Social Sciences in 2002. Yurdigül, who began his doctorate at the Istanbul University Institute of Social Sciences in Radio, Television and Cinema in the same year, completed his PhD in sociology of communication in 2007. He worked as a faculty member at the Istanbul University Faculty of Communication (2000-2008), the Atatürk University Faculty of Communication (2008-) and the Manas University Faculty of Communication (2016-). Yurdigül, who also worked as a reporter and director on various national channels in Turkey, has numerous short films and documentaries. He is married with two children.

Aslı Yurdigül is professor in the Faculty of Communication at Atatürk University, Turkey. She graduated from Marmara University, in Istanbul. Then, she worked many different televisions channel as co-director and director. And she worked international company as public relation officer. She started her academic career at Atatürk University, Faculty of Communication, Radio, Television and Cinema Department in 2005. She has published books, book chapters, articles, papers on media and communication studies. She wrote Cinema and News and Media and Culture of Fear; edited Basic Concepts in Public Relations. Also, she directed many documentary films. Her film named "Sixty Minds and Seventy Ideas" was supported by the Ministry of Culture and Tourism of the Republic of Turkey in 2020. She teaches courses such as Sociology of the News, Scientific Research Methods, Intercultural Communication, Media Politics at undergraduate, graduate and doctoral levels. She has recently been conducting different research projects such as media politics and e-sports.

Index

A

B

C

D

E

S

T

U

V

www.ingramcontent.com/pod-product-compliance
Lightning Source LLC
LaVergne TN
LVHW081529300926
849161LV00006B/1357